Symmetry with Operator Theory and Equations

Symmetry with Operator Theory and Equations

Special Issue Editor
Ioannis Argyros

MDPI • Basel • Beijing • Wuhan • Barcelona • Belgrade

Special Issue Editor
Ioannis Argyros
Cameron University
USA

Editorial Office
MDPI
St. Alban-Anlage 66
4052 Basel, Switzerland

This is a reprint of articles from the Special Issue published online in the open access journal *Symmetry* (ISSN 2073-8994) in 2019 (available at: https://www.mdpi.com/journal/symmetry/special_issues/Symmetry_Operator_Theory_Equations).

For citation purposes, cite each article independently as indicated on the article page online and as indicated below:

LastName, A.A.; LastName, B.B.; LastName, C.C. Article Title. *Journal Name* **Year**, *Article Number*, Page Range.

ISBN 978-3-03921-666-6 (Pbk)
ISBN 978-3-03921-667-3 (PDF)

© 2019 by the authors. Articles in this book are Open Access and distributed under the Creative Commons Attribution (CC BY) license, which allows users to download, copy and build upon published articles, as long as the author and publisher are properly credited, which ensures maximum dissemination and a wider impact of our publications.

The book as a whole is distributed by MDPI under the terms and conditions of the Creative Commons license CC BY-NC-ND.

Contents

About the Special Issue Editor vii

Preface to "Symmetry with Operator Theory and Equations" ix

Alicia Cordero, Jonathan Franceschi, Juan R. Torregrosa and Anna Chiara Zagati
A Convex Combination Approach for Mean-Based Variants of Newton's Method
Reprinted from: *Symmetry* 2019, 11, 1106, doi:10.3390/sym11091106 1

Alicia Cordero, Ivan Girona and Juan R. Torregrosa
A Variant of Chebyshev's Method with 3αth-Order of Convergence by Using FractionalDerivatives
Reprinted from: *Symmetry* 2019, 11, 1017, doi:10.3390/sym11081017 17

R.A. Alharbey, Ioannis K. Argyros and Ramandeep Behl
Ball Convergence for Combined Three-Step Methods Under Generalized Conditions in Banach Space
Reprinted from: *Symmetry* 2019, 11, 1002, doi:10.3390/sym11081002 28

Ioannis K. Argyros, Santhosh George, Chandhini Godavarma and Alberto A. Magreñán
Extended Convergence Analysis of the Newton–Hermitian and Skew–Hermitian Splitting Method
Reprinted from: *Symmetry* 2019, 11, 981, doi:10.3390/sym11080981 39

R. A. Alharbey, Munish Kansal, Ramandeep Behl and J.A. Tenreiro Machado
Efficient Three-Step Class of Eighth-Order Multiple Root Solvers and Their Dynamics
Reprinted from: *Symmetry* 2019, 11, 837, doi:10.3390/sym11070837 54

Janak Raj Sharma, Sunil Kumar and Ioannis K. Argyros
Development of Optimal Eighth Order Derivative-Free Methods for Multiple Roots of Nonlinear Equations
Reprinted from: *Symmetry*, 11, 766, doi:10.3390/sym11060766 84

Mujahid Abbas, Yusuf Ibrahim, Abdul Rahim Khan and Manuel de la Sen
Strong Convergence of a System of Generalized Mixed Equilibrium Problem, Split Variational Inclusion Problem and Fixed Point Problem in Banach Spaces
Reprinted from: *Symmetry* 2019, 11, 722, doi:10.3390/sym11050722 101

Mehdi Salimi and Ramandeep Behl
Sixteenth-Order Optimal Iterative Scheme Based on Inverse Interpolatory Rational Function for Nonlinear Equations
Reprinted from: *Symmetry* 2019, 11, 691, doi:10.3390/sym11050691 122

Munish Kansal, Ramandeep Behl, Mohammed Ali A. Mahnashi and Fouad Othman Mallawi
Modified Optimal Class of Newton-Like Fourth-Order Methods for Multiple Roots
Reprinted from: *Symmetry* 2019, 11, 526, doi:10.3390/sym11040526 133

Janak Raj Sharma, Deepak Kumar and Ioannis K. Argyros
An Efficient Class of Traub-Steffensen-Like Seventh Order Multiple-Root Solvers with Applications
Reprinted from: *Symmetry*, 11, 518, doi:10.3390/sym11040518 144

Ramandeep Behl, M. Salimi, M. Ferrara, S. Sharifi and Samaher Khalaf Alharbi
Some Real-Life Applications of a Newly Constructed DerivativeFree Iterative Scheme
Reprinted from: *Symmetry* **2019**, *11*, 239, doi:10.3390/sym11020239 . . . . **161**

Ioannis K. Argyros, Stepan Shakhno, Halyna Yarmola
Two-Step Solver for Nonlinear Equations
Reprinted from: *Symmetry* **2019**, *11*, 128, doi:10.3390/sym11020128 . . . . **175**

Ramandeep Behl, Ioannis K. Argyros, J.A. Tenreiro Machado and Ali Saleh Alshomrani
Local Convergence of a Family of Weighted-Newton Methods
Reprinted from: *Symmetry* **2019**, *11*, 103, doi:10.3390/sym11010103 . . . . **184**

About the Special Issue Editor

Ioannis Argyros, he is a professor at the Department of Mathematics Sciences, Cameron University. His research interests include: Applied mathematics, Operator theory, Computational mathematics and iterative methods especially on Banach spaces. He has published more than a thousand peer reviewed papers, thirty two books and seventeen chapters in books. He is an active reviewer of a plethora of papers and books, and has received several national and international awards. He has supervised two PhD students, several MSc. and undergraduate students, and has been the external evaluator for many PhD theses, tenure and promotion applicants.

Preface to "Symmetry with Operator Theory and Equations"

The development of iterative procedures for solving systems or nonlinear equations in abstract spaces is an important and challenging task. Recently such procedures have been extensively used in many diverse disciplines such as Applied Mathematics; Mathematical: Biology; Chemistry; Economics; Physics, and also Engineering to mention a few.

The main purpose of this special issue is to present new ideas in this area of research with applications. This issue gives an opportunity to researchers and practitioners to communicate their recent works.

Topics included in this issue are:

From the 35 articles received, this special issue includes 13 high-quality peer-reviewed papers reflecting recent trends in the aforementioned topics. We hope that the presented results would lead to new ideas in the future.

Ioannis Argyros
Special Issue Editor

Article

A Convex Combination Approach for Mean-Based Variants of Newton's Method

Alicia Cordero [1], Jonathan Franceschi [1,2,*], Juan R. Torregrosa [1] and Anna C. Zagati [1,2]

[1] Institute of Multidisciplinary Mathematics, Universitat Politècnica de València, Camino de Vera, s/n, 46022-Valencia, Spain
[2] Universitá di Ferrara, via Ludovico Ariosto, 35, 44121 Ferrara, Italy
* Correspondence: jofra1@posgrado.upv.es

Received: 31 July 2019; Accepted: 18 August 2019; Published: 2 September 2019

Abstract: Several authors have designed variants of Newton's method for solving nonlinear equations by using different means. This technique involves a symmetry in the corresponding fixed-point operator. In this paper, some known results about mean-based variants of Newton's method (MBN) are re-analyzed from the point of view of convex combinations. A new test is developed to study the order of convergence of general MBN. Furthermore, a generalization of the Lehmer mean is proposed and discussed. Numerical tests are provided to support the theoretical results obtained and to compare the different methods employed. Some dynamical planes of the analyzed methods on several equations are presented, revealing the great difference between the MBN when it comes to determining the set of starting points that ensure convergence and observing their symmetry in the complex plane.

Keywords: nonlinear equations; iterative methods; general means; basin of attraction

1. Introduction

We consider the problem of finding a simple zero α of a function $f\colon I \subset \mathbb{R} \to \mathbb{R}$, defined in an open interval I. This zero can be determined as a fixed point of some function g by means of the one-point iteration method:

$$x_{n+1} = g(x_n), \quad n = 0, 1, \ldots, \tag{1}$$

where x_0 is the starting point. The most widely-used example of these kinds of methods is the classical Newton's method given by:

$$x_{n+1} = x_n - \frac{f(x_n)}{f'(x_n)}, \quad n = 0, 1, \ldots. \tag{2}$$

It is well known that it converges quadratically to simple zeros and linearly to multiple zeros. In the literature, many modifications of Newton's scheme have been published in order to improve its order of convergence and stability. Interesting overviews about this area of research can be found in [1–3]. The works of Weerakoon and Fernando [4] and, later, Özban [5] have inspired a whole set of variants of Newton's method, whose main characteristic is the use of different means in the iterative expression.

It is known that if a sequence $\{x_n\}_{n\geq 0}$ tends to a limit α in such a way that there exist a constant $C > 0$ and a positive integer n_0 such that:

$$|x_{n+1} - \alpha| \leq C|x_n - \alpha|^p, \quad \forall n \geq n_0, \tag{3}$$

for $p \geq 1$, then p is called the order of convergence of the sequence and C is known as the asymptotic error constant. For $p = 1$, constant C satisfies $0 < C \leq 1$.

If we denote by $e_n = x_n - \alpha$ the exact error of the nth iterate, then the relation:

$$e_{n+1} = Ce_n^p + \mathcal{O}(e_n^{p+1}) \qquad (4)$$

is called the error equation for the method and p is the order of convergence.

Let us suppose that $f : I \subseteq \mathbb{R} \to \mathbb{R}$ is a sufficiently-differentiable function and α is a simple zero of f. It is plain that:

$$f(x) = f(x_n) + \int_{x_n}^{x} f'(t)\, dt. \qquad (5)$$

Weerakoon and Fernando in [4] approximated the definite integral (5) by using the trapezoidal rule and taking $x = \alpha$, getting:

$$0 \approx f(x_n) + 1/2(\alpha - x_n)(f'(x_n) + f'(\alpha)), \qquad (6)$$

and therefore, a new approximation x_{n+1} to α is given by:

$$x_{n+1} = x_n - \frac{f(x_n)}{(f'(x_n) + f'(z_n))/2}, \qquad z_n = x_n - \frac{f(x_n)}{f'(x_n)}, \qquad n = 0, 1, \ldots. \qquad (7)$$

Thus, this variant of Newton's scheme can be considered to be obtained by replacing the denominator $f'(x_n)$ of Newton's method (2) by the arithmetic mean of $f'(x_n)$ and $f'(z_n)$. Therefore, it is known as the arithmetic mean Newton method (AN).

In a similar way, the arithmetic mean can be replaced by other means. In particular, the harmonic mean $M_{Ha}(x,y) = 2xy/(x+y)$, where x and y are two nonnegative real numbers, from a different point of view:

$$M_{Ha}(x,y) = \frac{2xy}{x+y} = x\underbrace{\frac{y}{x+y}}_{\theta} + y\underbrace{\frac{x}{x+y}}_{1-\theta}, \qquad (8)$$

where since $0 \leq y \leq x + y$, then $0 \leq \theta \leq 1$, i.e., the harmonic mean can be seen as a convex combination between x and y, where every element is given the relevance of the other one in the sum. Now, let us switch the roles of x and y; we get:

$$x\frac{x}{x+y} + y\frac{y}{x+y} = \frac{x^2 + y^2}{x+y} = M_{Ch}(x,y), \qquad (9)$$

that is the contraharmonic mean between x and y.

Özban in [5] used the harmonic mean instead of the arithmetic one, which led to a new method:

$$x_{n+1} = x_n - \frac{f(x_n)(f'(x_n) + f'(z_n))}{2f'(x_n)f'(z_n)}, \qquad n = 0, 1, \ldots, \qquad (10)$$

being z_n a Newton step, which he called the harmonic mean Newton method (HN).

Ababneh in [6] designed an iterative method associated with this mean, called the contraharmonic mean Newton method (CHN), whose iterative expression is:

$$x_{n+1} = x_n - \frac{(f'(x_n) + f'(z_n))f(x_n)}{f'(x_n)^2 + f'(z_n)^2}, \qquad (11)$$

with third-order of convergence for simple roots of $f(x) = 0$, as well as the methods proposed by Weerakoon and Fernando [4] and Özban [5].

This idea has been used by different authors for designing iterative methods applying other means, generating symmetric fixed point operators. For example, Xiaojian in [7] employed the generalized mean of order $m \in \mathbb{R}$ between two values x and y defined as:

$$M_G(x,y) = \left(\frac{x^m + y^m}{2}\right)^{1/m}, \qquad (12)$$

to construct a third-order iterative method for solving nonlinear equations. Furthermore Singh et al. in [8] presented a third-order iterative scheme by using the Heronian mean between two values x and y, defined as:

$$M_{He}(x,y) = \frac{1}{3}(x + \sqrt{xy} + y). \qquad (13)$$

Finally, Verma in [9], following the same procedure, designed a third-order iterative method by using the centroidal mean between two values x and y, defined as:

$$M_{Ce}(x,y) = \frac{2(x^2 + xy + y^2)}{3(x + y)}. \qquad (14)$$

In this paper, we check that all these means are functional convex combinations means and develop a simple test to prove easily the third-order of the corresponding iterative methods, mentioned before. Moreover, we introduce a new method based on the Lehmer mean of order $m \in \mathbb{R}$, defined as:

$$M_{L_m}(x,y) = \frac{x^m + y^m}{x^{m-1} + y^{m-1}} \qquad (15)$$

and propose a generalization that also satisfies the previous test. Finally, all these schemes are numerically tested, and their dependence on initial estimations is studied by means of their basins of attraction. These basins are shown to be clearly symmetric.

The rest of the paper is organized as follows: Section 2 is devoted to designing a test that allows us to characterize the third-order convergence of the iterative method defined by a mean. This characterization is used in Section 3 for giving an alternative proof of the convergence of mean-based variants of Newton's (MBN) methods, including some new ones. In Section 4, we generalize the previous methods by using the concept of σ-means. Section 5 is devoted to numerical results and the use of basins of attraction in order to analyze the dependence of the iterative methods on the initial estimations used. With some conclusions, the manuscript is finished.

2. Convex Combination

In a similar way as has been stated in the Introduction for the arithmetic, harmonic, and contraharmonic means, the rest of the mentioned means can be also regarded as convex combinations. This is not coincidental: one of the most interesting properties that a mean satisfies is the averaging property:

$$\min(x,y) \leq M(x,y) \leq \max(x,y), \qquad (16)$$

where $M(x,y)$ is any mean function of x and y nonnegative. This implies that every mean that satisfies this property is a certain convex combination among its terms.

Indeed, there exists a unique $\theta(x,y) \in [0,1]$ such that:

$$\theta(x,y) = \begin{cases} \frac{M(x,y)-y}{x-y} & \text{if } x \neq y \\ 0 & \text{if } x = y \end{cases}. \tag{17}$$

This approach suggests that it is possible to generalize every mean-based variant of Newton's method (MBN), by studying their convex combination counterparts. As a matter of fact, every mean-based variant of Newton's method can be rewritten as:

$$x_{n+1} = x_n - \frac{f(x_n)}{\theta f'(x_n) + (1-\theta)f'(z_n)}, \tag{18}$$

where $\theta = \theta(f'(x_n), f'(z_n))$. This is a particular case of a family of iterative schemes constructed in [10].

We are interested in studying its order of convergence as a function of θ. Thus, we need to compute the approximated Taylor expansion of the convex combination at the denominator and then its inverse:

$$\begin{aligned}
\theta f'(x_n) + (1-\theta)f'(z_n) &= \theta f'(\alpha)[1 + 2c_2 e_n + 3c_3 e_n^2 + 4c_4 e_n^3 + \mathcal{O}(e_n^4)] + \\
&+ (1-\theta)f'(\alpha)[1 + 2c_2^2 e_n^2 + 4c_2(c_3 - c_2^2)e_n^3 + \mathcal{O}(e_n^4)] \\
&= f'(\alpha)[\theta + 2\theta c_2 e_n + 3\theta c_3 e_n^2 + 4\theta c_4 e_n^3 + \mathcal{O}(e_n^4)] + \\
&+ f'(\alpha)[1 + 2c_2^2 e_n^2 + 4c_2(c_3 - c_2^2)e_n^3 + \mathcal{O}(e_n^4)] + \\
&- f'(\alpha)[\theta + 2\theta c_2^2 e_n^2 + 4\theta c_2(c_3 - c_2^2)e_n^3 + \mathcal{O}(e_n^4)] \\
&= f'(\alpha)[1 + 2\theta c_2 e_n + (2c_2^2 + 3\theta c_3 - 2\theta c_2^2 + 3\theta c_3)e_n^2] + \\
&+ f'(\alpha)[(4\theta c_4 + (1-\theta)4c_2(c_3 - c_2^2))e_n^3 + \mathcal{O}(e_n^4)];
\end{aligned} \tag{19}$$

where $c_j = \frac{1}{j!}\frac{f^{(j)}(\alpha)}{f'(\alpha)}$, $j = 1, 2, \ldots$. Then, its inverse can be expressed as:

$$\begin{aligned}
&f'(\alpha)^{-1}\Big(1 - [2\theta c_2 e_n + (2c_2^2 + 3\theta c_3 - 2\theta c_2^2 + 3\theta c_3)e_n^2 + (4\theta c_4 + (1-\theta)4c_2(c_3 - c_2^2))e_n^3 + \mathcal{O}(e_n^4)] + \\
&+ [2\theta c_2 e_n + (2c_2^2 + 3\theta c_3 - 2\theta c_2^2 + 3\theta c_3)e_n^2 + (4\theta c_4 + (1-\theta)4c_2(c_3 - c_2^2))e_n^3 + \mathcal{O}(e_n^4)]^2 - \cdots\Big) \\
&= f'(\alpha)^{-1}[1 - 2\theta c_2 e_n + (2\theta c_2^2 - 2c_2^2 + 4\theta^2 c_2^2 - 3\theta c_3)e_n^2 - (4\theta c_4 + (1-\theta)4c_2(c_3 - c_2^2))e_n^3 + \mathcal{O}(e_n^4)].
\end{aligned} \tag{20}$$

Now,

$$\frac{f(x_n)}{\theta f'(x_n) + (1-\theta)f'(z_n)} = e_n + c_2(1-2\theta)e_n^2 + (4\theta^2 c_2^2 - 2c_2^2 + c_3 - 3\theta c_3)e_n^3 + \mathcal{O}(e_n^4), \tag{21}$$

and by replacing it in (18), it leads to the MBN error equation as a function of θ:

$$e_{n+1} = -c_2(1-2\theta)e_n^2 - (4\theta^2 c_2^2 - 2c_2^2 + c_3 - 3\theta c_3)e_n^3 + \mathcal{O}(e_n^4) =: \Phi(\theta). \tag{22}$$

Equation (22) can be used to re-discover the results of convergence: for example, for the contraharmonic mean, we have:

$$\theta(f'(x_n), f'(z_n)) = \frac{f'(x_n)}{f'(x_n) + f'(z_n)}, \tag{23}$$

where:

$$f'(x_n) + f'(z_n) = 2f'(\alpha)[1 + c_2 e_n(c_2^2 - 3/2c_3)e_n^2 + 2(c_2 c_3 - c_2^3 + c_4)e_n^3 + \mathcal{O}(e_n^4)], \tag{24}$$

so that:

$$\frac{1}{f'(x_n) + f'(z_n)} = (2f'(\alpha))^{-1}[1 - c_2 e_n - 3/2 c_3 e_n^2 + 4 c_2^3 e_n^3 - 2 c_4 e_n^3 + c_2 c_3 e_n^3 + \mathcal{O}(e_n^4)]$$
$$= (2f'(\alpha))^{-1}[1 - c_2 e_n - 3/2 c_3 e_n^2 + (4 c_2^3 - 2 c_4 + c_2 c_3) e_n^3 + \mathcal{O}(e_n^4)]. \quad (25)$$

Thus, we can obtain the θ associated with the contraharmonic mean:

$$\theta(f'(x_n), f'(z_n)) = [1/2 + c_2 e_n + 3/2 c_3 e_n^2 + 2 c_4 e_n^3 + \mathcal{O}(e_n^4)] \cdot$$
$$\cdot [1 - c_2 e_n - 3/2 c_3 e_n^2 + (4 c_2^3 + c_2 c_3 - 2 c_4) e_n^3 + \mathcal{O}(e_n^4)]$$
$$= 1/2 + 1/2 c_2 e_n - c_2^2 e_n^2 + 3/4 c_3 e_n^2 + 2 c_2^3 e_n^3 + c_4 e_n^3 - 5/2 c_2 c_3 e_n^3 + \mathcal{O}(e_n^4) \quad (26)$$
$$= 1/2 + 1/2 c_2 e_n + (3/4 c_3 - c_2^2) e_n^2 + (2 c_2^3 + c_4 - 5/2 c_2 c_3) e_n^3 + \mathcal{O}(e_n^4).$$

Finally, by replacing the previous expression in (22):

$$e_{n+1} = (1/2 c_3 + 2 c_2^2) e_n^3 + \mathcal{O}(e_n^4), \quad (27)$$

and we obtain again that the convergence for the contraharmonic mean Newton method is cubic.

Regarding the harmonic mean, it is straightforward that it is a functional convex combination, with:

$$\theta(f'(x_n), f'(z_n)) = 1 - \frac{f'(x_n)}{f'(x_n) + f'(z_n)} \quad (28)$$
$$= 1/2 + 1/2 c_2 e_n + (c_2^2 - 3/4 c_3) e_n^2 + (5/2 c_2 c_3 - 2 c_2^3 - c_4) e_n^3 + \mathcal{O}(e_n^4).$$

Replacing this expression in (22), we find the cubic convergence of the harmonic mean Newton method,

$$e_{n+1} = 1/2 c_3 e_n^3 + \mathcal{O}(e_n^4). \quad (29)$$

In both cases, the independent term of $\theta(f'(x_n), f'(z_n))$ was $1/2$; it was not a coincidence, but an instance of the following more general result.

Theorem 1. *Let $\theta = \theta(f'(x_n), f'(z_n))$ be associated with the mean-based variant of Newton's method (MBN):*

$$x_{n+1} = x_n - \frac{f(x_n)}{M(f'(x_n), f'(z_n))}, \qquad z_n = x_n - \frac{f(x_n)}{f'(x_n)}, \quad (30)$$

where M is a mean function of the variables $f'(x_n)$ and $f'(z_n)$. Then, MBN converges, at least, cubically if and only if the estimate:

$$\theta = 1/2 + \mathcal{O}(e_n). \quad (31)$$

holds.

Proof. We replace $\theta = 1/2 + \mathcal{O}(e_n)$ in the MBN error Equation (22), obtaining:

$$e_{n+1} = (4\theta^2 c_2^2 - 2 c_2^2 + c_3 - 3\theta c_3) e_n^3 + \mathcal{O}(e_n^4). \quad \Box \quad (32)$$

Now, some considerations follow.

Remark 1. *Generally speaking,*

$$\theta = a_0 + a_1 e_n + a_2 e_n^2 + a_3 e_n^3 + \mathcal{O}(e_n^4), \tag{33}$$

where a_i are real numbers. If we put (33) in (22), we have:

$$e_{n+1} = -c_2(1 - 2a_0)e_n^2 - (4a_0^2 c_2^2 - 3a_0 c_3 - 2a_1 c_2 - 2c_2^2 + c_3)e_n^3 + \mathcal{O}(e_n^4); \tag{34}$$

it follows that, in order to attain cubic convergence, the coefficient of e_n^2 must be zero. Therefore, $a_0(u) = 1/2$. On the other hand, to achieve a higher order (i.e., at least four), we need to solve the following system:

$$\begin{cases} 1 - 2a_0 &= 0 \\ 4a_0^2 c_2^2 - 3a_0 c_3 - 2a_1 c_2 - 2c_2^2 + c_3 &= 0 \end{cases}. \tag{35}$$

This gives us that $a_0(u) = 1/2, a_1(u) = -1/4(2c_2^2 + c_3)/(c_2)$ assure at least a fourth-order convergence of the method. However, none of the MBN methods under analysis satisfy these conditions simultaneously.

Remark 2. *The only convex combination involving a constant θ that converges cubically is $\theta = 1/2$, i.e., the arithmetic mean.*

The most useful aspect of Theorem 1 is synthesized in the following corollary, which we call the "θ-test".

Corollary 1 (θ-test). *With the same hypothesis of Theorem 1, an MBN converges at least cubically if and only if the Taylor expansion of the mean holds:*

$$M(f'(x_n), f'(z_n)) = f'(\alpha)\left[1 + \frac{1}{2}c_2 e_n\right] + \mathcal{O}(e_n^2). \tag{36}$$

Let us notice that Corollary 1 provides a test to analyze the convergence of an MBN without having to find out the inherent θ, therefore sensibly reducing the overall complexity of the analysis.

Re-Proving Known Results for MBN

In this section, we apply Corollary 1 to prove the cubic convergence of known MBN via a convex combination approach.

(i) Arithmetic mean:

$$\begin{aligned} M_A(f'(x_n), f'(z_n)) &= \frac{f'(x_n) + f'(z_n)}{2} \\ &= \frac{1}{2}(f'(\alpha)[1 + 2c_2 e_n + \mathcal{O}(e_n^2)] + f'(\alpha)[1 + \mathcal{O}(e_n^2)]) \\ &= f'(\alpha)[1 + c_2 e_n + \mathcal{O}(e_n^2)]. \end{aligned} \tag{37}$$

(ii) Heronian mean: In this case, the associated θ-test is:

$$\begin{aligned} M_{He}f'(x_n), f'(z_n) &= \frac{1}{3}(f'(\alpha)[1 + 2c_2 e_n + \mathcal{O}(e_n^2)] + f'(\alpha)[1 + c_2 e_n + \mathcal{O}(e_n^2)] + f'(\alpha)[1 + \mathcal{O}(e_n^2)]) \\ &= \frac{f'(\alpha)}{3}[3 + 2c_2 e_n + c_2 e_n + \mathcal{O}(e_n^2)]. \end{aligned} \tag{38}$$

(iii) Generalized mean:

$$\begin{aligned}
M_G(f'(x_n), f'(z_n)) &= \left(\frac{f'(x_n)^m + f'(z_n)^m}{2}\right)^{1/m} \\
&= \left(\frac{f'(\alpha)^m[1 + 2c_2 e_n + \mathcal{O}(e_n^2)]^m + f'(\alpha)^m[1 + \mathcal{O}(e_n^2)]^m}{2}\right)^{1/m} \\
&= f'(\alpha)\left([1 + c_2 e_n + \mathcal{O}(e_n^2)]^m\right)^{1/m} \\
&= f'(\alpha)[1 + c_2 e_n + \mathcal{O}(e_n^2)].
\end{aligned} \qquad (39)$$

(iv) Centroidal mean:

$$\begin{aligned}
M_{Ce}(f'(x_n), f'(z_n)) &= \frac{2(f'(x_n)^2 + f'(x_n)f'(z_n) + f'(z_n))}{3(f'(x_n) + f'(z_n))} \\
&= \frac{2(f'(\alpha)^2[1 + 2c_2 e_n + \mathcal{O}(e_n^2)] + f'(\alpha)^2[2 + 4c_2 e_n + \mathcal{O}(e_n^2)])}{3(f'(\alpha)[2 + 2c_2 e_n + \mathcal{O}(e_n^2)])} \\
&= \frac{2(f'(\alpha)^2[3 + 6c_2 e_n + \mathcal{O}(e_n^2)])}{3(f'(\alpha)[2 + 2c_2 e_n + \mathcal{O}(e_n^2)])} \\
&= f'(\alpha)[1 + 2c_2 e_n + \mathcal{O}(e_n^2)][1 + c_2 e_n + \mathcal{O}(e_n^2)] \\
&= f'(\alpha)[1 + c_2 e_n + \mathcal{O}(e_n^2)].
\end{aligned} \qquad (40)$$

3. New MBN by Using the Lehmer Mean and Its Generalization

The iterative expression of the scheme based on the Lehmer mean of order $m \in \mathbb{R}$ is:

$$x_{n+1} = x_n - \frac{f(x_n)}{M_{L_m}(f'(x_n), f'(z_n))},$$

where $z_n = x_n - \frac{f(x_n)}{f'(x_n)}$ and:

$$M_{L_m}(f'(x_n), f'(z_n)) = \frac{f'(x_n)^m + f'(z_n)^m}{f'(x_n)^{m-1} + f'(z_n)^{m-1}}. \qquad (41)$$

Indeed, there are suitable values of parameter p such that the associated Lehmer mean equals the arithmetic one and the geometric one, but also the harmonic and the contraharmonic ones. In what follows, we will find it again, this time in a more general context.

By analyzing the associated θ-test, we conclude that the iterative scheme designed with this mean has order of convergence three.

$$\begin{aligned}
M_{L_m}(f'(x_n), f'(z_n)) &= \frac{f'(x_n)^m + f'(z_n)^m}{f'(x_n)^{m-1} + f'(z_n)^{m-1}} \\
&= \frac{f'(\alpha)^m[1 + 2c_2 e_n + \mathcal{O}(e_n^2)]^m + f'(\alpha)^m[1 + \mathcal{O}(e_n^2)]^m}{f'(\alpha)^{m-1}[1 + 2c_2 e_n + \mathcal{O}(e_n^2)]^{m-1} + f'(\alpha)^{m-1}[1 + \mathcal{O}(e_n^2)]^{m-1}} \\
&= f'(\alpha)[1 + mc_2 e_n + \mathcal{O}(e_n^2)] \cdot [1 - ((m-1)c_2 e_n + \mathcal{O}(e_n^2)) + ((m-1)c_2 e_n + \mathcal{O}(e_n^2))^2 + \ldots] \\
&= f'(\alpha)[1 + mc_2 e_n + \mathcal{O}(e_n^2)] \cdot [1 - (m-1)c_2 e_n + \mathcal{O}(e_n^2)] \\
&= f'(\alpha)[1 + c_2 e_n + \mathcal{O}(e_n^2)].
\end{aligned} \qquad (42)$$

σ-Means

Now, we propose a new family of means of n variables, starting again from convex combinations. The core idea in this work is that, in the end, two distinct means only differ in their corresponding weights θ and $1 - \theta$. In particular, we can regard the harmonic mean as an "opposite-weighted" mean, while the contraharmonic one is a "self-weighted" mean.

This behavior can be generalized to n variables:

$$M_{CH}(x_1,\ldots,x_n) = \frac{\sum_{i=1}^{n} x_i^2}{\sum_{i=1}^{n} x_i} \qquad (43)$$

is the contraharmonic mean among n numbers. Equation (43) is just a particular case of what we call σ-mean.

Definition 1 (σ-mean). *Given $x = (x_1,\ldots,x_n) \in \mathbb{R}^n$ a vector of n real numbers and a bijective map $\sigma\colon \{1,\ldots,n\} \to \{1,\ldots,n\}$ (i.e., $\sigma(x)$ is a permutation of $x_1,\ldots,x_n$), we call the σ-mean of order $m \in \mathbb{R}$ the real number given by:*

$$M_\sigma(x_1,\ldots,x_n) := \frac{\sum_{i=1}^{n} x_i \cdot x_{\sigma(i)}^m}{\sum_{j=1}^{n} x_j^m}. \qquad (44)$$

Indeed, it is easy to see that, in an σ-mean, the weight assigned to each node x_i is:

$$\frac{x_{\sigma(i)}^m}{\sum_{j=1}^{n} x_{\sigma(j)}^m} = \frac{x_{\sigma(i)}^m}{\sum_{j=1}^{n} x_j^m} \in [0,1], \qquad (45)$$

where the equality holds because σ is a permutation of the indices. We are, therefore, still dealing with a convex combination, which implies that Definition 1 is well posed.

We remark that if we take $\sigma = \mathbb{1}$, i.e., the identical permutation, in (44), we find the Lehmer mean of order m. Actually, the Lehmer mean is a very special case of the σ-mean, as the following result proves.

Proposition 1. *Given $m \in \mathbb{R}$, the Lehmer mean of order m is the maximum σ-mean of order m.*

Proof. We recall the rearrangement inequality:

$$x_n y_1 + \cdots + x_1 y_n \leq x_{\sigma(1)} y_1 + \cdots + x_{\sigma(n)} y_n \leq x_1 y_1 + \cdots + x_n y_n, \qquad (46)$$

which holds for every choice of $x_1,\ldots,x_n$ and $y_1,\ldots,y_n$ regardless of the signs, assuming that both x_i and y_j are sorted in increasing order. In particular, $x_1 < x_2 < \cdots < x_n$ and $y_1 < y_2 < \cdots < y_n$ imply that the upper bound is attained only for the identical permutation.

Then, to prove the result, it is enough to replace every y_i with the corresponding weight defined in (45). □

The Lehmer mean and σ-mean are deeply related: if $n = 2$, as is the case of MBN, there are only two possible permutations, the identical one and the one that swaps one and two. We have already observed that the identical permutation leads to the Lehmer mean; however, if we express σ in standard cycle notation as $\bar{\sigma} = (1, 2)$, we have that:

$$M_{\bar{\sigma}}(x_1, x_2) = \frac{x_1 x_2 (x_1^m + x_2^m)}{x_1^{m+1} + x_2^{m+1}} = \frac{x_1^{-m} + x_2^{-m}}{x_1^{-m-1} + x_2^{-m-1}} = M_{L_{-m}}(x_1, x_2). \tag{47}$$

We conclude this section proving another property of σ-means, which is that the arithmetic mean of all possible σ-means of n numbers equals the arithmetic mean of the numbers themselves.

Proposition 2. *Given n real numbers $x_1, \ldots, x_n$ and Σ_n denoting the set of all possible permutations of $\{1 \ldots, n\}$, we have:*

$$\frac{1}{n!} \sum_{\sigma \in \Sigma_n} M_\sigma(x_1, \ldots, x_n) = \frac{1}{n} \sum_{i=1}^{n} x_i \tag{48}$$

for all $m \in \mathbb{R}$.

Proof. Let us rewrite Equation (48); by definition, we have:

$$\frac{1}{n!} \sum_{\sigma \in \Sigma_n} M_\sigma(x_1, \ldots, x_n) = \frac{1}{n!} \sum_{\sigma \in \Sigma_n} \left(\frac{\sum_{i=1}^{n} x_i x_{\sigma(i)}^m}{\sum_{j=1}^{n} x_j^m} \right) = \frac{1}{n} \sum_{i=1}^{n} x_i \tag{49}$$

and we claim that the last equality holds. Indeed, we notice that every term in the sum of the σ-means on the left side of the last equality involves a constant denominator, so we can multiply both sides by it and also by $n!$ to get:

$$\sum_{\sigma \in \Sigma_n} \left(\sum_{i=1}^{n} x_i x_{\sigma(i)}^m \right) = (n-1)! \left(\sum_{j=1}^{n} x_j^m \right) \left(\sum_{i=1}^{n} x_i \right). \tag{50}$$

Now, it is just a matter of distributing the product on the right in a careful way:

$$(n-1)! \left(\sum_{j=1}^{n} x_j^m \right) \left(\sum_{i=1}^{n} x_i \right) = \sum_{i=1}^{n} \left(x_i \cdot \sum_{k=1}^{n} ((n-1)!) x_k^m \right), \tag{51}$$

If we fix $i \in \{1, \ldots, n\}$, in Σ_n, there are exactly $(n-1)!$ permutations σ such that $\sigma(i) = i$. Therefore, the equality in (50) follows straightforwardly. □

4. Numerical Results and Dependence on Initial Estimations

Now, we present the results of some numerical computations, in which the following test functions have been used.

(a) $f_1(x) = x^3 + 4x^2 - 10$,
(b) $f_2(x) = \sin(x)^2 - x^2 + 1$,
(c) $f_3(x) = x^2 - e^x - 3x + 2$,
(d) $f_4(x) = \cos(x) - x$,
(e) $f_5(x) = (x-1)^3 - 1$.

The numerical tests were carried out by using MATLAB with double precision arithmetics in a computer with processor i7-8750H @2.20 GHz, 16 Gb of RAM, and the stopping criterion used was $|x_{n+1} - x_n| + |f(x_{n+1})| < 10^{-14}$.

We used the harmonic mean Newton method (HN), the contraharmonic mean Newton method (CHN), the Lehmer mean Newton method (LN(m)), a variant of Newton's method where the mean is a convex combination with $\theta = 1/3, 1/3N$, and the classic Newton method (CN). The main goals of these calculations are to confirm the theoretical results stated in the preceding sections and to compare the different methods, with CN as a control benchmark. In Table 1, we show the number of iterations that each method needs for satisfying the stopping criterion and also the approximated computational order of convergence, defined in [11], with the expression:

$$ACOC = \frac{\ln\left(|x_{n+1} - x_n|/|x_n - x_{n-1}|\right)}{\ln\left(|x_n - x_{n-1}|/|x_{n-1} - x_{n-2}|\right)}, \quad n = 2, 3, \ldots,$$

which is considered as a numerical approximation of the theoretical order of convergence p.

Table 1. Numerical results. HN, the harmonic mean Newton method; CHN, the contraharmonic mean Newton method; LN, the Lehmer–Newton method; CN, the classic Newton method.

Function	x_0	Number of Iterations					ACOC				
		HN	CHN	LN(−7)	1/3 N	CN	HN	CHN	LN(−7)	1/3 N	CN
(a)	−0.5	50	18	55	6	132	3.10	3.03	2.97	1.99	2.00
	1	4	5	5	5	6	2.94	3.01	2.96	2.02	2.00
	2	4	5	5	5	6	3.10	2.99	3.02	2.00	2.00
(b)	1	4	5	6	6	7	3.06	3.16	3.01	2.01	2.00
	3	4	5	7	6	7	3.01	2.95	3.02	2.01	2.00
(c)	2	5	5	5	5	6	3.01	2.99	3.11	2.01	2.00
	3	5	6	5	6	7	3.10	3.00	3.10	2.01	2.00
(d)	−0.3	5	5	5	6	6	2.99	3.14	3.02	2.01	1.99
	1	4	4	4	5	5	2.99	2.87	2.88	2.01	2.00
	1.7	4	4	5	5	5	3.00	2.72	3.02	2.01	1.99
(e)	0	6	>1000	7	7	10	3.06	3.00	3.02	2.01	2.00
	1.5	5	7	7	7	8	3.04	3.01	2.99	2.01	2.00
	2.5	4	5	5	5	7	3.07	2.96	3.01	1.99	2.00
	3.0	5	6	6	6	7	3.04	2.99	2.98	2.00	2.00
	3.5	5	6	6	6	8	3.07	2.95	2.99	2.00	2.00

Regarding the efficiency of the MBN, we used the efficiency index defined by Ostrowski in [12] as $EI = p^{\frac{1}{d}}$, where p is the order of convergence of the method and d is the number of functional evaluations per iteration. In this sense, all the MBN had the same $EI_{MBN} = 3^{\frac{1}{3}}$; meanwhile, Newton's scheme had the index $EI_{CN} = 2^{\frac{1}{2}}$. Therefore, all MBN were more efficient than the classical Newton method.

The presented numerical tests showed the performance of the different iterative methods to solve specific problems with fixed initial estimations and a stringent stopping criterion. However, it is useful to know their dependence on the initial estimation used. Although the convergence of the methods has been proven for real functions, it is usual to analyze the sets of convergent initial guesses in the complex plane (the proof would be analogous by changing the condition on the function to be differentiable by being holomorphic). To get this aim, we plotted the dynamical planes of each one of the iterative methods on the nonlinear functions $f_i(x), i = 1, 2, \ldots, 5$, used in the numerical tests. In them, a mesh of 400×400 initial estimations was employed in the region of the complex plane $[−3, 3] \times [−3, 3]$.

We used the routines appearing in [13] to plot the dynamical planes corresponding to each method. In them, each point of the mesh was an initial estimation for the analyzed method on the specific problem. If the method reached the root in less than 40 iterations (closer than $10^{−3}$), then this point is painted in orange (green for the second, etc.) color; if the process converges to another attractor different from the roots, then the point is painted in black. The zeros of the nonlinear functions are presented in the different pictures by white stars.

In Figure 1, we observe that Harmonic and Lehmer (for $m = -7$) means showed the most stable performance, whose unique basins of attraction were those of the roots (plotted in orange, red, and green). In the rest of the cases, there existed black areas of no convergence to the zeros of the nonlinear function $f_1(x)$. Specially unstable were the cases of Heronian, convex combination ($\theta = \pm 2$), and generalized means, with wide black areas and very small basins of the complex roots.

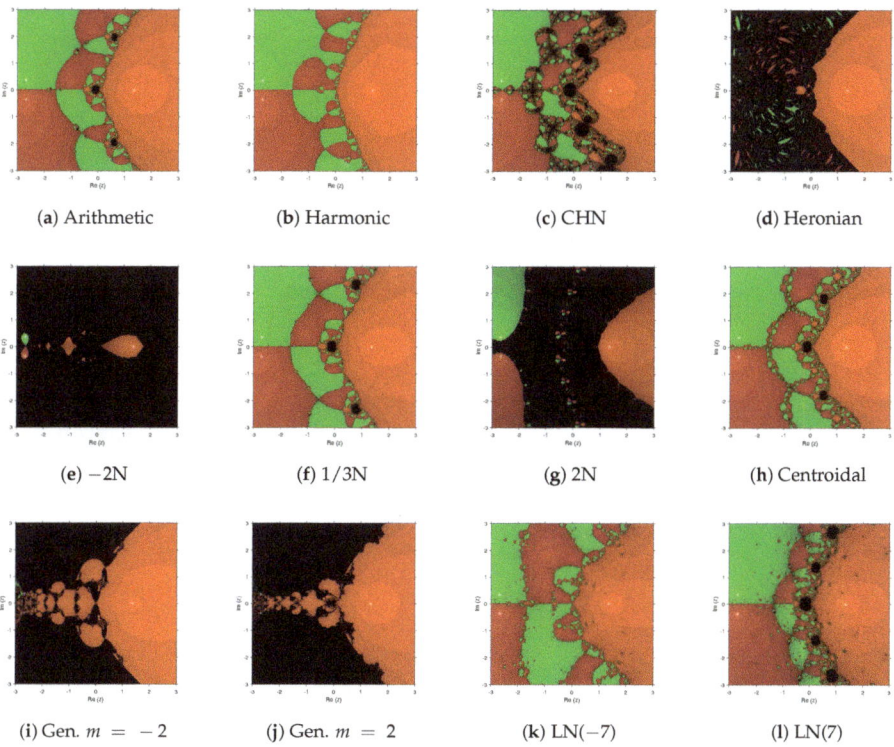

Figure 1. Dynamical planes of the mean-based methods on $f_1(x) = x^3 + 4x^2 - 10$.

Regarding Figure 2, again Heronian, convex combination ($\theta = -2$), and generalized means showed convergence only to one of the roots or very narrow basins of attraction. There existed black areas of no convergence to the roots in all cases, but the widest green and orange basins (corresponding to the zeros of $f_2(x)$) corresponded to harmonic, contra harmonic, centroidal, and Lehmer means.

Function $f_3(x)$ had only one zero at $x \approx 0.25753$, whose basin of attraction is painted in orange color in Figure 3. In general, most of the methods presented good performance; however, three methods did not converge to the root in the maximum of iterations required: Heronian and generalized means with $m = \pm 2$. Moreover, the basin of attraction was reduced when the parameter θ of the convex combination mean was used.

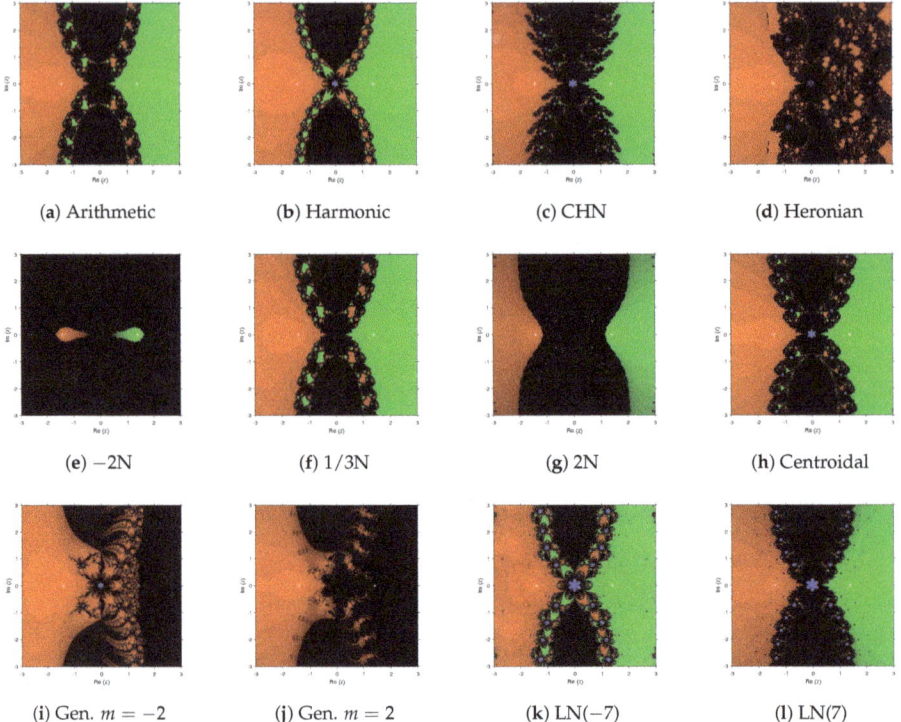

Figure 2. Dynamical planes of mean-based methods on $f_2(x) = sin(x)^2 - x^2 + 1$.

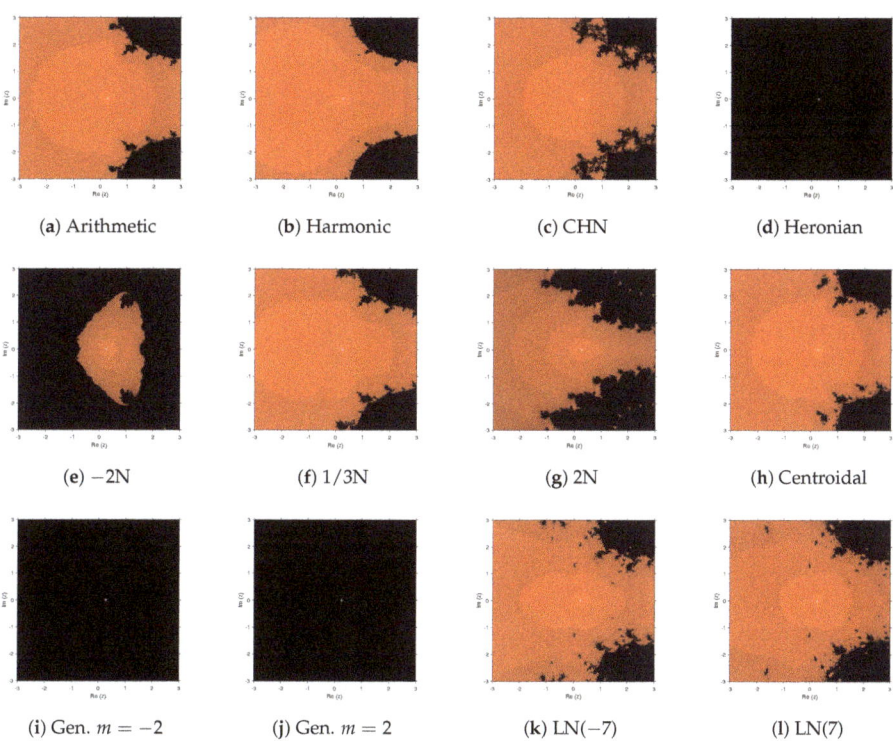

Figure 3. Dynamical planes of mean-based methods on $f_3(x) = x^2 - e^x - 3x + 2$.

A similar performance is observed in Figure 4, where Heronian and generalized means with $m = \pm 2$ showed no convergence to only the root of $f_4(x)$; meanwhile, the rest of the methods presented good behavior. Let us remark that in some cases, blue areas appear; this corresponded to initial estimations that, after 40 consecutive iterations, had an absolute value higher than 1000. In these cases, they and the surrounding black areas were identified as regions of divergence of the method. The best methods in this case were associated with the arithmetic and harmonic means.

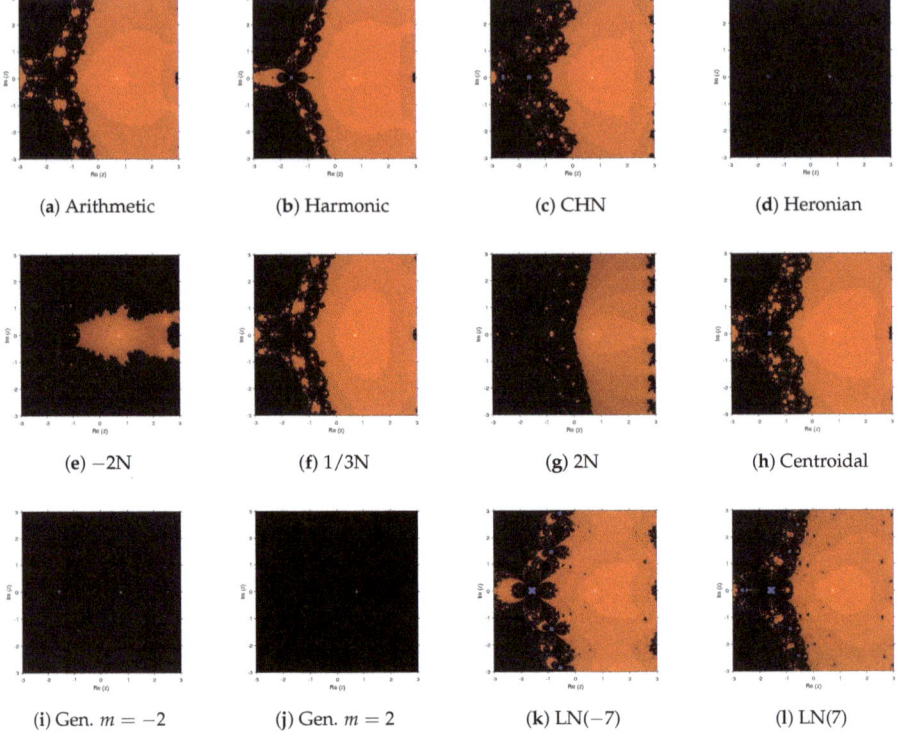

Figure 4. Dynamical planes of mean-based methods on $f_4(x) = cos(x) - x$.

In Figure 5, the best results in terms of the wideness of the basins of the attraction of the roots were for harmonic and Lehmer means, for $m = -7$. The biggest black areas corresponded to convex combination with $\theta = -2$, where the three basins of attraction of the roots were very narrow, and for Heronian and generalized means, there was only convergence to the real root.

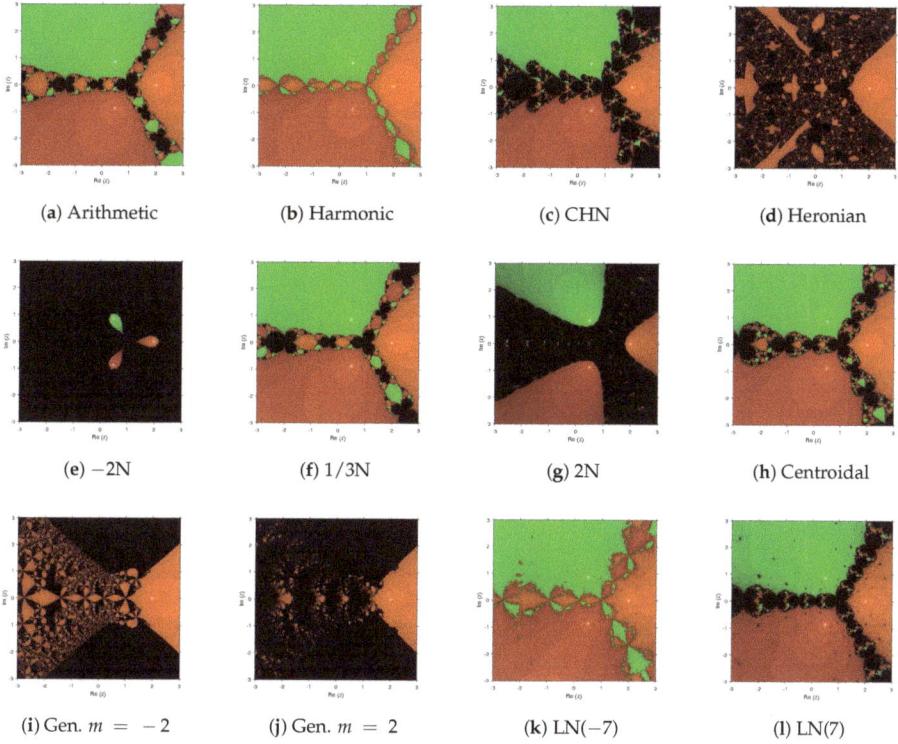

Figure 5. Dynamical planes of mean-based methods on $f_5(x) = (x-1)^3 - 1$.

5. Conclusions

The proposed θ-test (Corollary 1) has proven to be very useful to reduce the calculations of the analysis of convergence of any MBN. Moreover, though the employment of σ-means in the context of mean-based variants of Newton's method is probably not the best one to appreciate their flexibility, their use could still lead to interesting results due to their much greater capability of interpolating between numbers than already powerful means, such as the Lehmer one.

With regard to the numerical performance, Table 1 confirms that a convex combination with a constant coefficient could converge cubically if and only if it was the arithmetic mean; otherwise, as with this case, it converged quadratically, even if it may have done so with less iterations, generally speaking, than CN. Regarding the number of iterations, there were non-linear functions for which LN(m) converged with fewer iterations than HN. In our calculations, we set $m = -7$, but similar results were achieved also for different parameters. Regarding the dependence on initial estimations, the harmonic and Lehmer methods were proven to be very stable, with the widest areas of convergence in most of the nonlinear problems used in the tests.

Author Contributions: The individual contributions of the authors are as follows: conceptualization, J.R.T.; writing, original draft preparation, J.F. and A.C.Z.; validation, A.C. and J.R.T. formal analysis, A.C.; numerical experiments, J.F. and A.C.Z.

Funding: This research was partially funded by Spanish Ministerio de Ciencia, Innovación y Universidades PGC2018-095896-B-C22 and by Generalitat Valenciana PROMETEO/2016/089 (Spain).

Acknowledgments: The authors would like to thank the anonymous reviewers for their comments and suggestions, which improved the final version of this manuscript.

Conflicts of Interest: The authors declare that there is no conflict of interest regarding the publication of this paper.

References

1. Traub, J.F. *Iterative Methods for the Solution of Equations*; Prentice-Hall: Englewood Cliffs, NY, USA, 1964.
2. Petković, M.S.; Neta, B.; Petković, L.D.; Džunić, J. *Multipoint Methods for Solving Nonlinear Equations*; Academic Press: Cambridge, MA, USA, 2013.
3. Amat, S.; Busquier, S. *Advances in Iterative Methods for Nonlinear Equations*; SEMA SIMAI Springer Series; Springer: Berlin, Germany, 2016; Volume 10.
4. Weerakoon, S.; Fernando, T.A. A variant of Newton's method with accelerated third-order convergence. *Appl. Math. Lett.* **2000**, *13*, 87–93. [CrossRef]
5. Özban, A. Some new variants of Newton's method. *Appl. Math. Lett.* **2004**, *17*, 677–682. [CrossRef]
6. Ababneh, O.Y. New Newton's method with third order convergence for solving nonlinear equations. *World Acad. Sci. Eng. Technol.* **2012**, *61*, 1071–1073.
7. Xiaojian, Z. A class of Newton's methods with third-order convergence. *Appl. Math. Lett.* **2007**, *20*, 1026–1030.
8. Singh, M.K.; Singh, A.K. A new-mean type variant of Newton's method for simple and multiple roots. *Int. J. Math. Trends Technol.* **2017**, *49*, 174–177. [CrossRef]
9. Verma, K.L. On the centroidal mean Newton's method for simple and multiple roots of nonlinear equations. *Int. J. Comput. Sci. Math.* **2016**, *7*, 126–143. [CrossRef]
10. Zafar, F.; Mir, N.A. A generalized family of quadrature based iterative methods. *Gen. Math.* **2010**, *18*, 43–51.
11. Cordero, A; Torregrosa, J.R. Variants of Newton's method using fifth-order quadrature formulas. *Appl. Math. Comput.* **2007**, *190*, 686–698. [CrossRef]
12. Ostrowski, A.M. *Solution of Equations and Systems of Equatiuons*; Academic Press: New York, NY, USA, 1966.
13. Chicharro, F.I.; Cordero, A.; Torregrosa, J.R. Drawing dynamical and pa- rameters planes of iterative families and methods. *Sci. World J.* **2013**, *2013*, 780153. [CrossRef] [PubMed]

© 2019 by the authors. Licensee MDPI, Basel, Switzerland. This article is an open access article distributed under the terms and conditions of the Creative Commons Attribution (CC BY) license (http://creativecommons.org/licenses/by/4.0/).

Article

A Variant of Chebyshev's Method with 3αth-Order of Convergence by Using Fractional Derivatives

Alicia Cordero [1,*], Ivan Girona [2] and Juan R. Torregrosa [1]

[1] Institute of Multidisciplinary Mathematics, Universitat Politècnica de València, Camino de Vera, s/n, 46022 Valencia, Spain
[2] Facultat de Matemàtiques, Universitat de València, 46010 València, Spain
* Correspondence: acordero@mat.upv.es

Received: 19 July 2019; Accepted: 3 August 2019; Published: 6 August 2019

Abstract: In this manuscript, we propose several iterative methods for solving nonlinear equations whose common origin is the classical Chebyshev's method, using fractional derivatives in their iterative expressions. Due to the symmetric duality of left and right derivatives, we work with right-hand side Caputo and Riemann–Liouville fractional derivatives. To increase as much as possible the order of convergence of the iterative scheme, some improvements are made, resulting in one of them being of 3α-th order. Some numerical examples are provided, along with an study of the dependence on initial estimations on several test problems. This results in a robust performance for values of α close to one and almost any initial estimation.

Keywords: nonlinear equations; Chebyshev's iterative method; fractional derivative; basin of attraction

1. Introduction

The concept of fractional calculus was introduced simultaneously with the development of classical one. The first references date back to 1695, the year in which Leibniz and L'Hospital came up with the concept of semi-derivative. Other researchers of the time were also interested in this idea, such as Riemann, Liouville, or Euler.

Since their early development in the XIX-th century until nowadays, fractional calculus has evolved from theoretical aspects to the appearance in many real world applications: mechanical engineering, medicine, economy, and others. They are frequently modeled by differential equations with derivatives of fractional order (see, for example [1–4] and the references therein).

Nowadays, fractional calculus has numerous applications in science and engineering. The fundamental reason for this is the greater degree of freedom of fractional calculation tools compared to classical calculation ones. This makes it the most suitable procedure for modeling problems whose hereditary properties must be preserved. In this sense, one of the most significant tools of fractional calculation is the fractional (integral) derivative.

Many times, these kinds of problems are related with systems of equations, that can be nonlinear if it is the differential equation. So, it is not strange to adapt iterative techniques for solving nonlinear equations by means of fractional derivatives of different orders, and see which is the resulting effect on the convergence. This has been studied in some previous works by Brambila et al. [5] holding the original expression of Newton's iterative method and without proving the order of convergence. In [6], a fractional Newton's method was deduced to achieve 2α-th order of convergence and showing good numerical properties. However, it is known (see for example the text of Traub [7]) that in point-to-point methods to increase the order of the iterative methods implies to add functional evaluations of higher-order derivatives of the nonlinear function. Our starting question is: how affects this higher-order derivative when it is replaced by a fractional one to the global order of convergence of the iterative method?

The aim of this work is to use the Chebyshev's method with fractional derivative to solve $f(x) = 0$, $f : D \subseteq \mathbb{R} \to \mathbb{R}$. Let us consider $\bar{x} \in \mathbb{R}$ as the solution of the equation $f(x) = 0$, such that $f'(\bar{x}) \neq 0$. First of all, we remind the standard Chebyshev's method:

$$x_{k+1} = x_k - \left(1 + \frac{1}{2} L_f(x_k)\right) \frac{f(x_k)}{f'(x_k)}, \quad k = 0, 1, \ldots \tag{1}$$

being $L_f(x_k) = \dfrac{f(x_k) f''(x_k)}{f'(x_k)^2}$, known as logarithmic convexity degree. Then, we will change the first and second order integer derivatives by the notion of fractional derivative and see if is the order of convergence of the original method is held.

Now, we set up some definitions, properties and results that will be helpful in this work (for more information, see [8] and the references therein).

Definition 1. *The gamma function is defined as*

$$\Gamma(x) = \int_0^{+\infty} u^{x-1} e^{-u} du,$$

whenever $x > 0$.

The gamma function is known as a generalization of the factorial function, due to $\Gamma(1) = 1$ and $\Gamma(n+1) = n!$, when $n \in \mathbb{N}$. Let us now see the notion of fractional Riemann-Liouville and Caputo derivatives.

Definition 2. *Let $f : \mathbb{R} \to \mathbb{R}$ be an element of $L^1([a,x])$ ($-\infty < a < x < +\infty$), with $\alpha \geq 0$ and $n = [\alpha] + 1$, being $[\alpha]$ the integer part of α. Then, the Riemann–Liouville fractional derivative of order α of $f(x)$ is defined as follows:*

$$\left(D_{a^+}^\alpha\right) f(x) = \begin{cases} \dfrac{1}{\Gamma(n-\alpha)} \dfrac{d^n}{dx^n} \displaystyle\int_a^x \dfrac{f(t)}{(x-t)^{\alpha-n+1}} dt, & \alpha \notin \mathbb{N}, \\ \dfrac{d^{n-1} f(x)}{dx^{n-1}}, & \alpha = n - 1 \in \mathbb{N} \cup \{0\}. \end{cases} \tag{2}$$

Let us remark that definition (2) is consistent if the integral of the first identity in (2) is n-times derivable or, in another case, f is $(n-1)$-times derivable.

Definition 3. *Let $f : \mathbb{R} \to \mathbb{R}$ be an element of $C^{+\infty}([a,x])$ ($-\infty < a < x < +\infty$), $\alpha \geq 0$ and $n = [\alpha] + 1$. Thus, the Caputo fractional derivative of order α of $f(x)$ is defined as follows:*

$$\left({}^C D_a^\alpha\right) f(x) = \begin{cases} \dfrac{1}{\Gamma(n-\alpha)} \displaystyle\int_a^x \dfrac{d^n f(t)}{dt^n} \dfrac{dt}{(x-t)^{\alpha-n+1}}, & \alpha \notin \mathbb{N}, \\ \dfrac{d^{n-1} f(x)}{dx^{n-1}}, & \alpha = n - 1 \in \mathbb{N} \cup \{0\}. \end{cases} \tag{3}$$

In [9], Caputo and Torres generated a duality theory for left and right fractional derivatives, called symmetric duality, using it to relate left and right fractional integrals and left and right fractional Riemann–Liouville and Caputo derivatives.

The following result will be useful to prove Theorem 4.

Theorem 1 ([8], Proposition 26). *Let $\alpha \geq 0$, $n = [\alpha] + 1$, and $\beta \in \mathbb{R}$. Thus, the following identity holds:*

$$D_{a+}^{\alpha}(x-a)^{\beta} = \frac{\Gamma(\beta+1)}{\Gamma(\beta+1-\alpha)}(x-a)^{\beta-\alpha}. \tag{4}$$

The following result shows a relationship between Caputo and Riemann–Liouville fractional derivatives.

Theorem 2 ([8], Proposition 31). *Let $\alpha \notin \mathbb{N}$ such that $\alpha \geq 0$, $n = [\alpha] + 1$ and $f \in L^1([a,b])$ a function whose Caputo and Riemann-Liouville fractional derivatives exist. Thus, the following identity holds:*

$$^C D_a^{\alpha} f(x) = D_{a+}^{\alpha} f(x) - \sum_{k=0}^{n-1} \frac{f^{(k)}(a)}{\Gamma(k+1-\alpha)} (x-a)^{k-\alpha}, \quad x > a. \tag{5}$$

As a consequence of the two previous results, we obtain that

$$^C D_{x_0}^{\alpha} (x-x_0)^k = D_{x_0}^{\alpha} (x-x_0)^k, \quad k = 1, 2, \ldots$$

Remark 1. *In what follows, due to previous results and consequences, we work with Caputo fractional derivative, since all our conclusions are also valid for Riemann–Liouville fractional derivative at the same extent.*

The next result shows a generalization of the classical Taylor's theorem by using derivatives of fractional order.

Theorem 3 ([10], Theorem 3). *Let us assume that $^C D_a^{j\alpha} g(x) \in C([a,b])$, for $j = 1, 2, \ldots, n+1$, where $0 < \alpha \leq 1$. Then, we have*

$$g(x) = \sum_{i=0}^{n} {}^C D_a^{i\alpha} g(a) \frac{(x-a)^{i\alpha}}{\Gamma(i\alpha+1)} + {}^C D^{(n+1)\alpha} g(\xi) \frac{(x-a)^{(n+1)\alpha}}{\Gamma((n+1)\alpha+1)} \tag{6}$$

being $a \leq \xi \leq x$, for all $x \in (a,b]$, where $^C D_a^{n\alpha} = {}^C D_a^{\alpha} \cdot \ldots \cdot {}^C D_a^{\alpha}$ (n-times).

When the assumptions of Theorem 3 are satisfied, the Taylor development of $f(x)$ around $\bar{x}$, by using Caputo-fractional derivatives, is

$$f(x) = \frac{^C D_{\bar{x}}^{\alpha} f(\bar{x})}{\Gamma(\alpha+1)} \left[(x-\bar{x})^{\alpha} + C_2(x-\bar{x})^{2\alpha} + C_3(x-\bar{x})^{3\alpha} \right] + \mathcal{O}((x-\bar{x})^{4\alpha}), \tag{7}$$

being $C_j = \frac{\Gamma(\alpha+1)}{\Gamma(j\alpha+1)} \frac{{}^C D_{\bar{x}}^{j\alpha} f(\bar{x})}{{}^C D_{\bar{x}}^{\alpha} f(\bar{x})}$, for $j \geq 2$.

The rest of the manuscript is organized as follows: Section 2 deals with the design of high-order one-point fractional iterative methods and their analysis of convergence. In Section 3, some numerical tests are made in order to check the theoretical results and we show the corresponding convergence planes in order to study the dependence on the initial estimations of the proposed schemes. Finally, some conclusions are stated.

2. Proposed Methods and Their Convergence Analysis

In order to extend Chebyshev's iterative method to fractional calculus, let us define

$$^C L_f^{\alpha}(x) = \frac{f(x) \, {}^C D_a^{2\alpha} f(x)}{(D_a^{\alpha} f(x))^2},$$

that we call fractional logarithmic convexity degree of Caputo-type.

Then, an iterative Chebyshev-type method using Caputo derivatives can be constructed. The following result show the convergence conditions of this new method.derivative with 2α.

Theorem 4. *Let $f : D \subset \mathbb{R} \longrightarrow \mathbb{R}$ be a continuous function with k-order fractional derivatives, $k \in \mathbb{N}$ and any α, $0 < \alpha \leq 1$, in the interval D. If x_0 is close enough to the zero $\bar{x}$ of $f(x)$ and ${}^C D_{\bar{x}}^\alpha f(x)$ is continuous and non zero in $\bar{x}$, then the local order of convergence of the Chebyshev's fractional method using Caputo-type derivatives*

$$x_{k+1} = x_k - \Gamma(\alpha+1)\left(1 + \frac{1}{2}{}^C L_f^\alpha(x_k)\right) \frac{f(x_k)}{{}^C D_a^\alpha f(x_k)}, \qquad (8)$$

that we denote by CFC1, is at least 2α and the error equation is

$$e_{k+1}^\alpha = C_2 \left(\frac{2\Gamma^2(\alpha+1) - \Gamma(2\alpha+1)}{2\Gamma^2(\alpha+1)}\right) e_k^{2\alpha} + \mathcal{O}(e_k^{3\alpha}), \qquad (9)$$

being $e_k = x_k - \bar{x}$.

Proof. According to Theorems 1 and 3, we get that the Taylor expansion in fractional derivatives of ${}^C D^\alpha f(x_k)$ around $\bar{x}$ is

$$^C D^\alpha f(x_k) = \frac{{}^C D_{\bar{x}}^\alpha f(x_0)}{\Gamma(\alpha+1)} \left[\Gamma(\alpha+1) + \frac{\Gamma(2\alpha+1)}{\Gamma(\alpha+1)} C_2 e_k^\alpha + \frac{\Gamma(3\alpha+1)}{\Gamma(2\alpha+1)} C_3 e_k^{2\alpha}\right] + \mathcal{O}(e_k^{3\alpha}), \qquad (10)$$

where $C_j = \dfrac{\Gamma(\alpha+1)}{\Gamma(\alpha j+1)} \dfrac{{}^C D_{\bar{x}}^{j\alpha} f(\bar{x})}{{}^C D_{\bar{x}}^\alpha f(\bar{x})}$, for $j \geq 2$.

Then,

$$\frac{f(x_k)}{{}^C D_a^\alpha f(x_k)} = \frac{1}{\Gamma(\alpha+1)} e_k^\alpha + \frac{(\Gamma^2(\alpha+1)) - \Gamma(2\alpha+1)}{(\Gamma^3(\alpha+1))} C_2 e_k^{2\alpha} + \mathcal{O}(e_k^{3\alpha}). \qquad (11)$$

On the other hand, it is clear, by identity (10), that

$$^C D_{\bar{x}}^{2\alpha} f(x_k) = \frac{{}^C D_{\bar{x}}^\alpha f(\bar{x})}{\Gamma(\alpha+1)} \left[\Gamma(2\alpha+1) C_2 + \frac{\Gamma(3\alpha+1)}{\Gamma(\alpha+1)} C_3 e_k^\alpha\right] + O(e_k^{2\alpha}). \qquad (12)$$

Therefore,

$$f(x_k){}^C D_{\bar{x}}^{2\alpha} f(x_k) = \left(\frac{{}^C D_{\bar{x}}^\alpha f(\bar{x})}{\Gamma(\alpha+1)}\right)^2 \left[\Gamma(2\alpha+1) C_2 e_k^\alpha + \left(C_2^2 \Gamma(2\alpha+1) + \frac{\Gamma(3\alpha+1)}{\Gamma(\alpha+1)} C_3\right) e_k^{2\alpha}\right] + \mathcal{O}(e_k^{3\alpha})$$

and

$$\left({}^C D_a^\alpha f(x_k)\right)^2 = \left(\frac{({}^C D_{\bar{x}}^\alpha f)(\bar{x})}{\Gamma(\alpha+1)}\right)^2 \left[(\Gamma(\alpha+1))^2 + 2\Gamma(2\alpha+1) C_2 e_k^\alpha \right.$$
$$\left. + \left(\frac{(\Gamma(2\alpha+1))^2}{(\Gamma(\alpha+1))^2} C_2^2 + 2\frac{\Gamma(\alpha+1)\Gamma(3\alpha+1)}{\Gamma(2\alpha+1)} C_3\right) e_k^{2\alpha} \right.$$
$$\left. + 2\frac{\Gamma(2\alpha+1)\Gamma(3\alpha+1)}{\Gamma(\alpha+1)\Gamma(2\alpha+1)} C_2 C_3 e_k^{3\alpha}\right] + \mathcal{O}(e_k^{4\alpha}).$$

Let us now calculate the Taylor expansion of ${}^C L_f^\alpha(x_k)$ around $\bar{x}$,

$$^C L_f^\alpha(x_k) = \frac{\Gamma(2\alpha+1)}{\Gamma(\alpha+1)^2} C_2 e_k^\alpha$$
$$+ \frac{1}{\Gamma(\alpha+1)^2}\left[C_2^2 \Gamma(2\alpha+1) + \frac{\Gamma(3\alpha+1)}{\Gamma(\alpha+1)} C_3 - 2\frac{1}{\Gamma(\alpha+1)^2}\Gamma(2\alpha+1)^2 C_2^2\right] e_k^{2\alpha} + \mathcal{O}(e_k^{3\alpha}).$$

As a consequence, we obtain that)

$$1 + \frac{1}{2}{}^C L_f^\alpha(x_k) = 1 + \frac{\Gamma(2\alpha+1)}{2\Gamma(\alpha+1)^2}C_2 e_k^\alpha$$
$$+ \frac{1}{2\Gamma(\alpha+1)^2}\left[C_2^2\Gamma(2\alpha+1) + \frac{\Gamma(3\alpha+1)}{\Gamma(\alpha+1)}C_3 - 2\frac{1}{\Gamma(\alpha+1)^2}\Gamma(2\alpha+1)^2 C_2^2\right]e_k^{2\alpha} + \mathcal{O}(e$$

Accordingly, a Chebyshev-like quotient can be obtained, and written in terms of the error at the k-th iterate $e_k = x_k - \bar{x}$.

$$\frac{f(x_k)}{({}^C D_a^\alpha f)(x_k)}\left(1 + \frac{1}{2}{}^C L_f^\alpha(x_k)\right) = \frac{1}{\Gamma(\alpha+1)}e_k^\alpha + C_2\left(\frac{2\Gamma(\alpha+1)^2 - \Gamma(2\alpha+1)}{2\Gamma(\alpha+1)^3}\right)e_k^{2\alpha} + \mathcal{O}(e_k^{3\alpha}). \quad (13)$$

Then, it is clear that, to make null the term of e_k^α, a Caputo-fractional Chebyshev's method should include as a factor $\Gamma(\alpha+1)$ to conclude that the error expression is (9). □

Let us remark that, when $\alpha = 1$, we get the classical Chebyshev method, whose order is 3. However, the iterative expression defined in (8) does not achieve the required maximum order of convergence 3α. The following theorem presents another Caputo-fractional variant of Chebyshev's method, defined by replacing the second order derivative with a fractional one, with order $\alpha + 1$, $0 < \alpha \leq 1$. Its proof is omitted, as it is similar to that of Theorem 4. We denote this Caputo-fractional Chebyshev variant by CFC2.

Theorem 5. *Let $f : D \subset \mathbb{R} \longrightarrow \mathbb{R}$ be a continuous function with k-order fractional derivatives, $k \in \mathbb{N}$ and any α, $0 < \alpha \leq 1$, in the interval D. If x_0 is close enough to the zero $\bar{x}$ of $f(x)$ and ${}^C D_{\bar{x}}^\alpha f(x)$ is continuous and non zero in $\bar{x}$, then the local order of convergence of the Chebyshev's fractional method using Caputo-type derivatives (CFC2)*

$$x_{k+1} = x_k - \Gamma(\alpha+1)\frac{f(x_k)}{{}^C D_a^\alpha f(x_k)}\left(1 + \frac{1}{2}\frac{{}^C D_a^{\alpha+1} f(x_k) f(x_k)}{{}^C D_a^\alpha f(x_k){}^C D_a^\alpha f(x_k)}\right),$$

is at least 2α, being $0 < \alpha < 1$. On the one hand, if $0 < \alpha \leq \frac{2}{3}$, the error equation is

$$e_{k+1}^\alpha = \left(\frac{\Gamma(\alpha+1)^2 - \Gamma(2\alpha+1)}{\Gamma(\alpha+1)^3}\right)C_2 + \frac{1}{2}\frac{1}{\Gamma(\alpha+1)^2}\frac{{}^C D_{\bar{x}}^{\alpha+1} f(\bar{x})}{{}^C D^\alpha f(\bar{x})}\right)e_k^{2\alpha} + \mathcal{O}(e_k^{3\alpha+1}).$$

On the other hand, if $\frac{2}{3} \leq \alpha < 1$, then

$$e_{k+1}^\alpha = \left(\frac{\Gamma(\alpha+1)^2 - \Gamma(2\alpha+1)}{\Gamma(\alpha+1)^3}\right)C_2 + \frac{1}{2}\frac{1}{\Gamma(\alpha+1)^2}\frac{{}^C D_{\bar{x}}^{\alpha+1} f(\bar{x})}{{}^C D^\alpha f(\bar{x})}\right)e_k^{2\alpha} + \mathcal{O}(e_k^{3\alpha}).$$

Can the order of fractional Chebyshev's method be higher than 2α? In the following result it is shown that it is possible if the coefficients in the iterative expression are changed. The resulting fractional iterative scheme is denoted by CFC3. Again, the proof of the following result is omitted as it is similar to that of Theorem 4.

Theorem 6. *Let $f : D \subset \mathbb{R} \longrightarrow \mathbb{R}$ be a continuous function with k-order fractional derivatives, $k \in \mathbb{N}$ and any α, $0 < \alpha \leq 1$, in the interval D. If x_0 is close enough to the zero $\bar{x}$ of $f(x)$ and ${}^C D_{\bar{x}}^\alpha f(x)$ is* continuous and non zero in $\bar{x}$, then the local order of convergence of the Chebyshev's fractional method using Caputo-type derivatives (CFC3)

$$x_{k+1} = x_k - \Gamma(\alpha+1)\frac{f(x_k)}{^C D_a^\alpha f(x_k)}\left(A + B\,^C L_f^\alpha(x_k)\right), \tag{14}$$

is at least 3α only if $A = 1$ and $B = \dfrac{\Gamma(2\alpha+1) - \Gamma(\alpha+1)^2}{\Gamma(2\alpha+1)}$, being $0 < \alpha < 1$, and the error equation

$$\begin{aligned}e_{k+1}^\alpha &= \left[-\Gamma(2\alpha+1)\left(1 - \frac{\Gamma(2\alpha+1)}{\Gamma^4(\alpha+1)}\right)C_2 + \frac{B\Gamma(2\alpha+1)}{\Gamma^3(\alpha+1)}\left(2 - 3\frac{\Gamma(2\alpha+1)}{\Gamma^2(\alpha+1)}\right)C_2^2 \right.\\ &\quad \left. + \frac{1}{\Gamma(\alpha+1)}\left(\frac{B\Gamma(3\alpha+1)}{\Gamma^3(\alpha+1)} - \frac{\Gamma(3\alpha+1)}{\Gamma(2\alpha+1)\Gamma(\alpha+1)} + 1\right)C_3\right]e_k^{3\alpha} + \mathcal{O}(e_k^{4\alpha}),\end{aligned} \tag{15}$$

being $e_k = x_k - \bar{x}$.

According to the efficiency index defined by Ostrowski in [11], in Figure 1 we show that, with the same number of functional evaluations per iteration than CFC2 and CFC1 but with higher order of convergence, CFC3 has the best efficiency index, even compared with the fractional Newton's method CFN defined in [6]. In it, let us remark that incides of CFC1 and CFC2 coincide, as they have the same order of convergence and number of functional evaluations iteration.

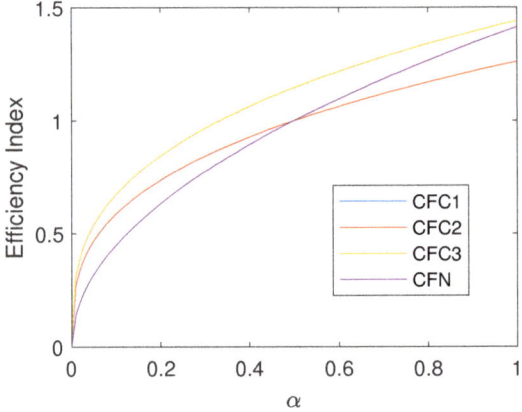

Figure 1. Efficiency indices of used methods.

In the following section, we analyze the dependence on the initial guess of the different Chebyshev-type fractional methods.

3. Numerical Performance of Proposed Schemes

In this section, we use *Matlab* R2018b with double precision for solving different kind of nonlinear equations. The stopping criterium used is $|x_{k+1} - x_k| < 10^{-6}$ with at most 250 iterations. The Gamma function is calculated by means of the routine made by Paul Godfrey (15 digits of accuracy along real axis and 13 elsewhere in $\mathbb{C}$. On the other hand, we use the program *mlf* of Mathworks for computing Mittag-Lefler function that has a precission of 9 significant digits.

The first test function is $f(x) = x^3 + x$, whose roots are $\bar{x}_1 = 0$, $\bar{x}_2 = i$ and $\bar{x}_3 = -i$. In Tables 1–6, we show the different solutions, the number of iterations, and the residual errors of the difference between the two last iterations and the value of function f at the last iteration. In Tables 1 and 2 we observe the performance of Caputo-fractional Chebyshev's method CFC2 and CFC3 estimating different roots with the initial guess $x_0 = 1.1$.

Table 1. CFC2 results for $f(x) = x^3 + x$ with $x_0 = 1.1$.

α	$\bar{x}$	$|x_{k+1} - x_k|$	$|f(x_{k+1})|$	Iterations
0.90	$-8.1274\text{e-}05 + i4.6093\text{e-}04$	9.2186e-04	4.6804e-04	250
0.91	$-3.2556\text{e-}05 + i2.0787\text{e-}04$	4.1575e-04	2.1041e-04	250
0.92	$-1.0632\text{e-}05 + i7.7354\text{e-}05$	1.5471e-04	7.8081e-05	250
0.93	$-2.5978\text{e-}06 + i2.1869\text{e-}05$	4.3739e-05	2.2023e-05	250
0.94	$-4.1185\text{e-}07 + i4.0939\text{e-}06$	8.1878e-06	4.1145e-06	250
0.95	$-1.6155\text{e-}08 - i9.1923\text{e-}07$	1.9214e-06	9.1937e-07	23
0.96	$1.4985\text{e-}07 - i7.7007\text{e-}07$	1.9468e-06	7.8451e-07	13
0.97	$3.4521\text{e-}07 - i7.6249\text{e-}07$	2.6824e-06	8.3699e-07	9
0.98	$2.7084\text{e-}07 - i3.4599\text{e-}07$	1.9608e-06	4.3939e-07	7
0.99	$-6.3910\text{e-}07 + i2.1637\text{e-}07$	6.4318e-06	6.7474e-07	5
1.00	$-2.9769\text{e-}08 + i0.0000\text{e+}00$	3.0994e-03	2.9769e-08	3

Table 2. CFC3 results for $f(x) = x^3 + x$ with $x_0 = 1.1$.

α	$\bar{x}$	$|x_{k+1} - x_k|$	$|f(x_{k+1})|$	Iterations
0.90	$-8.1270\text{e-}05 + i4.6090\text{e-}04$	9.2190e-04	4.6800e-04	250
0.91	$-3.2560\text{e-}05 + i2.0790\text{e-}04$	4.1570e-04	2.1040e-04	250
0.92	$-1.0630\text{e-}05 + i7.7350\text{e-}05$	1.5470e-04	7.8080e-05	250
0.93	$-2.5980\text{e-}06 + i2.1870\text{e-}05$	4.3740e-05	2.2020e-05	250
0.94	$-4.1180\text{e-}07 + i4.0940\text{e-}06$	8.1880e-06	4.1150e-06	250
0.95	$-1.6680\text{e-}07 + i9.2200\text{e-}07$	1.9640e-06	9.3690e-07	22
0.96	$-3.4850\text{e-}07 + i7.2840\text{e-}07$	2.0220e-06	8.0750e-07	12
0.97	$5.5470\text{e-}07 - i6.3310\text{e-}07$	2.6850e-06	8.4180e-07	7
0.98	$-3.1400\text{e-}07 + i2.0370\text{e-}07$	1.6820e-06	3.7430e-07	7
0.99	$1.2990\text{e-}07 - i8.4600\text{e-}08$	1.2680e-06	1.5500e-07	6
1.00	$-2.9770\text{e-}08 + i0.0000\text{e+}00$	3.0990e-03	2.9770e-08	3

As we know, when α is near to 1 the method needs less iterations and the evaluating of the last iteration is smaller than in first values of the parameter α. Let us now compare our proposed schemes with fractional Newton's method designed in [6] (see Table 3).

Table 3. CFN results for $f(x)$ with $x_0 = 1.1$.

α	$\bar{x}$	$|x_{k+1} - x_k|$	$|f(x_{k+1})|$	Iterations
0.90	$-8.1275\text{e-}05 - i4.6093\text{e-}04$	9.2187e-04	4.6804e-04	250
0.91	$-3.2556\text{e-}05 - i2.0787\text{e-}04$	4.1575e-04	2.1041e-04	250
0.92	$-1.0632\text{e-}05 - i7.7354\text{e-}05$	1.5471e-04	7.8081e-05	250
0.93	$-2.5978\text{e-}06 - i2.1869\text{e-}05$	4.3739e-05	2.2023e-05	250
0.94	$-4.1185\text{e-}07 + i4.0939\text{e-}06$	8.1878e-06	4.1145e-06	250
0.95	$2.8656\text{e-}08 - i9.4754\text{e-}07$	1.9837e-06	9.4797e-07	23
0.96	$-3.2851\text{e-}07 + i6.6774\text{e-}07$	1.8538e-06	7.4418e-07	14
0.97	$2.3413\text{e-}07 - i4.2738\text{e-}07$	1.5022e-06	4.8731e-07	11
0.98	$2.0677\text{e-}07 - i2.4623\text{e-}07$	1.4017e-06	3.2154e-07	9
0.99	$3.0668\text{e-}07 - i2.2801\text{e-}07$	3.3615e-06	3.8216e-07	7
1.00	$1.2192\text{e-}16 + i0.0000\text{e+}00$	3.9356e-06	1.2192e-16	5

As we can see in Tables 1–3, there are no big differences in terms of convergence to the real root but in the case of Caputo-fractional Newton's method CFN, the number of iterations needed to converge is higher than for CFC2 and CFC3.

To end, we show the convergence plane (see [12]) in Figures 2 and 3 where the abscissa axis corresponds to the initial approximations and α appears in the ordinate axis. We use a mesh of

400×400 points. Points painted in orange correspond to initial estimations that converge to $\bar{x}_1$ with a tolerance of 10^{-3}, a point is painted in blue if it converges to $\bar{x}_2$ and in green if it converges to $\bar{x}_3$. In any other case, points are painted in black, showing that no root is found in a maximum of 250 iterations. The estimations point are located in $[-5, 5]$, although convergence to the real root is found in $[-50, 50]$ paragraph has been moved, so that it can been seen before the figure 2 shows. please check and confirm.

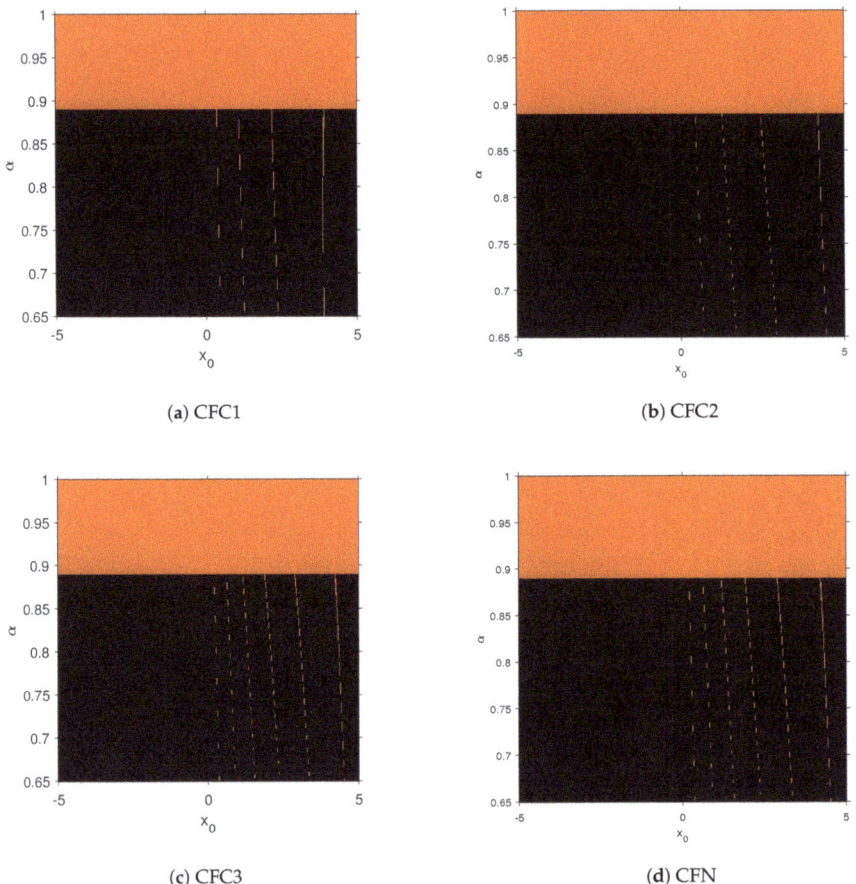

Figure 2. Convergence plane of proposed methods and CFN on $f(x)$.

We show the convergence plane of Chebyshev' and Newton's fractional methods. In Figure 2, it can be observed that for any real initial estimation in the interval used, if $\alpha \geq 0.89$, both methods converge to one of the zeros of $f(x)$ and if $\alpha < 0.89$, Newton's and Chebyshev's fractional methods do not converge to any solution. However, the higher order of convergence of Chebyshev's scheme can make the difference.

However, we have got only convergence to the real root, by using real initial guesses. In what follows, we use complex initial estimations, of equal real and imaginary parts, $x_0 = \lambda + i\lambda$ $\lambda \in \mathbb{R}$, in order to calculate the corresponding convergence plane.

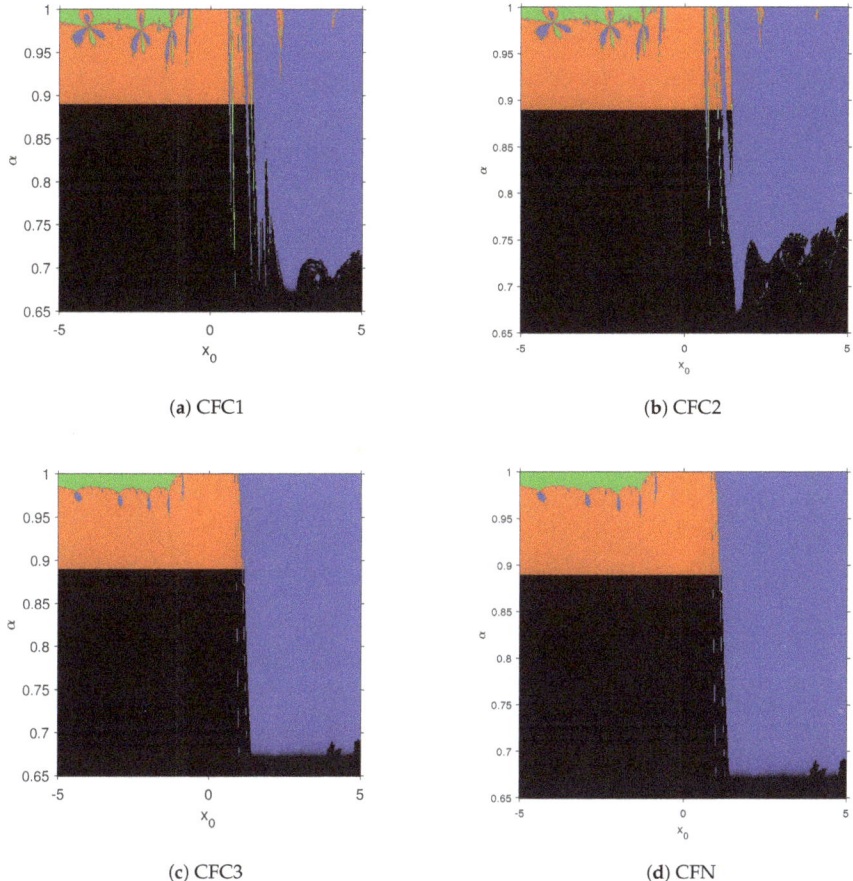

Figure 3. Convergence plane of proposed methods and CFN on $f(x)$ with complex initial estimation.

In Figure 3, we can observe that starting with complex initial values is more efficient to find all the roots of the nonlinear equation. In it, orange color means convergence to the real root $\bar{x}_1$, blue color is convergence to $\bar{x}_2$ and (x_0, α) in green color converge to $\bar{x}_3$. It is observed that it is possible to converge to $\bar{x}_3$ with lower values of α with CFC3 than using CFC1 or CFC2. Moreover, the methods converge mostly to $\bar{x}_3$ when the real and complex part of the initial estimation is positive, meanwhile it is possible to converge to any of the roots when the real and complex part of x_0 is negative.

Iterations

Also in Tables 4–6 we see that, with the same initial estimation, it is possible to approximate all the roots of the function by changing the value of α.

Table 4. CFC2 results for $f(x)$ with $x_0 = -1.3 - i1.3$.

α	$\bar{x}$	$|x_{k+1} - x_k|$	$|f(x_{k+1})|$	Iterations
0.90	−8.1274e-05-i4.6093e-04	9.2186e-04	4.6804e-04	250
0.91	−3.2556e-05-i2.0787e-04	4.1575e-04	2.1041e-04	250
0.92	−1.0632e-05-i7.7354e-05	1.5471e-04	7.8081e-05	250
0.93	−2.5978e-06-i2.1869e-05	4.3739e-05	2.2023e-05	250
0.94	−4.1185e-07+i4.0939e-06	8.1878e-06	4.1145e-06	250
0.95	−1.1203e-07+i9.8331e-07	2.0799e-06	9.8967e-07	26
0.96	−3.2473e-09-i7.0333e-07	1.7375e-06	7.0333e-07	19
0.97	5.9619e-08-i7.4110e-07	2.3750e-06	7.4349e-07	18
0.98	4.7121e-07+i7.4115e-07	4.1416e-06	8.7826e-07	19
0.99	3.9274e-08-i3.9674e-07	3.5758e-06	3.9868e-07	14
1.00	3.9559e-08+i4.8445e-07	7.8597e-03	4.8606e-07	8

Table 5. CFC3 results for $f(x)$ with $x_0 = -1.3 - i1.3$.

α	$\bar{x}$	$|x_{k+1} - x_k|$	$|f(x_{k+1})|$	Iterations
0.90	−8.1308e-05-i4.6100e-04	9.2200e-04	4.6811e-04	250
0.91	−3.2562e-05-i2.0789e-04	4.1577e-04	2.1042e-04	250
0.92	−1.0633e-05-i7.7355e-05	1.5471e-04	7.8082e-05	250
0.93	−2.5978e-06-i2.1870e-05	4.3739e-05	2.2023e-05	250
0.94	−4.1185e-07+i4.0939e-06	8.1878e-06	4.1145e-06	250
0.95	−9.6563e-08+i9.3217e-07	1.9628e-06	9.3716e-07	28
0.96	1.3446e-08-i7.0728e-07	1.7477e-06	7.0741e-07	22
0.97	−9.7497e-08+i1.0000e+00	1.7666e-06	2.1081e-07	15
0.98	−1.8598e-07-i1.0000e+00	5.5924e-06	4.4631e-07	15
0.99	−1.5051e-07+i5.1262e-07	4.9442e-06	5.3426e-07	13
1.00	3.9559e-08+i4.8445e-07	7.8597e-03	4.8606e-07	8

Table 6. Fractional Newton results for $f(x)$ with $x_0 = -1.3 - i1.3$.

α	$\bar{x}$	$|x_{k+1} - x_k|$	$|f(x_{k+1})|$	Iterations
0.90	−8.1275e-05+i4.6093e-04	9.2187e-04	4.6804e-04	250
0.91	−3.2556e-05+i2.0787e-04	4.1575e-04	2.1041e-04	250
0.92	−1.0632e-05+i7.7354e-05	1.5471e-04	7.8081e-05	250
0.93	−2.5978e-06+i2.1869e-05	4.3739e-05	2.2023e-05	250
0.94	−4.1185e-07+i4.0939e-06	8.1878e-06	4.1145e-06	250
0.95	−9.1749e-08-i9.2392e-07	1.9434e-06	9.2846e-07	28
0.96	1.5946e-08+i1.0000e+00	6.4777e-07	1.0272e-07	12
0.97	1.2679e-07+i1.0000e+00	4.3336e-06	5.1715e-07	16
0.98	−5.1142e-07+i7.8442e-07	4.5155e-06	9.3641e-07	16
0.99	9.3887e-08-i1.0000e+00	4.7305e-06	1.8942e-07	11
1.00	−2.9297e-10-i1.0000e+00	1.4107e-05	5.9703e-10	9

4. Conclusions

In this manuscript, we have designed several Chebyshev-type fractional iterative methods, by using Caputo's fractional derivative. We have shown that the order of convergence can reach 3α, $0 < \alpha \leq 1$, by means of an appropriate design in the iterative expression, including a Gamma function as a dumping parameter, but also an specific treatment of the high-order derivative derivative. It has been proven that the replacement of high-order integer derivatives by fractional ones must be carefully done, as it is not obvious that the order of convergence will be preserved. The theoretical results have been checked in the numerical section, with special emphasis on the dependence on the initial guess (that is shown in the convergence planes), comparing Newton' and Chebyshev's fractional methods performances. CFC3 method is not only the most efficient scheme (see Figure 1), but also converges to any of the searched roots for values of α lower than those needed by CFC1 and CFC2.

Author Contributions: the contribution of the authors to this manuscript can be defined as: Conceptualization and Validation, A.C. and J.R.T.; investigation, I.G.; writing—original draft preparation, I.G.; writing—review and editing, A.C. and J.R.T.

Funding: This research was partially supported by Ministerio de Ciencia, Innovación y Universidades under grants PGC2018-095896-B-C22 and by Generalitat Valenciana PROMETEO/2016/089.

Acknowledgments: The authors would like to thank the anonymous reviewers for their comments and suggestions that have improved the final version of this manuscript.

Conflicts of Interest: The authors declare no conflict of interest.

References

1. Mathai, A.M.; Haubold, H.J. Fractional and Multivariable Calculus. In *Model Building and Optimization Problems*; Springer Optimization and Its Applications 122; Springer: Berlin, Germany, 2017.
2. Ross, B. A brief history and exposition of the fundamental theory of fractional calculus. In *Fractional Calculus and Its Applications*; Ross, B., Ed.; Lecture Notes in Mathematics; Springer: Berlin/Heidelberg, Germany, 1975; Volume 457, pp. 1–36.
3. Atanackovic, T.M.; Pilipovic, S.; Stankovic, B.; Zorica, D. *Fractional Calculus with Applications in Mechanics: Wave Propagation, Impact and Variational Principles*; Wiley: London, UK, 2014.
4. Khan, M.A.; Ullah, S.; Farhan, M. The dynamics of Zika virus with Caputo fractional derivative. *AIMS Math.* **2019**, *4*, 134–146. [CrossRef]
5. Brambila, F.; Torres, A. Fractional Newton-Raphson Method. *arXiv* **2017**, arXiv:1710.07634v3.
6. Akgül, A.; Cordero, A.; Torregrosa, J.R. A fractional Newton method with 2th-order of convergence and its stability. *Appl. Math. Lett.* **2019**, *98*, 344–351. [CrossRef]
7. Traub, J.F. *Iterative Methods for the Solution of Equations*; Prentice-Hall: Englewood Cliffs, NJ, USA, 1964.
8. Lombardero, A. Cálculo Fraccionario y dinámica newtoniana. *Pensam. Mat.* **2014**, *IV*, 77–106.
9. Caputo, M.C.; Torres, D.F.M. Duality for the left and right fractional derivatives. *Signal Process* **2015**, *107*, 265–271. [CrossRef]
10. Odibat, Z.M.; Shawagfeh, N.T. Generalized Taylor's formula. *Appl. Math. Comput.* **2007**, *186*, 286–293. [CrossRef]
11. Ostrowski, A.M. *Solution of Equations and Systems of Equations*; Prentice-Hall: Englewood Cliffs, NJ, USA, 1964.
12. Magreñan, A.Á. A new tool to study real dynamics: The convergence plane. *Appl. Math. Comput.* **2014**, *248*, 215–224. [CrossRef]

© 2019 by the authors. Licensee MDPI, Basel, Switzerland. This article is an open access article distributed under the terms and conditions of the Creative Commons Attribution (CC BY) license (http://creativecommons.org/licenses/by/4.0/).

Article

Ball Convergence for Combined Three-Step Methods Under Generalized Conditions in Banach Space

R. A. Alharbey [1], Ioannis K. Argyros [2] and Ramandeep Behl [1,*]

[1] Department of Mathematics, King Abdulaziz University, Jeddah 21589, Saudi Arabia
[2] Department of Mathematics Sciences, Cameron University, Lawton, OK 73505, USA
* Correspondence: ramanbehl87@yahoo.in

Received: 21 June 2019; Accepted: 24 July 2019; Published: 3 August 2019

Abstract: Problems from numerous disciplines such as applied sciences, scientific computing, applied mathematics, engineering to mention some can be converted to solving an equation. That is why, we suggest higher-order iterative method to solve equations with Banach space valued operators. Researchers used the suppositions involving seventh-order derivative by Chen, S.P. and Qian, Y.H. But, here, we only use suppositions on the first-order derivative and Lipschitz constrains. In addition, we do not only enlarge the applicability region of them but also suggest computable radii. Finally, we consider a good mixture of numerical examples in order to demonstrate the applicability of our results in cases not covered before.

Keywords: local convergence; convergence order; Banach space; iterative method

PACS: 65G99; 65H10; 47J25; 47J05; 65D10; 65D99

1. Introduction

One of the most useful task in numerical analysis concerns finding a solution κ of

$$\Theta(x) = 0, \tag{1}$$

where $\Theta : \mathbb{D} \subset \mathbb{X} \to \mathbb{Y}$ is a Fréchet-differentiable operator, $\mathbb{X}$, $\mathbb{Y}$ are Banach spaces and $\mathbb{D}$ is a convex subset of $\mathbb{X}$. The $L(\mathbb{X}, \mathbb{Y})$ is the space of bounded linear operators from $\mathbb{X}$ to $\mathbb{Y}$.

Consider, a three step higher-order convergent method defined for each $l = 0, 1, 2, \ldots$ by

$$
\begin{aligned}
y_l &= x_l - \Theta'(x_l)^{-1}\Theta(x_l), \\
z_l &= \phi\big(x_l, \Theta(x_l), \Theta'(x_l), \Theta'(y_l)\big), \\
x_{l+1} &= z_l - \beta A_l^{-1}\Theta(z_l),
\end{aligned} \tag{2}
$$

where $\alpha, \beta \in \mathbb{S}$, $A_l = (\beta - \alpha)\Theta'(x_l) + \alpha\Theta'(y_l)$, ($\mathbb{S} = \mathbb{R}$ or $\mathbb{S} = \mathbb{C}$) and the second sub step represents any iterative method, in which the order of convergence is at least $m = 1, 2, 3, \ldots$. If $\mathbb{X} = \mathbb{Y} = \mathbb{R}$, then it was shown in [1]. The proof uses Taylor series expansions and the conditions on function Θ is up to the seventh differentiable. These suppositions of derivatives on the considered function Θ hamper the applicability of (2). Consider, a function μ on $\mathbb{X} = \mathbb{Y} = \mathbb{R}$, $\mathbb{D} = [-0.5, 1.5]$ by

$$\mu(t) = \begin{cases} 0, & t = 0 \\ t^3 \ln t^2 + t^5 - t^4, & t \neq 0 \end{cases}.$$

Then, we have that

$$\mu'(t) = 3t^2 \ln t^2 + 5t^4 - 4t^3 + 2t^2,$$

$$\mu''(t) = 6t \ln t^2 + 20t^3 - 12t^2 + 10t$$

and

$$\mu'''(t) = 6 \ln t^2 + 60t^2 - 24t + 22.$$

Then, obviously the third-order derivative $\mu'''(t)$ is not bounded on $\mathbb{D}$. Method (2) studied in [1], for $\mathbb{X} = \mathbb{Y} = \mathbb{R}$ suffers from several following defects:

(i) Applicable only on the real line.
(ii) Range of initial guesses for granted convergence is not discussed.
(iii) Higher than first order derivatives and Taylor series expansions were used limiting the applicability.
(iv) No computable error bounds on $\|\Omega_l\|$ (where $\Omega_l = x_l - \kappa$) were given.
(v) No uniqueness result was addressed.
(vi) The convergence order claim by them is also not correct, e.g., see the following method 43 [1]

$$\begin{aligned} y_l &= x_l - \frac{\Theta(x_l)}{\Theta'(x_l)}, \\ z_l &= x_l - \frac{2\Theta(x_l)}{\Theta'(y_l) + \Theta'(x_l)}, \\ x_{l+1} &= z_l - \frac{\beta \Theta(z_l)}{\alpha \Theta'(y_l) + (\beta - \alpha)\Theta'(x_l)}. \end{aligned} \quad (3)$$

It has fifth-order of convergence for $\alpha = \beta$ but $\alpha \neq \beta \in \mathbb{R}$ provides fourth-order convergence. But, authors claimed sixth-order convergence for every $\alpha, \beta \in \mathbb{R}$ that is not correct. The new proof is given in Section 2.

(vii) They can't choose special cases like methods 41, 47 and 49 (numbering from their paper [1]) because Chen and Qian [1], consider $y_l = x_l - \frac{f(x_l)}{f'(x_l)}$ in the proof of theorem. Additionally, it is clearly mentioned in the expression of (21) (from their paper [1]).

To address all these problems, we first extend method (2) to Banach space valued operators. The order of convergence is computed by using COC or ACOC (see remark 2.2(d)). Our technique uses only the first derivative in the analysis of method (2), so we can solve classes of equations not possible before in [1].

The remaining material of the paper is ordered as proceeds: Section 2 suggest convergence study of scheme (2). The applicability of our technique appears in Section 3.

2. Convergence Analysis

We consider some scalars functions and constraints for convergence study. Therefore, we assume that functions v, w_0, w, $\tilde{g}_2 : [0, +\infty) \to [0, +\infty)$ are continuous and nondecreasing with $w_0(0) = w(0) = 0$ and $\alpha, \beta \in \mathbb{S}$. Assume equation

$$w_0(t) = 1 \quad (4)$$

has a minimal positive solution r_0.

Functions g_1, h_1, p and h_p defined on $[0, r_0)$ as follow:

$$g_1(t) = \frac{\int_0^1 w((1-\eta)t)d\eta}{1 - w_0(t)},$$

$$h_1(t) = g_1(t) - 1,$$

$$p(t) = |\beta|^{-1}\Big[|\beta - \alpha|w_0(t) + |\alpha|w_0(g_1(t)t)\Big], \beta \neq 0,$$

and

$$h_p(t) = p(t) - 1.$$

Notice, that $h_1(0) = h_p(0) = -1 < 0$ and $h_1(t) \to +\infty$, $h_q(t) \to +\infty$ as $t \to r_0^-$. Then, by the intermediate value theorem (IVT), the functions h_1 and h_p have roots in $(0, r_0)$. Let r_1 and r_p, stand respectively the smallest such roots of the function h_1 and h_p. Additionally, we consider two functions g_2 and h_2 on $(0, r_0)$ by

$$g_2(t) = \bar{g}_2(t)t^{m-1},$$

and

$$h_2(t) = g_2(t) - 1.$$

Suppose that

$$\bar{g}_2(0) < 1, \text{ if } m = 1 \tag{5}$$

and

$$g_2(t) \to a \text{ (a number greater than one or } +\infty) \tag{6}$$

as $t \to \bar{r}_0^-$ for some $\bar{r}_0 \leq r_0$. Then, again by adopting IVT that function h_2 has some roots $(0, \bar{r}_0)$. Let r_2 be the smallest such root. Notice that, if $m > 1$ condition (5) is not needed to show $h_2(0) < 0$, since in this case $h_2(0) = g_2(0) - 1 = 0 - 1 = -1 < 0$.

Finally, functions g_3 and h_3 on $[0, \bar{r}_p)$ by

$$g_3(t) = \left(1 + \frac{\int_0^1 v(\eta g_2(t)t)d\eta}{1 - p(t)}\right)g_2(t),$$

and

$$h_3(t) = g_3(t) - 1,$$

where $\bar{r}_p = \min\{r_p, r_2\}$. Suppose that

$$(1 + v(0))\bar{g}_2(0) < 1, \text{ if } m = 1, \tag{7}$$

we get by (7) that $h_3(0) = (1 + v(0))\bar{g}_2(0) - 1 < 0$ and $h_3(t) \to +\infty$ or positive number as $t \to \bar{r}_p^-$. Let r_3 stand for the smallest root of function h_3 in $(0, r_p)$. Consider a radius of convergence r as

$$r = \min\{r_1, r_3\}. \tag{8}$$

Then, it holds

$$0 \leq g_i(t) < 1, i = 1, 2, 3 \text{ for each } t \in [0, r). \tag{9}$$

Let us assume that we have center $z \in \mathbb{X}$ and radius $\rho > 0$ of $U(z, \rho)$ and $\bar{U}(z, \rho)$ open and closed ball, respectively, in the Banach space $\mathbb{X}$.

Theorem 1. *Let $\Theta : \mathbb{D} \subseteq \mathbb{X} \to \mathbb{Y}$ be a differentiable operator. Let v, w_0, w, $\bar{g}_2 : [0, \infty) \to [0, \infty)$ be nondecreasing continuous functions with $w_0(0) = w(0) = 0$. Additionally, we consider that $r_0 \in [0, \infty)$, $\alpha \in S$, $\beta \in S - \{0\}$ and $m \geq 1$. Assume that there exists $\kappa \in \mathbb{D}$ such that for every $\lambda_1 \in \mathbb{D}$*

$$\Theta(\kappa) = 0, \quad \Theta'(\kappa)^{-1} \in L(\mathbb{Y}, \mathbb{X}), \tag{10}$$

$$\|\Theta'(\kappa)^{-1}(\Theta'(\lambda_1) - \Theta'(\kappa))\| \leq w_0(\|\lambda_1 - \kappa\|). \tag{11}$$

and Equation (4) has a minimal solution r_0 and (5) holds.

Moreover, assume that for each $\lambda_1, \lambda_2 \in \mathbb{D}_0 := \mathbb{D} \cap U(\kappa, r_0)$

$$\|\Theta'(\kappa)^{-1}(\Theta'(\lambda_1) - \Theta'(\lambda_2))\| \leq w(\|\lambda_1 - \lambda_2\|), \tag{12}$$

$$\|\Theta'(\kappa)^{-1}\Theta'(\lambda_1)\| \leq v(\|\lambda_1 - \kappa\|), \tag{13}$$

$$\|\phi(\lambda_1, \Theta(\lambda_1), \Theta'(\lambda_1), \Theta'(\lambda_2))\| \leq \bar{g}_2(\|\lambda_1 - \kappa\|)\|\lambda_1 - \kappa\|^m \tag{14}$$

and

$$\bar{U}(\kappa, r) \subseteq \mathbb{D}. \tag{15}$$

Then, for $x_0 \in U(\kappa, r) - \{\kappa\}$, we have $\lim_{l \to \infty} x_l = \kappa$, where $\{x_l\} \subset U(\kappa, r)$ and the following assertions hold

$$\|y_l - \kappa\| \leq g_1(\|\Omega_l\|)\|\Omega_l\| \leq \|\Omega_l\| < r, \tag{16}$$

$$\|z_l - \kappa\| \leq g_2(\|\Omega_l\|)\|\Omega_l\| \leq \|\Omega_l\| \tag{17}$$

and

$$\|x_{l+1} - \kappa\| \leq g_3(\|\Omega_l\|)\|\Omega_l\| \leq \|\Omega_l\|, \tag{18}$$

where $x_l - \kappa = \Omega_l$ and functions g_i, $i = 1, 2, 3$ are given previously. Moreover, if $R \geq r$

$$\int_0^1 w_0(\eta R) d\eta < 1, \tag{19}$$

then κ is unique in $\mathbb{D}_1 := \mathbb{D} \cap \bar{U}(\kappa, R)$.

Proof. We demonstrate that the sequence $\{x_l\}$ is well-defined in $U(\kappa, r)$ and converges to κ by adopting mathematical induction. By the hypothesis $x_0 \in U(\kappa, r) - \{\kappa\}$, (4), (6) and (13), we yield

$$\|\Theta'(\kappa)^{-1}(\Theta'(x_0) - \Theta'(\kappa))\| \leq w_0(\|\Omega_0\|) < w_0(r) < 1, \tag{20}$$

where $\Omega_0 = x_0 - \kappa$ and $\Theta'(x_0)^{-1} \in L(\mathbb{Y}, \mathbb{X})$, y_0 exists by the first two sub steps of method (2) and

$$\|\Theta'(x_0)^{-1}\Theta'(\kappa)\| \leq \frac{1}{1 - w_0(\|\Omega_0\|)}. \tag{21}$$

From (4), (8), (9) (for $i = 1$), (10), (12), (21) and the first substep of (2), we have

$$
\begin{aligned}
\|y_0 - \kappa\| &= \|\Omega_0 - \Theta'(x_0)^{-1}\Theta(x_0) - \kappa\| \\
&= \left\|\Theta'(x_0)^{-1}[\Theta'(x_0)(\Omega_0 - \kappa) - (\Theta(x_0) - \Theta(\kappa))]\right\| \\
&= \left\|[\Theta'(x_0)^{-1}\Theta'(\kappa)]\left[\Theta'(\kappa)^{-1}\Big(\Theta'(x_0)(\Omega_0 - \kappa) - (\Theta(x_0) - \Theta(\kappa))\Big)\right]\right\| \\
&\leq \|\Theta'(x_0)^{-1}\Theta(\kappa)\| \left\|\int_0^1 \Big(\Theta'(\kappa)^{-1}(\Theta'(\kappa + \eta(\Omega_0 - \kappa)) - \Theta'(x_0))(\Omega_0)\Big)d\eta\right\| \\
&\leq \frac{\int_0^1 w((1-\eta)\|\Omega_0\|)d\eta\|\Omega_0\|}{1 - w_0(\|\Omega_0\|)} \\
&\leq g_1(\|\Omega_0\|)\|\Omega_0\| \leq \|\Omega_0\| < r,
\end{aligned}
\tag{22}
$$

which implies (16) for $l = 0$ and $y_0 \in U(\kappa, r)$.

By (8), (9) (for $i = 2$) and (14), we get

$$
\begin{aligned}
\|z_0 - \kappa\| &= \|\phi(x_0, \Theta(x_0), \Theta'(x_0), \Theta'(y_0))\| \\
&\leq \bar{g}_2(\|\Omega_0\|)\|\Omega_0\|^m \\
&= g_2(\|\Omega_0\|)\|\Omega_0\| \leq \|\Omega_0\| < r,
\end{aligned}
\tag{23}
$$

so (17) holds $l = 0$ and $z_0 \in U(\kappa, r)$.

Using expressions (4), (8) and (11), we obtain

$$
\begin{aligned}
\left\|(\beta\Theta'(\kappa))^{-1}\left[(\beta - \alpha)(\Theta'(x_0) - \Theta'(\kappa)) + \alpha(\Theta'(y_0) - \Theta'(\kappa))\right]\right\| \\
\leq |\beta|^{-1}\left[|\beta - \alpha|w_0(\|\Omega_0\|) + |\alpha|w_0(\|y_0 - \kappa\|)\right] \\
\leq |\beta|^{-1}\left[|\beta - \alpha|w_0(\|\Omega_0\|) + |\alpha|w_0(g_1(\|\Omega_0\|)\|\Omega_0\|)\right] \\
= p(\|\Omega_0\|) \leq p(r) < 1,
\end{aligned}
\tag{24}
$$

so

$$
\|((\beta - \alpha)\Theta'(x_0) + \alpha\Theta'(y_0))^{-1}\Theta'(\kappa)\| \leq \frac{1}{1 - p(\|\Omega_0\|)}.
\tag{25}
$$

and x_1 is well-defined.

In view of (4), (8), (9) (for $i = 3$), (13), (22), (23) and (24), we get in turn that

$$
\begin{aligned}
\|x_1 - \kappa\| &= \|z_0 - \kappa\| + |\beta|\int_0^1 v(\eta\|z_0 - \kappa\|)d\eta\|\Omega_0\| \\
&\leq \left(1 + \frac{|\beta|\int_0^1 v(\eta g_2(\|\Omega_0\|))d\eta}{|\beta|(1 - p(\|\Omega_0\|))}\right)g_2(\|\Omega_0\|)\|\Omega_0\| \\
&= g_3(\|\Omega_0\|)\|\Omega_0\| \leq \|\Omega_0\| < r,
\end{aligned}
\tag{26}
$$

that demonstrates (18) and $x_1 \in U(\kappa, r)$. If we substitute x_0, y_0, x_1 by x_l, y_l, x_{l+1}, we arrive at (18) and (19). By adopting the estimates

$$
\|x_{l+1} - \kappa\| \leq c\|\Omega_l\| < r, \; c = g_2(\|\Omega_0\|) \in [0, 1),
\tag{27}
$$

so $\lim_{l \to \infty} x_l = \kappa$ and $x_{l+1} \in U(\kappa, r)$.

Now, only the uniqueness part is missing, so we assume that $\kappa^* \in \mathbb{D}_1$ with $\Theta(\kappa^*) = 0$. Consider, $Q = \int_0^1 \Theta'(\kappa + \eta(\kappa - \kappa^*))d\eta$. From (8) and (15), we obtain

$$\|\Theta'(\kappa)^{-1}(Q - \Theta'(\kappa))\| \leq \|\int_0^1 w_0(\eta\|\kappa^* - \kappa\|)d\eta$$
$$\leq \int_0^1 w_0(\eta R)d\eta < 1, \tag{28}$$

and by

$$0 = \Theta(\kappa) - \Theta(\kappa^*) = Q(\kappa - \kappa^*), \tag{29}$$

we derive $\kappa = \kappa^*$. □

Remark 1.

(a) By expression (13) hypothesis (15) can be omitted, if we set

$$v(t) = 1 + w_0(t) \text{ or } v(t) = 1 + w_0(r_0), \tag{30}$$

since,

$$\|\Theta'(\kappa)^{-1}\left[(\Theta'(x) - \Theta'(\kappa)) + \Theta'(\kappa)\right]\| = 1 + \|\Theta'(\kappa)^{-1}(\Theta'(x) - \Theta'(\kappa))\|$$
$$\leq 1 + w_0(\|x - \kappa\|) \tag{31}$$
$$= 1 + w_0(t) \text{ for } \|x - \kappa\| \leq r_0.$$

(b) Consider w_0 to be strictly increasing, so we have

$$r_0 = w_0^{-1}(1) \tag{32}$$

for (4).

(c) If w_0 and w are constants, then

$$r_1 = \frac{2}{2w_0 + w} \tag{33}$$

and

$$r \leq r_1, \tag{34}$$

where r_1 is the convergence radius for well-known Newton's method

$$x_{l+1} = x_l - \Theta'(x_l)^{-1}\Theta(x_l), \tag{35}$$

given in [2].

On the other hand, Rheindoldt [3] and Traub [4] suggested

$$r_{TR} = \frac{2}{3w_1}, \tag{36}$$

where as Argyros [2,5]

$$r_A = \frac{2}{2w_0 + w_1}, \tag{37}$$

where w_1 is the Lipschitz constant for (9) on $\mathbb{D}$. Then,

$$w \leq w_1, \, w_0 \leq w_1, \tag{38}$$

so

$$r_{TR} \leq r_A \leq r_1 \tag{39}$$

and

$$\frac{r_{TR}}{r_A} \to \frac{1}{3} \quad \text{as} \quad \frac{w_0}{w} \to 0. \tag{40}$$

(d) We use the following rule for COC

$$\zeta = \frac{\ln \frac{\|x_{l+2}-\kappa\|}{\|x_{l+1}-\kappa\|}}{\ln \frac{\|x_{l+1}-\kappa\|}{\|x_l-\kappa\|}}, \quad \text{for each } l = 0,1,2,\ldots \tag{41}$$

or ACOC [6], defined as

$$\zeta^* = \frac{\ln \frac{\|x_{l+2}-x_{l+1}\|}{\|x_{l+1}-x_l\|}}{\ln \frac{\|x_{l+1}-x_n\|}{\|x_n-x_{l-1}\|}}, \quad \text{for each } l = 1,2,\ldots \tag{42}$$

not requiring derivatives and ζ^* does not depend on κ.

(e) Our results can be adopted for operators Θ that satisfy [2,5]

$$\Theta'(x) = P(\Theta(x)), \tag{43}$$

for a continuous operator P. The beauty of our study is that we can use the results without prior knowledge of solution κ, since $\Theta'(\kappa) = P(\Theta(\kappa)) = P(0)$. As an example $\Theta(x) = e^x - 1$, so we assume $P(x) = x + 1$.

(f) Let us show how to consider functions ϕ, $\bar{g}_2$, g_2 and m. Define function ϕ by

$$\phi(x_l, \Theta(x_l), \Theta'(x_l), \Theta'(y_l)) = y_l - \Theta'(y_l)^{-1}\Theta(y_l). \tag{44}$$

Then, we can choose

$$g_2(t) = \frac{\int_0^1 w((1-\eta)g_1(t)t)d\eta}{1 - w_0(g_1(t)t)} g_1(t). \tag{45}$$

If w_0, w, v are given in particular by $w_0(t) = L_0 t$, $w(t) = Lt$ and $v(t) = M$ for some $L_0, L > 0$, and $M \geq 1$, then we have that

$$\bar{g}_2(t) = \frac{\frac{L^2}{8(1-L_0 t)^2}}{1 - \frac{L_0 L t^2}{2(1-L_0 t)}}, \tag{46}$$

$$g_2(t) = \bar{g}_2(t)t^3 \text{ and } m = 4.$$

(g) If $\beta = 0$, we can obtain the results for the two-step method

$$\begin{aligned} y_l &= x_l - \Theta'(x_l)^{-1}\Theta(x_l), \\ x_{l+1} &= \phi(x_l, \Theta(x_l), \Theta'(x_l), \Theta'(y_l)) \end{aligned} \tag{47}$$

by setting $z_l = x_{l+1}$ in Theorem 1.

Convergence Order of Expression (3) from [1]

Theorem 2. *Let $\Theta : \mathbb{R} \to \mathbb{R}$ has a simple zero ζ being a sufficiently many times differentiable function in an interval containing ζ. Further, we consider that initial guess $x = x_0$ is sufficiently close to ζ. Then, the iterative scheme defined by (3) from [1] has minimum fourth-order convergence and satisfy the following error equation*

$$\begin{aligned} e_{l+1} = &-\frac{c_2\left(2c_2^2 + c_3\right)(\alpha - \beta)}{\beta}e_l^4 + \frac{1}{2\beta^2}\Big[4\beta c_4 c_2(\beta - \alpha) - 4c_2^4(2\alpha^2 - 8\alpha\beta + 5\beta^2) \\ &- 2c_3 c_2^2(2\alpha^2 + \alpha\beta - 4\beta^2) + 3\beta c_3^2(\beta - \alpha)\Big]e_l^5 + O(e_l^6), \end{aligned} \tag{48}$$

where $\alpha, \beta \in \mathbb{R}$, $e_l = x_l - \xi$ and $c_j = \frac{\Theta^{(j)}(\xi)}{j!\Theta'(\xi)}$ for $j = 1, 2, \ldots 6$.

Proof. The Taylor's series expansion of function $\Theta(x_l)$ and its first order derivative $\Theta'(x_l)$ around $x = \xi$ with the assumption $\Theta'(\xi) \neq 0$ leads us to:

$$\Theta(x_l) = \Theta'(\xi) \left[\sum_{j=1}^{6} c_j e_l^j + O(e_l^7) \right], \tag{49}$$

and

$$\Theta'(x_l) = \Theta'(\xi) \left[\sum_{j=1}^{6} j c_j e_l^j + O(e_l^7) \right], \tag{50}$$

respectively.

By using the Equations (49) and (50), we get

$$y_l - \xi = c_2 e_l^2 - 2(c_2^2 - c_3) e_l^3 + (4c_2^3 - 7c_3c_2 + 3c_4) e_l^4 + (-8c_2^4 + 20c_3c_2^2 - 10c_4c_2 - 6c_3^2 + 4c_5) e_l^5 \\
+ (16c_2^5 - 52c_3c_2^3 + 28c_4c_2^2 + (33c_3^2 - 13c_5)c_2 - 17c_3c_4 + 5c_6) e_l^6 + O(e_l^7). \tag{51}$$

The following expansion of $\Theta(y_l)$ about ξ

$$\Theta'(y_l) = \Theta'(\xi) \Big[1 + 2c_2^2 e_l^2 + (4c_2c_3 - 4c_2^3) e_l^3 + c_2(8c_2^3 - 11c_3c_2 + 6c_4) e_l^4 - 4c_2(4c_2^4 - 7c_3c_2^2 + 5c_4c_2 - 2c_5) e_l^5 \\
+ 2(16c_2^6 - 34c_3c_2^4 + 30c_4c_2^3 - 13c_5c_2^2 + (5c_6 - 8c_3c_4)c_2 + 6c_3^3) e_l^6 \Big]. \tag{52}$$

From Equations (50)–(52) in the second substep of (3), we have

$$z_l - \xi = \left(c_2^2 + \frac{c_3}{2}\right) e_l^3 + \left(-3c_2^3 + \frac{3c_3c_2}{2} + c_4\right) e_l^4 + \left(6c_2^4 - 9c_3c_2^2 + 2c_4c_2 - \frac{3}{4}(c_3^2 - 2c_5)\right) e_l^5 \\
+ \frac{1}{2}\left(-18c_2^5 + 50c_3c_2^3 - 30c_4c_2^2 - 5\left(c_3^2 - c_5\right)c_2 - 5c_3c_4 + 4c_6\right) e_l^6 + O(e_l^7). \tag{53}$$

Similarly, we can expand function $f(z_l)$ about ξ with the help of Taylor series expansion, which is defined as follows:

$$\Theta(z_l) = \Theta'(\xi) \Big[\left(c_2^2 + \frac{c_3}{2}\right) e_l^3 + \left(-3c_2^3 + \frac{3c_3c_2}{2} + c_4\right) e_l^4 + \left(6c_2^4 - 9c_3c_2^2 + 2c_4c_2 - \frac{3}{4}(c_3^2 - 2c_5)\right) e_l^5 \\
+ \left\{ c_2 \left(c_2^2 + \frac{c_3}{2}\right)^2 + \frac{1}{2}\left(-18c_2^5 + 50c_3c_2^3 - 30c_4c_2^2 - 5(c_3^2 - c_5)c_2 - 5c_3c_4 + 4c_6\right)\right\} e_l^6 + O(e_l^7) \Big]. \tag{54}$$

Adopting expressions (49)–(54), in the last sub-step of method (3), we have

$$e_{l+1} = -\frac{c_2 \left(2c_2^2 + c_3\right)(\alpha - \beta)}{\beta} e_l^4 + \frac{1}{2\beta^2} \Big[4\beta c_4 c_2 (\beta - \alpha) - 4c_2^4 (2\alpha^2 - 8\alpha\beta + 5\beta^2) \\
- 2c_3 c_2^2 (2\alpha^2 + \alpha\beta - 4\beta^2) + 3\beta c_3^2 (\beta - \alpha) \Big] e_l^5 + O(e_l^6). \tag{55}$$

For choosing $\alpha = \beta$ in (55), we obtain

$$e_{l+1} = \left(2c_2^4 + c_3 c_2^2\right) e_l^5 + O(e_l^6). \tag{56}$$

The expression (55) confirms that the scheme (3) have maximum fifth-order convergence for $\alpha = \beta$ (that can be seen in (56)). This completes the proof and also contradict the claim of authors [1]. □

This type of proof and theme are close to work on generalization of the fixed point theorem [2,5,7,8]. We recall a standard definition.

Definition 2. Let $\{x_l\}$ be a sequence in $\mathbb{X}$ which converges to κ. Then, the convergence is of order $\lambda \geq 1$ if there exist $\lambda > 0$, abd $l_0 \in \mathbb{N}$ such that

$$\|x_{l+1} - \kappa\| \leq \lambda \|x_l - \kappa\|^\lambda \text{ for each } l \geq l_0.$$

3. Examples with Applications

Here, we test theoretical results on four numerical examples. In the whole section, we consider $\phi(x_l, \Theta(x_l), \Theta'(x_l), \Theta'(y_l)) = x_l - \frac{2f(x_l)}{f'(y_l) + f'(x_l)}$, that means $m = 2$ for the computational point of view, called by $(M1)$.

Example 1. Set $\mathbb{X} = \mathbb{Y} = C[0, 1]$. Consider an integral equation [9], defined by

$$x(\beta) = 1 + \int_0^1 T(\beta, \alpha) \left(x(\alpha)^{\frac{3}{2}} + \frac{x(\alpha)^2}{2} \right) d\alpha \tag{57}$$

where

$$T(\beta, \alpha) = \begin{cases} (1 - \beta)\alpha, & \alpha \leq s, \\ \beta(1 - \alpha), & s \leq \alpha. \end{cases} \tag{58}$$

Consider corresponding operator $\Theta : C[0, 1] \to C[0, 1]$ as

$$\Theta(x)(\beta) = x(\beta) - \int_0^\alpha T(\beta, \alpha) \left(x(\alpha)^{\frac{3}{2}} + \frac{x(\alpha)^2}{2} \right) d\alpha. \tag{59}$$

But

$$\left\| \int_0^\alpha T(\beta, \alpha) d\alpha \right\| \leq \frac{1}{8}, \tag{60}$$

and

$$\Theta'(x)y(\beta) = y(\beta) - \int_0^\alpha T(\beta, \alpha) \left(\frac{3}{2} x(\alpha)^{\frac{1}{2}} + x(\alpha) \right) d\alpha.$$

Using $\kappa(s) = 0$, we obtain

$$\left\| \Theta'(\kappa)^{-1} \left(\Theta'(x) - \Theta'(y) \right) \right\| \leq \frac{1}{8} \left(\frac{3}{2} \|x - y\|^{\frac{1}{2}} + \|x - y\| \right), \tag{61}$$

So, we can set

$$w_0(\alpha) = w(\alpha) = \frac{1}{8} \left(\frac{3}{2} \alpha^{\frac{1}{2}} + \alpha \right).$$

Hence, by adopting Remark 2.2(a), we have

$$v(\alpha) = 1 + w_0(\alpha) \text{ or } v_0(\alpha) = M,$$

The results in [1] are not applicable, since Θ' is not Lipschitz. But, our results can be used. The radii of convergence of method (2) for example (1) are described in Table 1.

Table 1. Radii of convergence for problem (1).

α	β	m	r_1	r_p	r_2	r_3	r	Methods
1	1	2	2.6303	3.13475	2.6303	2.1546	2.1546	M1
1	2	2	2.6303	3.35124	2.6303	2.0157	2.0157	M1

Example 2. Consider a system of differential equations

$$\theta_1'(x) - \theta_1(x) - 1 = 0$$
$$\theta_2'(y) - (e-1)y - 1 = 0 \qquad (62)$$
$$\theta_3'(z) - 1 = 0$$

that model for the motion of an object for $\theta_1(0) = \theta_2(0) = \theta_3(0) = 0$. Then, for $v = (x, y, z)^T$ consider $\Theta := (\theta_1, \theta_2, \theta_3) : \mathbb{D} \to \mathbb{R}^3$ defined by

$$\Theta(v) = \left(e^x - 1, \frac{e-1}{2}y^2 + y, z\right)^T. \qquad (63)$$

We have

$$\Theta'(v) = \begin{bmatrix} e^x & 0 & 0 \\ 0 & (e-1)y + 1 & 0 \\ 0 & 0 & 1 \end{bmatrix}.$$

Then, we get $w_0(t) = L_0 t$, $w(t) = Lt$, $w_1(t) = L_1 t$ and $v(t) = M$, where $L_0 = e - 1 < L = e^{\frac{1}{L_0}} = 1.789572397$, $L_1 = e$ and $M = e^{\frac{1}{L_0}} = 1.7896$. The convergence radii of scheme (2) for example (2) are depicted in Table 2.

Table 2. Radii of convergence for problem (2).

α	β	r_1	r_p	r_2	r_3	r	Methods	x_0	n	ρ
1	1	0.382692	0.422359	0.321733	0.218933	0.218933	M1	0.15	3	4.9963
1	2	0.382692	0.441487	0.321733	0.218933	0.218933	M1	0.11	4	4.0000

We follow the stopping criteria for computer programming (i) $\|F(X_l)\|$ and (ii) $\|X_{l+1} - X_l\| < 10^{-100}$ in all the examples.

Example 3. Set $\mathbb{X} = \mathbb{Y} = C[0, 1]$ and $\mathbb{D} = \bar{U}(0, 1)$. Consider Θ on $\mathbb{D}$ as

$$\Theta(\varphi)(x) = \phi(x) - 5\int_0^1 x\eta\varphi(\eta)^3 d\eta. \qquad (64)$$

We have that

$$\Theta'(\varphi(\xi))(x) = \xi(x) - 15\int_0^1 x\eta\varphi(\eta)^2 \xi(\eta) d\eta, \text{ for each } \xi \in \mathbb{D}. \qquad (65)$$

Then, we get $\kappa = 0$, $L_0 = 7.5$, $L_1 = L = 15$ and $M = 2$. leading to $w_0(t) = L_0 t$, $v(t) = 2 = M$, $w(t) = Lt$, $w_1(t) = L_1 t$. The radii of convergence of scheme (2) for problem (3) are described in the Table 3.

Table 3. Radii of convergence for problem (3).

α	β	m	r_1	r_p	r_2	r_3	r	Methods
1	1	2	0.0666667	0.0824045	0.0233123	0.00819825	0.00819825	M1
1	2	2	0.0666667	0.0888889	0.0233123	0.00819825	0.00819825	M1

Example 4. We get $L = L_0 = 96.662907$ and $M = 2$ for example at introduction. Then, we can set $w_0(t) = L_0 t$, $v(t) = M = 2$, $w(t) = Lt$, $w_1(t) = Lt$. The convergence radii of the iterative method (2) for example (4) are mentioned in the Table 4.

Table 4. Radii of convergence for problem (4).

α	β	m	r_1	r_p	r_2	r_3	r	Methods	x_0	n	ρ
1	1	2	0.0102914	0.0102917	0.00995072	0.00958025	0.00958025	M1	1.008	3	5.0000
1	2	2	0.0102914	0.010292	0.00995072	0.00958025	0.00958025	M1	1.007	4	3.0000

4. Conclusions

A major problem in the development of iterative methods is the convergence conditions. In the case of especially high order methods, such as (2), the operator involved must be seventh times differentiable according to the earlier study [1] which do not appear in the methods, limiting the applicability. Moreover, no error bounds or uniqueness of the solution that can be computed are given. That is why we address these problems based only on the first order derivative which actually appears in the method. The convergence order is determined using *COC* or *ACOC* that do not require higher than first order derivatives. Our technique can be used to expand the applicability of other iterative methods [1–13] along the same lines.

Author Contributions: All the authors have equal contribution for this paper.

Funding: This project was funded by the Deanship of Scientific Research (DSR), King Abdulaziz University, Jeddah, under grant No. (D-253-247-1440). The authors, therefore, acknowledge, with thanks, the DSR technical and financial support.

Conflicts of Interest: The authors declare no conflict of interest.

References

1. Chen, S.P.; Qian, Y.H. A family of combined iterative methods for solving nonlinear equations. *Am. J. Appl. Math. Stat.* **2017**, *5*, 22–32. [CrossRef]
2. Argyros, I.K. *Convergence and Application of Newton-Type Iterations*; Springer: Berlin, Germany, 2008.
3. Rheinboldt, W.C. *An Adaptive Continuation Process for Solving Systems of Nonlinear Equations*; Polish Academy of Science, Banach Center Publications: Warsaw, Poland, 1978; Volume 3, pp. 129–142.
4. Traub, J.F. *Iterative Methods for the Solution of Equations*; Prentice-Hall Series in Automatic Computation; Prentice-Hall: Englewood Cliffs, NJ, USA, 1964.
5. Argyros, I.K.; Hilout, S. *Computational Methods in Nonlinear Analysis*; World Scientific Publishing Company: New Jersey, NJ, USA, 2013.
6. Kou, J. A third-order modification of Newton method for systems of nonlinear equations. *Appl. Math. Comput.* **2007**, *191*, 117–121.
7. Petkovic, M.S.; Neta, B.; Petkovic, L.; Džunič, J. *Multipoint Methods for Solving Nonlinear Equations*; Elsevier: Amsterdam, The Netherlands, 2013.
8. Sharma, J.R.; Ghua, R.K.; Sharma, R. An efficient fourth-order weighted-Newton method for system of nonlinear equations. *Numer. Algor.* **2013**, *62*, 307–323. [CrossRef]
9. Ezquerro, J.A.; Hernández, M.A. New iterations of R-order four with reduced computational cost. *BIT Numer. Math.* **2009**, *49*, 325–342. [CrossRef]
10. Amat, S.; Hernández, M.A.; Romero, N. A modified Chebyshev's iterative method with at least sixth order of convergence. *Appl. Math. Comput.* **2008**, *206*, 164–174. [CrossRef]
11. Argyros, I.K.; Magreñán, Á.A. Ball convergence theorems and the convergence planes of an iterative methods for nonlinear equations. *SeMA* **2015**, *71*, 39–55.
12. Cordero, A.; Torregrosa, J.R.; Vassileva, M.P. Increasing the order of convergence of iterative schemes for solving nonlinear system. *J. Comput. Appl. Math.* **2012**, *252*, 86–94. [CrossRef]
13. Potra, F.A.; Pták, V. Nondiscrete Introduction and Iterative Process. In *Research Notes in Mathematics*; Pitman Advanced Publishing Program: Boston, MA, USA, 1984; Volume 103.

© 2019 by the authors. Licensee MDPI, Basel, Switzerland. This article is an open access article distributed under the terms and conditions of the Creative Commons Attribution (CC BY) license (http://creativecommons.org/licenses/by/4.0/).

Article

Extended Convergence Analysis of the Newton–Hermitian and Skew–Hermitian Splitting Method

Ioannis K Argyros [1,*], Santhosh George [2], Chandhini Godavarma [2] and Alberto A Magreñán [3]

[1] Department of Mathematical Sciences, Cameron University, Lawton, OK 73505, USA
[2] Department of Mathematical and Computational Sciences, National Institute of Technology, Karnataka 575 025, India
[3] Departamento de Matemáticas y Computación, Universidad de la Rioja, 26006 Logroño, Spain
* Correspondence: iargyros@cameron.edu

Received: 24 June 2019; Accepted: 25 July 2019; Published: 2 August 2019

Abstract: Many problems in diverse disciplines such as applied mathematics, mathematical biology, chemistry, economics, and engineering, to mention a few, reduce to solving a nonlinear equation or a system of nonlinear equations. Then various iterative methods are considered to generate a sequence of approximations converging to a solution of such problems. The goal of this article is two-fold: On the one hand, we present a correct convergence criterion for Newton–Hermitian splitting (NHSS) method under the Kantorovich theory, since the criterion given in Numer. Linear Algebra Appl., 2011, 18, 299–315 is not correct. Indeed, the radius of convergence cannot be defined under the given criterion, since the discriminant of the quadratic polynomial from which this radius is derived is negative (See Remark 1 and the conclusions of the present article for more details). On the other hand, we have extended the corrected convergence criterion using our idea of recurrent functions. Numerical examples involving convection–diffusion equations further validate the theoretical results.

Keywords: Newton–HSS method; systems of nonlinear equations; semi-local convergence

1. Introduction

Numerous problems in computational disciplines can be reduced to solving a system of nonlinear equations with n equations in n variables like

$$F(x) = 0 \tag{1}$$

using Mathematical Modelling [1–11]. Here, F is a continuously differentiable nonlinear mapping defined on a convex subset Ω of the $n-$dimensional complex linear space $\mathbb{C}^n$ into $\mathbb{C}^n$. In general, the corresponding Jacobian matrix $F'(x)$ is sparse, non-symmetric and positive definite. The solution methods for the nonlinear problem $F(x) = 0$ are iterative in nature, since an exact solution x^* could be obtained only for a few special cases. In the rest of the article, some of the well established and standard results and notations are used to establish our results (See [3–6,10–14] and the references there in). Undoubtedly, some of the well known methods for generating a sequence to approximate x^* are the inexact Newton (IN) methods [1–3,5–14]. The IN algorithm involves the steps as given in the following:

Algorithm IN [6]

- *Step 1:* Choose initial guess x_0, tolerance value *tol*; Set $k = 0$
- *Step 2:* While $F(x_k) > tol \times F(x_0)$, Do

  1. Choose $\eta_k \in [0,1)$. Find d_k so that $\|F(x_k) + F'(x_k)d_k\| \leq \eta_k \|F(x_k)\|$.
  2. Set $x_{k+1} = x_k + d_k$; $k = k + 1$

Furthermore, if A is sparse, non-Hermitian and positive definite, the Hermitian and skew-Hermitian splitting (HSS) algorithm [4] for solving the linear system $Ax = b$ is given by,

Algorithm HSS [4]

- *Step 1:* Choose initial guess x_0, tolerance value *tol* and $\alpha > 0$; Set $l = 0$
- *Step 2:* Set $H = \frac{1}{2}(A + A^*)$ and $S = \frac{1}{2}(A - A^*)$, where H is Hermitian and S is skew-Hermitian parts of A.
- *Step 3:* While $\|b - Ax_l\| > tol \times \|b - Ax_0\|$, Do

  1. Solve $(\alpha I + H)x_{l+1/2} = (\alpha I - S)x_l + b$
  2. Solve $(\alpha I + S)x_l = (\alpha I - H)x_{l+1/2} + b$
  3. Set $l = l + 1$

Newton–HSS [5] algorithm combines appropriately both IN and HSS methods for the solution of the large nonlinear system of equations with positive definite Jacobian matrix. The algorithm is as follows:

Algorithm NHSS (The Newton–HSS method [5])

- *Step 1:* Choose initial guess x_0, positive constants α and *tol*; Set $k = 0$
- *Step 2:* While $\|F(x_k)\| > tol \times \|F(x_0)\|$

  - Compute Jacobian $J_k = F'(x_k)$
  - Set

  $$H_k(x_k) = \frac{1}{2}(J_k + J_k^*) \text{ and } S_k(x_k) = \frac{1}{2}(J_k - J_k^*), \qquad (2)$$

  where H_k is Hermitian and S_k is skew-Hermitian parts of J_k.
  - Set $d_{k,0} = 0$; $l = 0$
  - While

  $$\|F(x_k) + J_k d_{k,l}\| \geq \eta_k \times \|F(x_k)\| \quad (\eta_k \in [0,1)) \qquad (3)$$

  Do
  {
  1. Solve sequentially:

  $$(\alpha I + H_k)d_{k,l+1/2} = (\alpha I - S_k)d_{k,l} + b \qquad (4)$$
  $$(\alpha I + S_k)d_{k,l} = (\alpha I - H_k)d_{k,l+1/2} + b \qquad (5)$$

  2. Set $l = l + 1$

  }
  - Set

  $$x_{k+1} = x_k + d_{k,l}; \quad k = k + 1 \qquad (6)$$

- Compute J_k, H_k and S_k for new x_k

Please note that η_k is varying in each iterative step, unlike a fixed positive constant value in used in [5]. Further observe that if d_{k,ℓ_k} in (6) is given in terms of $d_{k,0}$, we get

$$d_{k,\ell_k} = (I - T_k^\ell)(I - T_k)^{-1} B_k^{-1} F(x_k) \tag{7}$$

where $T_k := T(\alpha, k)$, $B_k := B(\alpha, k)$ and

$$\begin{aligned} T(\alpha, x) &= B(\alpha, x)^{-1} C(\alpha, x) \\ B(\alpha, x) &= \frac{1}{2\alpha}(\alpha I + H(x))(\alpha I + S(x)) \\ C(\alpha, x) &= \frac{1}{2\alpha}(\alpha I - H(x))(\alpha I - S(x)). \end{aligned} \tag{8}$$

Using the above expressions for T_k and d_{k,ℓ_k}, we can write the Newton–HSS in (6) as

$$x_{k+1} = x_k - (I - T_k^\ell)^{-1} F'(x_k)^{-1} F(x_k). \tag{9}$$

A Kantorovich-type semi-local convergence analysis was presented in [7] for NHSS. However, there are shortcomings:

(i) The semi-local sufficient convergence criterion provided in (15) of [7] is false. The details are given in Remark 1. Accordingly, Theorem 3.2 in [7] as well as all the followings results based on (15) in [7] are inaccurate. Further, the upper bound function g_3 (to be defined later) on the norm of the initial point is not the best that can be used under the conditions given in [7].

(ii) The convergence domain of NHSS is small in general, even if we use the corrected sufficient convergence criterion (12). That is why, using our technique of recurrent functions, we present a new semi-local convergence criterion for NHSS, which improves the corrected convergence criterion (12) (see also Section 3 and Section 4, Example 4.4).

(iii) Example 4.5 taken from [7] is provided to show as in [7] that convergence can be attained even if these criteria are not checked or not satisfied, since these criteria are not sufficient too. The convergence criteria presented here are only sufficient.

Moreover, we refer the reader to [3–11,13,14] and the references therein to avoid repetitions for the importance of these methods for solving large systems of equations.

The rest of the note is organized as follows. Section 2 contains the semi-local convergence analysis of NHSS under the Kantorovich theory. In Section 3, we present the semi-local convergence analysis using our idea of recurrent functions. Numerical examples are discussed in Section 4. The article ends with a few concluding remarks.

2. Semi-Local Convergence Analysis

To make the paper as self-contained as possible we present some results from [3] (see also [7]). The semi-local convergence of NHSS is based on the conditions ($\mathcal{A}$). Let $x_0 \in \mathbb{C}^n$ and $F : \Omega \subset \mathbb{C}^n \longrightarrow \mathbb{C}^n$ be G–differentiable on an open neighborhood $\Omega_0 \subset \Omega$ on which $F'(x)$ is continuous and positive definite. Suppose $F'(x) = H(x) + S(x)$ where $H(x)$ and $S(x)$ are as in (2) with $x_k = x$.

($\mathcal{A}_1$) There exist positive constants β, γ and δ such that

$$\max\{\|H(x_0)\|, \|S(x_0)\|\} \le \beta, \ \|F'(x_0)^{-1}\| \le \gamma, \ \|F(x_0)\| \le \delta, \tag{10}$$

($\mathcal{A}_2$) There exist nonnegative constants L_h and L_s such that for all $x, y \in U(x_0, r) \subset \Omega_0$,

$$\|H(x) - H(y)\| \leq L_h \|x - y\|$$
$$\|S(x) - S(y)\| \leq L_s \|x - y\|. \tag{11}$$

Next, we present the corrected version of Theorem 3.2 in [7].

Theorem 1. *Assume that conditions ($\mathcal{A}$) hold with the constants satisfying*

$$\delta \gamma^2 L \leq \bar{g}_3(\eta) \tag{12}$$

where $\bar{g}_3(t) := \frac{(1-t)^2}{2(2+t+2t^2-t^3)}$, $\eta = \max\{\eta_k\} < 1$, $r = \max\{r_1, r_2\}$ with

$$r_1 = \frac{\alpha + \beta}{L}\left(\sqrt{1 + \frac{2\alpha\tau\theta}{(2\gamma + \gamma\tau\theta)(\alpha+\beta)^2}} - 1\right)$$
$$r_2 = \frac{b - \sqrt{b^2 - 2ac}}{a} \tag{13}$$
$$a = \frac{\gamma L(1+\eta)}{1 + 2\gamma^2 \delta L \eta}, \; b = 1 - \eta, \; c = 2\gamma\delta,$$

and with $\ell_* = \liminf_{k \to \infty} \ell_k$ satisfying $\ell_* > \lfloor \frac{\ln \eta}{\ln((\tau+1)\theta)} \rfloor$, (Here $\lfloor . \rfloor$ represents the largest integer less than or equal to the corresponding real number) $\tau \in (0, \frac{1-\theta}{\theta})$ and

$$\theta \equiv \theta(\alpha, x_0) = \|T(\alpha, x_0)\| < 1. \tag{14}$$

Then, the iteration sequence $\{x_k\}_{k=0}^{\infty}$ generated by Algorithm NHSS is well defined and converges to x_*, so that $F(x_*) = 0$.

Proof. We simply follow the proof of Theorem 3.2 in [7] but use the correct function $\bar{g}_3$ instead of the incorrect function g_3 defined in the following remark. □

Remark 1. *The corresponding result in [7] used the function bound*

$$g_3(t) = \frac{1-t}{2(1+t^2)} \tag{15}$$

instead of $\bar{g}_3$ in (12) (simply looking at the bottom of first page of the proof in Theorem 3.2 in [7]), i.e., the inequality they have considered is,

$$\delta\gamma^2 L \leq g_3(\eta). \tag{16}$$

However, condition (16) does not necessarily imply $b^2 - 4ac \geq 0$, which means that r_2 does not necessarily exist (see (13) where $b^2 - 2ac \geq 0$ is needed) and the proof of Theorem 3.2 in [7] breaks down. As an example, choose $\eta = \frac{1}{2}$, then $g_3(\frac{1}{2}) = \frac{1}{5}, \bar{g}_3(\frac{1}{2}) = \frac{1}{23}$ and for $\bar{g}_3(\frac{1}{2}) = \delta\gamma^2 L < g_3(\frac{1}{2})$, we have $b^2 - 4ac < 0$. Notice that our condition (12) is equivalent to $b^2 - 4ac \geq 0$. Hence, our version of Theorem 3.2 is correct. Notice also that

$$\bar{g}_3(t) < g_3(t) \text{ for each } t \geq 0, \tag{17}$$

so (12) implies (16) but not necessarily vice versa.

3. Semi-Local Convergence Analysis II

We need to define some parameters and a sequence needed for the semi-local convergence of NHSS using recurrent functions.

Let $\beta, \gamma, \delta, L_0, L$ be positive constants and $\eta \in [0,1)$. Then, there exists $\mu \geq 0$ such that $L = \mu L_0$. Set $c = 2\gamma\delta$. Define parameters p, q, η_0 and δ_0 by

$$p = \frac{(1+\eta)\mu\gamma L_0}{2}, \quad q = \frac{-p + \sqrt{p^2 + 4\gamma L_0 p}}{2\gamma L_0}, \tag{18}$$

$$\eta_0 = \sqrt{\frac{\mu}{\mu+2}} \tag{19}$$

and

$$\xi = \frac{\mu}{2} \min\left\{\frac{2(q-\eta)}{(1+\eta)\mu+2q}, \frac{(1+\eta)q - \eta - q^2}{(1+\eta)q - \eta}\right\}. \tag{20}$$

Moreover, define scalar sequence $\{s_k\}$ by

$$\begin{aligned} s_0 &= 0, s_1 = c = 2\gamma\delta \text{ and for each } k = 1, 2, \ldots \\ s_{k+1} &= s_k + \frac{1}{1 - \gamma L_0 s_k}[p(s_k - s_{k-1}) + \eta(1 - \gamma L_0 s_{k-1})](s_k - s_{k-1}). \end{aligned} \tag{21}$$

We need to show the following auxiliary result of majorizing sequences for NHSS using the aforementioned notation.

Lemma 1. *Let $\beta, \gamma, \delta, L_0, L$ be positive constants and $\eta \in [0,1)$. Suppose that*

$$\gamma^2 L \delta \leq \xi \tag{22}$$

and

$$\eta \leq \eta_0, \tag{23}$$

where η_0, ξ are given by (19) and (20), respectively. Then, sequence $\{s_k\}$ defined in (21) is nondecreasing, bounded from above by

$$s^{**} = \frac{c}{1-q} \tag{24}$$

and converges to its unique least upper bounds s^* which satisfies

$$c \leq s^* \leq s^{**}. \tag{25}$$

Proof. Notice that by (18)–(23) $q \in (0,1), q > \eta, \eta_0 \in [\frac{\sqrt{3}}{3}, 1), c > 0, (1+\eta)q - \eta > 0, (1+\eta)q - \eta - q^2 > 0$ and $\xi > 0$. We shall show using induction on k that

$$0 < s_{k+1} - s_k \leq q(s_k - s_{k-1}) \tag{26}$$

or equivalently by (21)

$$0 \leq \frac{1}{1 - \gamma L_0 s_k}[p(s_k - s_{k-1}) + \eta(1 - \gamma L_0 s_{k-1})] \leq q. \tag{27}$$

Estimate (27) holds true for $k = 1$ by the initial data and since it reduces to showing $\delta \leq \frac{\eta}{\gamma^2 L} \frac{q-\eta}{(1+\eta)\mu+2q}$, which is true by (20). Then, by (21) and (27), we have

$$0 < s_2 - s_1 \leq q(s_1 - s_0), \quad \gamma L_0 s_1 < 1$$

and

$$s_2 \leq s_1 + q(s_1 - s_0) = \frac{1-q^2}{1-q}(s_1 - s_0) < \frac{s_1 - s_0}{1-q} = s^{**}.$$

Suppose that (26),
$$\gamma L_0 s_k < 1 \tag{28}$$
and
$$s_{k+1} \leq \frac{1-q^{k+1}}{1-q}(s_1 - s_0) < s^{**} \tag{29}$$
hold true. Next, we shall show that they are true for k replaced by $k+1$. It suffices to show that
$$0 \leq \frac{1}{1-\gamma L_0 s_{k+1}}(p(s_{k+1} - s_k) + \eta(1-\gamma L_0 s_k)) \leq q$$
or
$$p(s_{k+1} - s_k) + \eta(1-\gamma L_0 s_k) \leq q(1-\gamma L_0 s_{k+1})$$
or
$$p(s_{k+1} - s_k) + \eta(1-\gamma L_0 s_k) - q(1-\gamma L_0 s_{k+1}) \leq 0$$
or
$$p(s_{k+1} - s_k) + \eta(1-\gamma L_0 s_1) + \gamma q L_0 s_{k+1}) - q \leq 0$$
(since $s_1 \leq s_k$) or
$$2\gamma\delta p q^k + 2\gamma^2 q L_0 \delta(1 + q + \ldots + q^k) + \eta(1 - 2\gamma^2 L_0 \delta) - q \leq 0. \tag{30}$$

Estimate (30) motivates us to introduce recurrent functions f_k defined on the interval $[0,1)$ by
$$f_k(t) = 2\gamma\delta p t^k + 2\gamma^2 L_0 \delta(1 + t + \ldots + t^k)t - t + \eta(1 - 2\gamma^2 L_0 \delta). \tag{31}$$

Then, we must show instead of (30) that
$$f_k(q) \leq 0. \tag{32}$$

We need a relationship between two consecutive functions f_k:
$$\begin{aligned} f_{k+1}(t) &= f_{k+1}(t) - f_k(t) + f_k(t) \\ &= 2\gamma\delta p t^{k+1} + 2\gamma^2 L_0 \delta(1 + t + \ldots t^{k+1})t - t \\ &\quad + \eta(1 - 2\gamma^2 L_0 \delta) - 2\gamma\delta p t^k - 2\gamma^2 L_0 \delta(1 + t + \ldots + t^k)t \\ &\quad + t - \eta(1 - 2\gamma^2 L_0 \delta) + f_k(t) \\ &= f_k(t) + 2\gamma\delta g(t) t^k, \end{aligned} \tag{33}$$

where
$$g(t) = \gamma L_0 t^2 + pt - p. \tag{34}$$

Notice that $g(q) = 0$. It follows from (32) and (34) that
$$f_{k+1}(q) = f_k(q) \text{ for each } k. \tag{35}$$

Then, since
$$f_\infty(q) = \lim_{k \to \infty} f_k(q), \tag{36}$$
it suffices to show
$$f_\infty(q) \leq 0 \tag{37}$$

instead of (32). We get by (31) that

$$f_\infty(q) = \frac{2\gamma^2 L_0 \delta q}{1-q} - q + \eta(1 - 2\gamma^2 L_0 \delta) \qquad (38)$$

so, we must show that

$$\frac{2\gamma^2 L_0 \delta q}{1-q} - q + \eta(1 - 2\gamma^2 L_0 \delta) \leq 0, \qquad (39)$$

which reduces to showing that

$$\delta \leq \frac{\mu}{2\gamma^2 L} \frac{(1+\eta)q - \eta - q^2}{(1+\eta)q - \eta}, \qquad (40)$$

which is true by (22). Hence, the induction for (26), (28) and (29) is completed. It follows that sequence $\{s_k\}$ is nondecreasing, bounded above by s^{**} and as such it converges to its unique least upper bound s^* which satisfies (25). □

We need the following result.

Lemma 2 ([14]). *Suppose that conditions (A) hold. Then, the following assertions also hold:*

(i) $\|F'(x) - F'(y)\| \leq L\|x - y\|$
(ii) $\|F'(x)\| \leq L\|x - y\| + 2\beta$
(iii) *If* $r < \frac{1}{\gamma L}$, *then* $F'(x)$ *is nonsingular and satisfies*

$$\|F'(x)^{-1}\| \leq \frac{\gamma}{1 - \gamma L\|x - x_0\|}, \qquad (41)$$

where $L = L_h + L_s$.

Next, we show how to improve Lemma 2 and the rest of the results in [3,7]. Notice that it follows from (i) in Lemma 2 that there exists $L_0 > 0$ such that

$$\|F'(x) - F'(x_0)\| \leq L_0\|x - x_0\| \text{ for each } x \in \Omega. \qquad (42)$$

We have that

$$L_0 \leq L \qquad (43)$$

holds true and $\frac{L}{L_0}$ can be arbitrarily large [2,12]. Then, we have the following improvement of Lemma 2.

Lemma 3. *Suppose that conditions (A) hold. Then, the following assertions also hold:*

(i) $\|F'(x) - F'(y)\| \leq L\|x - y\|$
(ii) $\|F'(x)\| \leq L_0\|x - y\| + 2\beta$
(iii) *If* $r < \frac{1}{\gamma L_0}$, *then* $F'(x)$ *is nonsingular and satisfies*

$$\|F'(x)^{-1}\| \leq \frac{\gamma}{1 - \gamma L_0\|x - x_0\|}. \qquad (44)$$

Proof. (ii) We have

$$\begin{aligned}
\|F'(x)\| &= \|F'(x) - F'(x_0) + F'(x_0)\| \\
&\leq \|F'(x) - F'(x_0)\| + \|F'(x_0)\| \\
&\leq L_0\|x - x_0\| + \|F'(x_0)\| \leq L_0\|x - x_0\| + 2\beta.
\end{aligned}$$

(iii)
$$\gamma \|F'(x) - F'(x_0)\| \leq \gamma L_0 \|x - x_0\| < 1. \tag{45}$$

It follows from the Banach lemma on invertible operators [1] that $F'(x)$ is nonsingular, so that (44) holds. □

Remark 2. *The new estimates (ii) and (iii) are more precise than the corresponding ones in Lemma 2, if $L_0 < L$.*

Next, we present the semi-local convergence of NHSS using the majorizing sequence $\{s_n\}$ introduced in Lemma 1.

Theorem 2. *Assume that conditions (A), (22) and (23) hold. Let $\eta = \max\{\eta_k\} < 1$, $r = \max\{r_1, t^*\}$ with*

$$r_1 = \frac{\alpha + \beta}{L}\left(\sqrt{1 + \frac{2\alpha\tau\theta}{(2\gamma + \gamma\tau\theta)(\alpha + \beta)^2}} - 1\right)$$

and s^* is as in Lemma 1 and with $\ell_* = \liminf_{k \to \infty} \ell_k$ satisfying $\ell_* > \lfloor \frac{\ln \eta}{\ln((\tau+1)\theta)} \rfloor$, (Here $\lfloor . \rfloor$ represents the largest integer less than or equal to the corresponding real number) $\tau \in (0, \frac{1-\theta}{\theta})$ and

$$\theta \equiv \theta(\alpha, x_0) = \|T(\alpha, x_0)\| < 1. \tag{46}$$

Then, the sequence $\{x_k\}_{k=0}^{\infty}$ generated by Algorithm NHSS is well defined and converges to x_*, so that $F(x_*) = 0$.

Proof. If we follow the proof of Theorem 3.2 in [3,7] but use (44) instead of (41) for the upper bound on the norms $\|F'(x_k)^{-1}\|$ we arrive at

$$\|x_{k+1} - x_k\| \leq \frac{(1+\eta)\gamma}{1 - \gamma L_0 s_k}\|F(x_k)\|, \tag{47}$$

where

$$\|F(x_k)\| \leq \frac{L}{2}(s_k - s_{k-1})^2 + \eta\frac{1 - \gamma L_0 s_{k-1}}{\gamma(1+\eta)}(s_k - s_{k-1}), \tag{48}$$

so by (21)

$$\|x_{k+1} - x_k\| \leq (1+\eta)\frac{\gamma}{1 - \gamma L_0 s_k}[\frac{L}{2}(s_k - s_{k-1}) + \eta\frac{1 - \gamma L_0 s_{k-1}}{\gamma(1+\eta)}](s_k - s_{k-1}) = s_{k+1} - s_k. \tag{49}$$

We also have that $\|x_{k+1} - x_0\| \leq \|x_{k+1} - x_k\| + \|x_k - x_{k-1}\| + \ldots + \|x_1 - x_0\| \leq s_{k+1} - s_k + s_k - s_{k-1} + \ldots + s_1 - s_0 = s_{k+1} - s_0 < s^*$. It follows from Lemma 1 and (49) that sequence $\{x_k\}$ is complete in a Banach space $\mathbb{R}^n$ and as such it converges to some $x_* \in \bar{U}(x_0, r)$ (since $\bar{U}(x_0, r)$ is a closed set). However, $\|T(\alpha; x_*)\| < 1$ [4] and NHSS, we deduce that $F(x_*) = 0$. □

Remark 3. *(a) The point s^* can be replaced by s^{**} (given in closed form by (24)) in Theorem 2.*
(b) Suppose there exist nonnegative constants L_h^0, L_s^0 such that for all $x \in U(x_0, r) \subset \Omega_0$

$$\|H(x) - H(x_0)\| \leq L_h^0 \|x - x_0\|$$

and

$$\|S(x) - S(x_0)\| \leq L_s^0 \|x - x_0\|.$$

Set $L_0 = L_h^0 + L_s^0$. Define $\Omega_0^1 = \Omega_0 \cap U(x_0, \frac{1}{\gamma L_0})$. Replace condition $(\mathcal{A}_2)$ by
$(\mathcal{A}_2')$ There exist nonnegative constants L_h' and L_s' such that for all $x, y \in U(x_0, r) \subset \Omega_0^1$

$$\|H(x) - H(y)\| \le L_h'\|x - y\|$$

$$\|S(x) - S(y)\| \le L_s'\|x - y\|.$$

Set $L' = L_h' + L_s'$. Notice that
$$L_h' \le L_h, \ L_s' \le L_s \text{ and } L' \le L, \tag{50}$$

since $\Omega_0^1 \subseteq \Omega_0$. Denote the conditions $(\mathcal{A}_1)$ and $(\mathcal{A}_2')$ by $(\mathcal{A}')$. Then, clearly the results of Theorem 2 hold with conditions $(\mathcal{A}')$, Ω_0^1, L' replacing conditions $(\mathcal{A})$, Ω_0 and L, respectively (since the iterates $\{x_k\}$ remain in Ω_0^1 which is a more precise location than Ω_0). Moreover, the results can be improved even further, if we use the more accurate set Ω_0^2 containing iterates $\{x_k\}$ defined by $\Omega_0^2 := \Omega \cap U(x_1, \frac{1}{\gamma L_0} - \gamma \delta)$. Denote corresponding to L' constant by L'' and corresponding conditions to $(\mathcal{A}')$ by $(\mathcal{A}'')$. Notice that (see also the numerical examples) $\Omega_0^2 \subseteq \Omega_0^1 \subseteq \Omega_0$. In view of (50), the results of Theorem 2 are improved and under the same computational cost.

(c) The same improvements as in (b) can be made in the case of Theorem 1.

The majorizing sequence $\{t_n\}$ in [3,7] is defined by

$$\begin{aligned} t_0 &= 0, t_1 = c = 2\gamma\delta \\ t_{k+1} &= t_k + \frac{1}{1 - \gamma L t_k}[p(t_k - t_{k-1}) + \eta(1 - \gamma L t_{k-1})](t_k - t_{k-1}). \end{aligned} \tag{51}$$

Next, we show that our sequence $\{s_n\}$ is tighter than $\{t_n\}$.

Proposition 1. *Under the conditions of Theorems 1 and 2, the following items hold*

(i) $s_n \le t_n$
(ii) $s_{n+1} - s_n \le t_{n+1} - t_n$ and
(iii) $s^* \le t^* = \lim_{k \to \infty} t_k \le r_2$.

Proof. We use a simple inductive argument, (21), (51) and (43). □

Remark 4. *Majorizing sequences using L' or L'' are even tighter than sequence $\{s_n\}$.*

4. Special Cases and Numerical Examples

Example 1. *The semi-local convergence of inexact Newton methods was presented in [14] under the conditions*

$$\begin{aligned} \|F'(x_0)^{-1}F(x_0)\| &\le \beta, \\ \|F'(x_0)^{-1}(F'(x) - F'(y))\| &\le \gamma\|x - y\|, \\ \frac{\|F'(x_0)^{-1}s_n\|}{\|F'(x_0)^{-1}F(x_n)\|} &\le \eta_n \end{aligned}$$

and
$$\beta\gamma \le g_1(\eta),$$
where
$$g_1(\eta) = \frac{\sqrt{(4\eta+5)^3} - (2\eta^3 + 14\eta + 11)}{(1+\eta)(1-\eta)^2}.$$

More recently, Shen and Li [11] substituted $g_1(\eta)$ with $g_2(\eta)$, where

$$g_2(\eta) = \frac{(1-\eta)^2}{(1+\eta)(2(1+\eta) - \eta(1-\eta)^2)}.$$

Estimate (22) can be replaced by a stronger one but directly comparable to (20). Indeed, let us define a scalar sequence $\{u_n\}$ (less tight than $\{s_n\}$) by

$$\begin{aligned} u_0 = 0, u_1 &= 2\gamma\delta, \\ u_{k+1} &= u_k + \frac{(\frac{1}{2}\rho(u_k - u_{k-1}) + \eta)}{1 - \rho u_k}(u_k - u_{k-1}), \end{aligned} \quad (52)$$

where $\rho = \gamma L_0(1+\eta)\mu$. Moreover, define recurrent functions f_k on the interval $[0,1)$ by

$$f_k(t) = \frac{1}{2}\rho c t^{k-1} + \rho c(1 + t + \ldots + t^{k-1})t + \eta - t$$

and function $g(t) = t^2 + \frac{t}{2} - \frac{1}{2}$. Set $q = \frac{1}{2}$. Moreover, define function g_4 on the interval $[0, \frac{1}{2})$ by

$$g_4(\eta) = \frac{1 - 2\eta}{4(1 + \eta)}. \quad (53)$$

Then, following the proof of Lemma 1, we obtain:

Lemma 4. Let $\beta, \gamma, \delta, L_0, L$ be positive constants and $\eta \in [0, \frac{1}{2})$. Suppose that

$$\gamma^2 L\delta \leq g_4(\eta) \quad (54)$$

Then, sequence $\{u_k\}$ defined by (52) is nondecreasing, bounded from above by

$$u^{**} = \frac{c}{1-q}$$

and converges to its unique least upper bound u^* which satisfies

$$c \leq u^* \leq u^{**}.$$

Proposition 2. Suppose that conditions (A) and (54) hold with $r = \min\{r_1, u^*\}$. Then, sequence $\{x_n\}$ generated by algorithm NHSS is well defined and converges to x_* which satisfies $F(x_*) = 0$.

These bound functions are used to obtain semi-local convergence results for the Newton–HSS method as a subclass of these techniques. In Figures 1 and 2, we can see the graphs of the four bound functions $g_1, g_2, \bar{g}_3$ and g_4. Clearly our bound function $\bar{g}_3$ improves all the earlier results. Moreover, as noted before, function g_3 cannot be used, since it is an incorrect bound function.

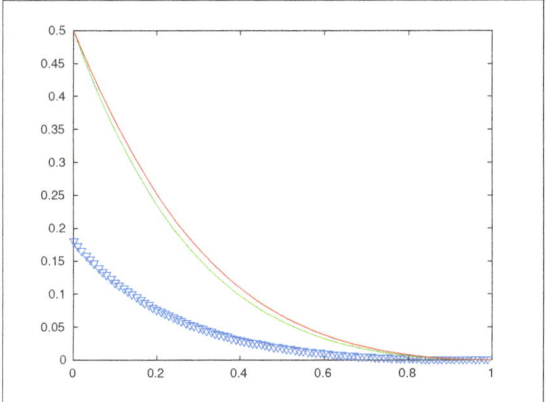

Figure 1. Graphs of $g_1(t)$ (Violet), $g_2(t)$ (Green), $\bar{g}_3$ (Red).

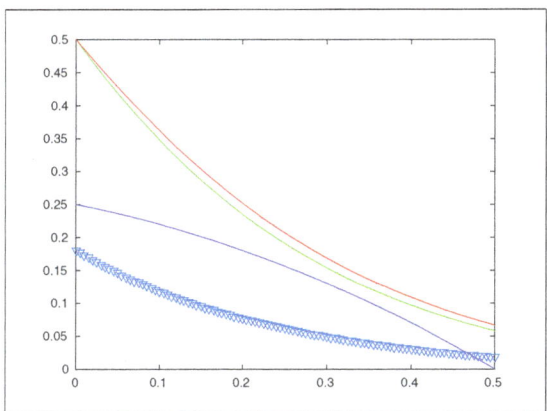

Figure 2. Graphs of $g_1(t)$ (Violet), $g_2(t)$ (Green), $\bar{g}_3$ (Red) and g_4 (Blue).

In the second example we compare the convergence criteria (22) and (12).

Example 2. Let $\eta = 1, \Omega_0 = \Omega = U(x_0, 1 - \lambda), x_0 = 1, \lambda \in [0, 1)$. Define function F on Ω by

$$F(x) = x^3 - \lambda. \tag{55}$$

Then, using (55) and the condition ($\mathcal{A}$), we get $\gamma = \frac{1}{3}, \delta = 1 - \lambda, L = 6(2 - \lambda), L_0 = 3(3 - \lambda)$ and $\mu = \frac{2(2-\lambda)}{3-\lambda}$. Choosing $\lambda = 0.8$., we get $L = 7.2, L_0 = 6.6, \delta = 0.2, \mu = 1.0909091, \eta_0 = 0.594088525, p = 1.392, q = 0.539681469, \gamma^2 L\delta = 0.16$. Let $\eta = 0.16 < \eta_0$, then, $\bar{g}_3(0.16) = 0.159847474$, $\xi = \min\{0.176715533, 0.20456064\} = 0.176715533$. Hence the old condition (12) is not satisfied, since $\gamma^2 L\delta > \bar{g}_3(0.16)$. However, the new condition (22) is satisfied, since $\gamma^2 L\delta < \xi$. Hence, the new results expand the applicability of NHSS method.

The next example is used for the reason already mentioned in (iii) of the introduction.

Example 3. *Consider the two-dimensional nonlinear convection–diffusion equation [7]*

$$-(u_{xx} + u_{yy}) + q(u_x + u_y) = -e^u, \quad (x,y) \in \Omega$$
$$u(x,y) = 0 \quad (x,y) \in \partial\Omega \tag{56}$$

where $\Omega = (0,1) \times (0,1)$ and $\partial\Omega$ is the boundary of Ω. Here $q > 0$ is a constant to control the magnitude of the convection terms (see [7,15,16]). As in [7], we use classical five-point finite difference scheme with second order central difference for both convection and diffusion terms. If N defines number of interior nodes along one co-ordinate direction, then $h = \frac{1}{N+1}$ and $Re = \frac{qh}{2}$ denotes the equidistant step-size and the mesh Reynolds number, respectively. Applying the above scheme to (56), we obtain the following system of nonlinear equations:

$$\tilde{A}u + h^2 e^u = 0$$
$$u = (u_1, u_2, \ldots, u_N)^T, \; u_i = (u_{i1}, u_{i2}, \ldots, u_{iN})^T, \; i = 1, 2, \ldots, N,$$

where the coefficient matrix $\tilde{A}$ is given by

$$\tilde{A} = T_x \otimes I + I \otimes T_y.$$

Here, $\otimes$ is the Kronecker product, T_x and T_y are the tridiagonal matrices

$$T_x = tridiag(-1 - Re, 4, -1 + Re), T_y = tridiag(-1 - Re, 0, -1 + Re).$$

In our computations, N is chosen as 99 so that the total number of nodes are 100×100. We use $\alpha = \frac{qh}{2}$ as in [7] and we consider two choices for η_k i.e., $\eta_k = 0.1$ and $\eta_k = 0.01$ for all k.

The results obtained in our computation is given in Figures 3–6. The total number of inner iterations is denoted by IT, the total number of outer iterations is denoted by OT and the total CPU time is denoted by t.

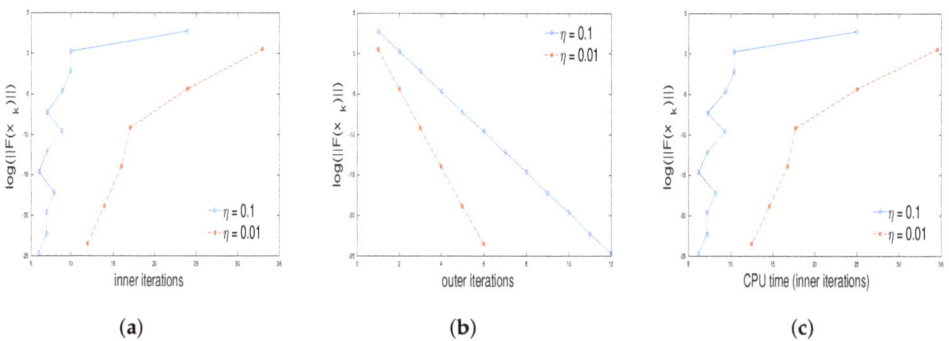

Figure 3. Plots of (**a**) inner iterations vs. $\log(\|F(x_k)\|)$, (**b**) outer iterations vs. $\log(\|F(x_k)\|)$, (**c**) CPU time vs. $\log(\|F(x_k)\|)$ for $q = 600$ and $x_0 = e$.

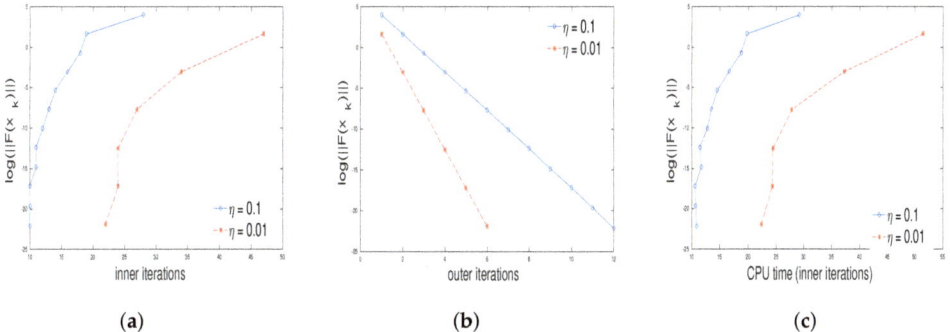

Figure 4. Plots of (**a**) inner iterations vs. $\log(\|F(x_k)\|)$, (**b**) outer iterations vs. $\log(\|F(x_k)\|)$, (**c**) CPU time vs. $\log(\|F(x_k)\|)$ for $q = 2000$ and $x_0 = e$.

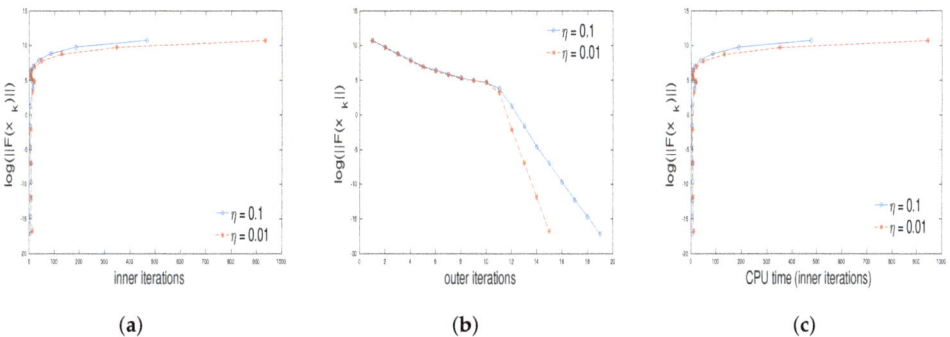

Figure 5. Plots of (**a**) inner iterations vs. $\log(\|F(x_k)\|)$, (**b**) outer iterations vs. $\log(\|F(x_k)\|)$, (**c**) CPU time vs. $\log(\|F(x_k)\|)$ for $q = 600$ and $x_0 = 6e$.

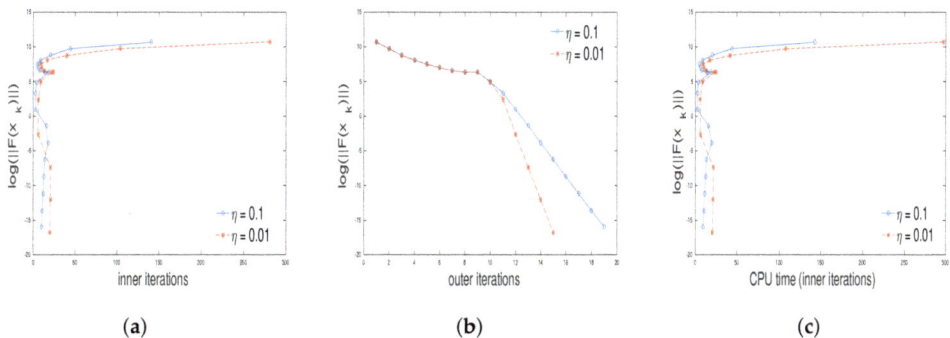

Figure 6. Plots of (**a**) inner iterations vs. $\log(\|F(x_k)\|)$, (**b**) outer iterations vs. $\log(\|F(x_k)\|)$, (**c**) CPU time vs. $\log(\|F(x_k)\|)$ for $q = 2000$ and $x_0 = 6e$.

5. Conclusions

A major problem for iterative methods is the fact that the convergence domain is small in general, limiting the applicability of these methods. Therefore, the same is true, in particular for Newton–Hermitian, skew-Hermitian and their variants such as the NHSS and other related methods [4–6,11,13,14]. Motivated by the work in [7] (see also [4–6,11,13,14]) we:

(a) Extend the convergence domain of NHSS method without additional hypotheses. This is done in Section 3 using our new idea of recurrent functions. Examples, where the new sufficient convergence criteria hold (but not previous ones), are given in Section 4 (see also the remarks in Section 3).

(b) The sufficient convergence criterion (16) given in [7] is false. Therefore, the rest of the results based on (16) do not hold. We have revisited the proofs to rectify this problem. Fortunately, the results can hold if (16) is replaced with (12). This can easily be observed in the proof of Theorem 3.2 in [7]. Notice that the issue related to the criteria (16) is not shown in Example 4.5, where convergence is established due to the fact that the validity of (16) is not checked. The convergence criteria obtained here are not necessary too. Along the same lines, our technique in Section 3 can be used to extend the applicability of other iterative methods discussed in [1–6,8,9,12–16].

Author Contributions: Conceptualization: I.K.A., S.G.; Editing: S.G., C.G.; Data curation: C.G. and A.A.M.

Funding: This research received no external funding.

Conflicts of Interest: The authors declare no conflict of interest.

References

1. Argyros, I.K.; Szidarovszky, F. *The Theory and Applications of Iteration Methods*; CRC Press: Boca Raton, FL, USA, 1993.
2. Argyros, I.K.; Magréñan, A.A. *A Contemporary Study of Iterative Methods*; Elsevier (Academic Press): New York, NY, USA, 2018.
3. Argyros, I.K.; George, S. Local convergence for an almost sixth order method for solving equations under weak conditions. *SeMA J.* **2018**, *75*, 163–171. [CrossRef]
4. Bai, Z.Z.; Golub, G.H.; Ng, M.K. Hermitian and skew-Hermitian splitting methods for non-Hermitian positive definite linear systems. *SIAM J. Matrix Anal. Appl.* **2003**, *24*, 603–626. [CrossRef]
5. Bai, Z.Z.; Guo, X.P. The Newton-HSS methods for systems of nonlinear equations with positive-definite Jacobian matrices. *J. Comput. Math.* **2010**, *28*, 235–260.
6. Dembo, R.S.; Eisenstat, S.C.; Steihaug, T. Inexact Newton methods. *SIAM J. Numer. Anal.* **1982**, *19*, 400–408. [CrossRef]
7. Guo, X.P.; Duff, I.S. Semi-local and global convergence of the Newton-HSS method for systems of nonlinear equations. *Numer. Linear Algebra Appl.* **2011**, *18*, 299–315. [CrossRef]
8. Magréñan, A.A. Different anomalies in a Jarratt family of iterative root finding methods. *Appl. Math. Comput.* **2014**, *233*, 29–38.
9. Magréñan, A.A. A new tool to study real dynamics: The convergence plane. *Appl. Math. Comput.* **2014**, *248*, 29–38. [CrossRef]
10. Ortega, J.M.; Rheinboldt, W.C. *Iterative Solution of Nonlinear Equations in Several Variables*; Academic Press: New York, NY, USA, 1970.
11. Shen, W.P.; Li, C. Kantorovich-type convergence criterion for inexact Newton methods. *Appl. Numer. Math.* **2009**, *59*, 1599–1611. [CrossRef]
12. Argyros, I.K. Local convergence of inexact Newton-like-iterative methods and applications. *Comput. Math. Appl.* **2000**, *39*, 69–75. [CrossRef]
13. Eisenstat, S.C.; Walker, H.F. Choosing the forcing terms in an inexact Newton method. *SIAM J. Sci. Comput.* **1996**, *17*, 16–32. [CrossRef]
14. Guo, X.P. On semilocal convergence of inexact Newton methods. *J. Comput. Math.* **2007**, *25*, 231–242.

15. Axelsson, O.; Catey, G.F. On the numerical solution of two-point singularly perturbed value problems, Computer Methods in Applied Mechanics and Engineering. *Comput. Methods Appl. Mech. Eng.* **1985**, *50*, 217–229. [CrossRef]
16. Axelsson, O.; Nikolova, M. Avoiding slave points in an adaptive refinement procedure for convection-diffusion problems in 2D. *Computing* **1998**, *61*, 331–357. [CrossRef]

© 2019 by the authors. Licensee MDPI, Basel, Switzerland. This article is an open access article distributed under the terms and conditions of the Creative Commons Attribution (CC BY) license (http://creativecommons.org/licenses/by/4.0/).

Article

Efficient Three-Step Class of Eighth-Order Multiple Root Solvers and Their Dynamics

R. A. Alharbey [1], Munish Kansal [2], Ramandeep Behl [1,*] and J. A. Tenreiro Machado [3]

1. Department of Mathematics, King Abdulaziz University, Jeddah 21589, Saudi Arabia
2. School of Mathematics, Thapar Institute of Engineering and Technology, Patiala 147004, India
3. ISEP-Institute of Engineering, Polytechnic of Porto Department of Electrical Engineering, 431, 4294-015 Porto, Portugal
* Correspondence: ramanbehl87@yahoo.in

Received: 27 May 2019; Accepted: 19 June 2019; Published: 26 June 2019

Abstract: This article proposes a wide general class of optimal eighth-order techniques for approximating multiple zeros of scalar nonlinear equations. The new strategy adopts a weight function with an approach involving the function-to-function ratio. An extensive convergence analysis is performed for the eighth-order convergence of the algorithm. It is verified that some of the existing techniques are special cases of the new scheme. The algorithms are tested in several real-life problems to check their accuracy and applicability. The results of the dynamical study confirm that the new methods are more stable and accurate than the existing schemes.

Keywords: multiple roots; optimal iterative methods; scalar equations; order of convergence

1. Introduction

Solving nonlinear problems is very important in numerical analysis and finds many applications in physics, engineering, and other applied sciences [1,2]. These problems occur in a variety of areas such as initial and boundary values, heat and fluid flow, electrostatics, or even in global positioning systems (GPS) [3–6]. It is difficult to find analytical solutions for nonlinear problems, but numerical techniques may be used to obtain approximate solutions. Therefore, iterative schemes provide an attractive alternative to solve such problems. When we discuss iterative solvers for finding multiple roots of nonlinear equations of the form $f(x) = 0$, where $f(x)$ is a real function defined in a domain $D \subseteq \mathbb{R}$, we recall the classical modified Newton's method [1,2,7] (also known as Rall's algorithm), given by:

$$x_{n+1} = x_n - m\frac{f(x_n)}{f'(x_n)}, \; n = 0, 1, 2, 3, \ldots, \qquad (1)$$

where m is the multiplicity of the required solution. Given the multiplicity $m \geq 1$, in advance, the algorithm converges quadratically for multiple roots. We find one-point iterative functions in the literature, but in the scope of the real world, they are not of practical interest because of their theoretical limitations regarding convergence order and the efficiency index. Moreover, most of the one-point techniques are computationally expensive and inefficient when they are tested on numerical examples. Therefore, multipoint iterative algorithms are better candidates to qualify as efficient solvers. The good thing about multipoint iterative schemes without memory for scalar nonlinear equations is that we can establish a conjecture about their convergence order. According to the Kung–Traub conjecture [1], any multipoint method without memory can reach its convergence order of at most 2^{n-1} for n functional evaluations. A number of researchers proposed various optimal fourth-order techniques (requiring three functional evaluations per iteration) [8–13] and non-optimal approaches [14,15] for approximating multiple zeros of nonlinear functions. Nonetheless, a limited number of multipoint point iterative algorithms having a sixth-order of convergence were formulated. Thukral [16] proposed the following sixth-order multipoint iteration scheme:

$$
\begin{aligned}
y_n &= x_n - m \frac{f(x_n)}{f'(x_n)}, \\
z_n &= x_n - m \frac{f(x_n)}{f'(x_n)} \sum_{i=1}^{3} \left(\frac{f(y_n)}{f(x_n)} \right)^{\frac{i}{m}}, \\
x_{n+1} &= z_n - m \frac{f(x_n)}{f'(x_n)} \left(\frac{f(z_n)}{f(x_n)} \right)^{\frac{1}{m}} \left[\sum_{i=1}^{3} \left(\frac{f(y_n)}{f(x_n)} \right)^{\frac{i}{m}} \right]^2.
\end{aligned} \qquad (2)
$$

Geum et al. [17] presented a non-optimal class of two-point sixth-order as follows:

$$
\begin{aligned}
y_n &= x_n - m \frac{f(x_n)}{f'(x_n)}, \quad m > 1, \\
x_{n+1} &= y_n - Q(u_n, s_n) \frac{f(y_n)}{f'(y_n)},
\end{aligned} \qquad (3)
$$

where $u_n = \sqrt[m]{\frac{f(y_n)}{f(x_n)}}$, $s_n = \sqrt[m-1]{\frac{f'(y_n)}{f'(x_n)}}$ and $Q : \mathbb{C} \to \mathbb{C}$ is a holomorphic function in the neighborhood of origin $(0,0)$. However, the main drawback of this algorithm is that it is not valid for simple zeros (i.e., for $m = 1$).

In 2016, Geum et al. [18] developed another non-optimal family of three-point sixth-order techniques for multiple zeros consisting of the steps:

$$
\begin{aligned}
y_n &= x_n - m \frac{f(x_n)}{f'(x_n)}, \quad m \geq 1, \\
w_n &= y_n - m G(u_n) \frac{f(x_n)}{f'(x_n)}, \\
x_{n+1} &= w_n - m K(u_n, v_n) \frac{f(x_n)}{f'(x_n)},
\end{aligned} \qquad (4)
$$

where $u_n = \sqrt[m]{\frac{f(y_n)}{f(x_n)}}$ and $v_n = \sqrt[m]{\frac{f(w_n)}{f(x_n)}}$. The weight functions $G : \mathbb{C} \to \mathbb{C}$ and $K : \mathbb{C}^2 \to \mathbb{C}$ are analytic in the neighborhood of zero and $(0,0)$, respectively. It can be seen that (3) and (4) require four function evaluations to achieve sixth-order convergence. Therefore, they are not optimal in the sense of the Kung and Traub conjecture [1]. It is needless to mention that several authors have tried to develop optimal eighth-order techniques for multiple zeros, but without success to the authors' best knowledge. Motivated by this fact, Behl et al. [19] introduced an optimal eighth-order iterative family for multiple roots given by:

$$
\begin{aligned}
y_n &= x_n - m \frac{f(x_n)}{f'(x_n)}, \\
z_n &= y_n - u_n Q(h_n) \frac{f(x_n)}{f'(x_n)}, \\
x_{n+1} &= z_n - u_n v_n G(h_n, v_n) \frac{f(x_n)}{f'(x_n)},
\end{aligned} \qquad (5)
$$

where $Q : \mathbb{C} \to \mathbb{C}$ is analytic in the neighborhood of (0) and $G : \mathbb{C}^2 \to \mathbb{C}$ is holomorphic in the neighborhood of $(0,0)$, with $u_n = \left(\frac{f(y_n)}{f(x_n)} \right)^{\frac{1}{m}}$, $h_n = \frac{u_n}{a_1 + a_2 u_n}$, and $v_n = \left(\frac{f(z_n)}{f(y_n)} \right)^{\frac{1}{m}}$. Moreover, a_1 and a_2 and free disposable real parameters.

Zafar et al. [20] presented an optimal eighth-order family using the weight function approach as follows:

$$y_n = x_n - m\frac{f(x_n)}{f'(x_n)},$$
$$z_n = y_n - mu_n H(u_n)\frac{f(x_n)}{f'(x_n)}, \quad (6)$$
$$x_{n+1} = z_n - u_n v_n (A_2 + A_3 u_n) P(v_n) G(w_n)\frac{f(x_n)}{f'(x_n)},$$

where A_2 and A_3 are real parameters and the weight functions $H, P, G : \mathbb{C} \to \mathbb{C}$ are analytic in the neighborhood of zero, with $u_n = \left(\frac{f(y_n)}{f(x_n)}\right)^{\frac{1}{m}}$, $v_n = \left(\frac{f(z_n)}{f(y_n)}\right)^{\frac{1}{m}}$, and $w_n = \left(\frac{f(z_n)}{f(x_n)}\right)^{\frac{1}{m}}$.

It is clear from the above review of the state-of-the-art that we have a very small number of optimal eighth-order techniques that can handle the case of multiple zeros. Moreover, these types of methods have not been discussed in depth to date. Therefore, the main motivation of the current research work is to present a new optimal class of iterative functions having eighth-order convergence, exploiting the weight function technique for computing multiple zeros. The new scheme requires only four function evaluations (i.e., $f(x_n)$, $f'(x_n)$, $f(y_n)$ and $f(z_n)$) per iteration, which is in accordance with the classical Kung–Traub conjecture. It is also interesting to note that the optimal eighth-order family (5) proposed by Behl et al. [19] can be considered as a special case of (7) for some particular values of the free parameters. In fact, the Artidiello et al. [21] family can be obtained as a special case of (7) in the case of simple roots. Therefore, the new algorithm can be treated as a more general family for approximating multiple zeros of nonlinear functions.

The rest of the paper is organized as follows. Section 2 presents the new eighth-order scheme and its convergence analysis. Section 2.1 discuss some special cases based on the different choices of weight functions employed in the second and third substeps of (7). Section 3 is devoted to the numerical experiments and the analysis of the dynamical behavior, which illustrate the efficiency, accuracy, and stability of (7). Section 4 presents the conclusions.

2. Construction of the Family

In this section, we develop a new optimal eighth-order scheme for multiple roots with known multiplicity $m \geq 1$. Here, we establish the main theorem describing the convergence analysis of the proposed family with the three steps as follows:

$$y_n = x_n - m\frac{f(x_n)}{f'(x_n)},$$
$$z_n = y_n - mu_n\frac{f(x_n)}{f'(x_n)}H(t_n), \quad (7)$$
$$x_{n+1} = z_n - u_n v_n \frac{f(x_n)}{f'(x_n)}G(t_n, s_n),$$

where the weight functions H and G are such that $H : \mathbb{C} \to \mathbb{C}$ is analytic in the neighborhood of origin and $G : \mathbb{C}^2 \to \mathbb{C}$ is holomorphic in the neighborhoods of $(0,0)$, with $u_n = \left(\frac{f(y_n)}{f(x_n)}\right)^{\frac{1}{m}}$, $t_n = \frac{u_n}{b_1+b_2 u_n}$, $v_n = \left(\frac{f(z_n)}{f(y_n)}\right)^{\frac{1}{m}}$, and $s_n = \frac{v_n}{b_3+b_4 v_n}$, $b_i \in \mathbb{R}$ (for $i = 1,2,3,4$) being arbitrary parameters.

In the following Theorem 1, we demonstrate how to construct weight functions H and G so that the algorithm arrives at the eighth order without requiring any additional functional evaluations.

Theorem 1. *Assume that $f : \mathbb{C} \to \mathbb{C}$ is an analytic function in the region enclosing the multiple zero $x = \alpha$ with multiplicity $m \geq 1$. The iterative Equation (7) has eighth-order convergence when it satisfies the conditions:*

$$\begin{cases} H(0) = 1, \quad H'(0) = 2b_1, \quad G_{00} = m, \quad G_{10} = 2mb_1, \quad G_{01} = mb_3, \\ G_{20} = m(H''(0) + 2b_1^2), \quad G_{11} = 4mb_1b_3, \quad G_{30} = m(H^{(3)}(0) + 6b_1 H''(0) - 24b_1^3 - 12b_1^2 b_2), \end{cases} \quad (8)$$

where $G_{ij} = \left. \dfrac{\partial^{i+j} G}{\partial t^i \partial s^j} \right|_{(t=0,s=0)}$, $i, j \in \{0, 1, 2, 3, 4\}$.

Proof. Let $x = \alpha$ be a multiple zero of $f(x)$. Using the Taylor series expansion of $f(x_n)$ and $f'(x_n)$ in the neighborhood of α, we obtain:

$$f(x_n) = \frac{f^{(m)}(\alpha)}{m!} e_n^m \left(1 + c_1 e_n + c_2 e_n^2 + c_3 e_n^3 + c_4 e_n^4 + c_5 e_n^5 + c_6 e_n^6 + c_7 e_n^7 + c_8 e_n^8 + O(e_n^9) \right) \quad (9)$$

and:

$$f'(x_n) = \frac{f^{m}(\alpha)}{m!} e_n^{m-1} \Big(m + c_1(m+1)e_n + c_2(m+2)e_n^2 + c_3(m+3)e_n^3 + c_4(m+4)e_n^4 + c_5(m+5)e_n^5 \\ + c_6(m+6)e_n^6 + c_7(m+7)e_n^7 + c_8(m+8)e_n^8 + O(e_n^9) \Big), \quad (10)$$

respectively, where $e_n = x_n - \alpha$, $c_k = \dfrac{1}{k!} \dfrac{f^{(k)}(\alpha)}{f'(\alpha)}$, and $k = 1, 2, 3, \ldots$.

Using the above Equations (9) and (10) in the first substep of (7), we get:

$$y_n - \alpha = \frac{c_1 e_n^2}{m} + \frac{(-(1+m)c_1^2 + 2mc_2)e_n^3}{m^2} + \sum_{j=1}^{5} \Gamma_j e_n^{j+3} + O(e_n^9), \quad (11)$$

where $\Gamma_j = \Gamma_j(m, c_1, c_2, \ldots, c_8)$ are given in terms of $m, c_1, c_2, c_3, \ldots, c_8$ for $1 \leq j \leq 5$. The explicit expressions for the first two terms Γ_1 and Γ_2 are given by $\Gamma_1 = \frac{1}{m^3}\{3m^2 c_3 + (m+1)^2 c_1^3 - m(3m+4)c_2 c_1\}$ and $\Gamma_2 = \frac{1}{m^4}\{(m+1)^3 c_1^4 - 2m(2m^2 + 5m + 3)c_2 c_1^2 + 2m^2(2m+3)c_3 c_1 + 2m^2(c_2^2(m+2) - 2c_4 m)\}$.

Using the Taylor series expansion again, we obtain:

$$f(y_n) = f^{(m)}(\alpha) e_n^{2m} \Bigg[\frac{\left(\frac{c_1}{m}\right)^m}{m!} + \frac{(2c_2 m - c_1^2(m+1))\left(\frac{c_1}{m}\right)^m e_n}{c_1 m!} + \left(\frac{c_1}{m}\right)^{1+m} \frac{1}{2m! c_1^3}\{(3 + 3m + 3m^2 + m^3)c_1^4 \\ - 2m(2 + 3m + 2m^2)c_1^2 c_2 + 4(-1+m)m^2 c_2^2 + 6m^2 c_1 c_3\}e_n^2 + \sum_{j=1}^{5} \tilde{\Gamma}_j e_n^{j+3} + O(e_n^9) \Bigg] \quad (12)$$

and:

$$u_n = \frac{c_1 e_n}{m} + \frac{(2c_2 m - c_1^2(m+2))e_n^2}{m^2} + \gamma_1 e_n^3 + \gamma_2 e_n^4 + \gamma_3 e_n^5 + O(e_n^6), \quad (13)$$

where:

$$\begin{cases} \gamma_1 = \dfrac{1}{2m^3}[c_1^3(2m^2+7m+7)+6c_3m^2-2c_2c_1m(3m+7)], \\ \gamma_2 = -\dfrac{1}{6m^4}[c_1^4(6m^3+29m^2+51m+34)-6c_2c_1^2m(4m^2+16m+17)+12c_3c_1m^2(2m+5)+12m^2(c_2^2(m+3) \\ \quad -2c_4m)], \\ \gamma_3 = \dfrac{1}{24m^5}\bigl[-24m^3(c_2c_3(5m+17)-5c_5m)+12c_3c_1^2m^2(10m^2+43m+49)+12c_1m^2\{c_2^2(10m^2+47m+53) \\ \quad -2c_4m(5m+13)\}-4c_2c_1^3m(30m^3+163m^2+306m+209)+c_1^5(24m^4+146m^3+355m^2+418m+209)\bigr]. \end{cases} \qquad (14)$$

Now, using the above Equation (14), we get:

$$t_n = \frac{c_1}{mb_1}e_n + \sum_{i=1}^{4}\Theta_j e_n^{j+1} + O(e_n^6), \qquad (15)$$

where $\Theta_j = \Theta_j(b_1,b_2,m,c_1,c_2,\ldots,c_8)$ are given in terms of $b_1,b_2,m,c_1,c_2,\ldots,c_8$, and the two coefficients Θ_1 and Θ_2 are written explicitly as $\Theta_1 = -\dfrac{b_2c_1^2+b_1((2+m)c_1^2-2mc_2)}{m^2b_1^2}$, $\Theta_2 = \dfrac{1}{2m^3b_1^3}[2b_2^2c_1^3 + 4b_1b_2c_1((2+m)c_1^2-2mc_2)+b_1^2\{(7+7m+2m^2)c_1^3-2m(7+3m)c_1c_2+6m^2c_3\}]$.

Since we have $t_n = \dfrac{u_n}{b_1+b_2u_n} = O(e_n)$, it suffices to expand weight function $H(t_n)$ in the neighborhood of origin by means of Taylor expansion up to the fifth-order term, yielding:

$$H(t_n) \approx H(0)+H'(0)t_n+\frac{1}{2!}H''(0)t_n^2+\frac{1}{3!}H^{(3)}(0)t_n^3+\frac{1}{4!}H^{(4)}(0)t_n^4+\frac{1}{5!}H^{(5)}(0)t_n^5, \qquad (16)$$

where $H^{(k)}$ represents the kth derivative. By inserting the Equations (9)–(16) in the second substep of (7), we have:

$$z_n - \alpha = \frac{(m-H(0))c_1}{m^2}e_n^2 + \frac{2m(m-H(0))b_1c_2-(H'(0)+(m+m^2-3H(0)-mH(0))b_1)c_1^2}{m^3b_1}e_n^3 \\ + \sum_{s=1}^{5}\Omega_s e_n^{s+3}+O(e_n^9), \qquad (17)$$

where $\Omega_s = \Omega_s(H(0),H'(0),H''(0),H^{(3)}(0),H^{(4)}(0),m,b_1,b_2,c_1,c_2,\ldots,c_8)$, $s=1,2,3,4,5$.

From the error Equation (17), it is clear that to obtain at least fourth-order convergence, the coefficients of e_n^2 and e_n^3 must vanish simultaneously. This result is possible only for the following values of $H(0)$ and $H'(0)$, namely:

$$H(0) = m, \quad H'(0) = 2mb_1, \qquad (18)$$

which can be calculated from the Equation (17).

Substituting the above values of $H(0)$ and $H'(0)$ in (17), we obtain:

$$z_n - \alpha = \frac{(m(9+m)b_1^2 - H''(0)+4mb_1b_2)c_1^3 - 2m^2b_1^2c_1c_2}{2m^4b_1^2}e_n^4 + \sum_{r=1}^{4}L_r e_n^{s+4}+O(e_n^9), \qquad (19)$$

where $L_r = L_r(H''(0),H^{(3)}(0),H^{(4)}(0),m,b_1,b_2,c_1,c_2,\ldots,c_8)$, $r=1,2,3,4$.

Using the Taylor series expansion again, we can write:

$$f(z_n) = f^{(m)}(\alpha)e_n^{4m}\left[\frac{2-m}{m!}\left(\frac{(m(9+m)b_1^2-H''(0)+4mb_1b_2)c_1^3-2m^2b_1^2c_1c_2}{m^4b_1^2}\right)^m + \sum_{s=1}^{5}\bar{P}_s e_n^s + O(e_n^6)\right], \qquad (20)$$

and:
$$v_n = \frac{(m(9+m)b_1^2 - H''(0) + 4mb_1b_2)c_1^2 - 2m^2b_1^2c_2}{2m^3b_1^2}e_n^2 + \Delta_1 e_n^3 + \Delta_2 e_n^4 + \Delta_3 e_n^5 + O(e_n^6), \qquad (21)$$

where $\Delta_1 = \frac{1}{3m^4 b_1^3}[3H''(0)b_2 c_1^3 - 12mb_1^2 b_2 c_1((3+m)c_1^2 - 2mc_2) + 3b_1(((3+m)H''(0) - 2mb_2^2)c_1^3 - 2mH''(0)c_1 c_2) - mb_1^3\{(49 + 27m + 2m^2)c_1^3 - 6m(9+m)c_1 c_2 + 6m^2 c_3\}]$.

Now, using the above Equation (21), we obtain:
$$s_n = \frac{v_n}{b_3 + b_4 v_n} = \frac{(-H''(0) + (9+m)b_1^2 + 4b_1 b_2)c_1^2 - 2mb_1^2 c_2}{2m^3 b_1^2 b_3}e_n^2 + \sigma_1 e_n^3 + \sigma_2 e_n^4 + \sigma_3 e_n^5 + O(e_n^6), \qquad (22)$$

where $\sigma_i = \sigma_i(m, b_1, b_2, b_3, b_4, H''(0), H^{(3)}(0), c_1, c_2, c_3, c_4, c_5)$ for $1 \leq i \leq 3$, with the explicit coefficient σ_1 written as:

$$\sigma_1 = \frac{1}{6m^3 b_1^3 b_3}\left[\left(H^{(3)}(0) + \left(98 + 54m + 4m^2\right)b_1^3 - 6H''(0)b_2 + 24(3+m)b_1^2 b_2 - 6b_1\left((3+m)H''(0) - 2b_2^2\right)\right)c_1^3 \right.$$
$$\left. -12mb_1\left(-H''(0) + (9+m)b_1^2 + 4b_1 b_2\right)c_1 c_2 + 12m^2 b_1^3 c_3\right)\right].$$

From Equations (15) and (22), we conclude that t_n and s_n are of orders e_n and e_n^2, respectively. We can expand the weight function $G(t, s)$ in the neighborhood of (0, 0) by Taylor series up to fourth-order terms:

$$G(t_n, s_n) \approx G_{00} + G_{10} t_n + G_{01} s_n + \frac{1}{2!}\left(G_{20} t_n^2 + 2G_{11} t_n s_n + G_{02} s_n^2\right) + \frac{1}{3!}\left(G_{30} t_n^3 + 3G_{21} t_n^2 s_n + 3G_{12} t_n s_n^2 + G_{03} s_n^3\right) \qquad (23)$$
$$+ \frac{1}{4!}\left(G_{40} t_n^4 + 4G_{31} t_n^3 s_n + 6G_{22} t_n^2 s_n^2 + 4G_{13} t_n s_n^3 + G_{04} s_n^4\right),$$

where $G_{ij} = \left.\frac{\partial^{i+j}}{\partial t^i \partial s^j} G(t, s)\right|_{(t=0, s=0)}, i, j \in \{0, 1, 2, 3, 4\}$.

Using the Equations (9)–(23) in (7), we have:

$$e_{n+1} = \frac{(G_{00} - m)c_1((H''(0) - (m+9)b_1^2 - 4b_1 b_2)c_1^2 + 2mb_1^2 c_2)}{2b_1^2 m^4}e_n^4 + \sum_{i=1}^{4} R_i e_n^{i+4} + O(e_n^9), \qquad (24)$$

where $R_i = R_i(m, b_1, b_2, b_3, b_4, H(0), H'(0), H''(0), H^{(3)}(0), c_1, c_2, \ldots, c_8), i = 1, 2, 3, 4$.

To obtain at least sixth-order convergence, we need to adopt $G_{00} = m$. Furthermore, substituting $G_{00} = m$ in $R_1 = 0$, one obtains:
$$G_{10} = 2b_1 m. \qquad (25)$$

Inserting $G_{00} = m$ and $G_{10} = 2b_1 m$ in $R_2 = 0$, we obtain the following relations:
$$G_{01} - mb_3 = 0, \quad G_{20} - mH''(0) - 2mb_1^2 = 0, \qquad (26)$$

which further yield:
$$G_{01} = mb_3, \quad G_{20} = mH''(0) + 2mb_1^2. \qquad (27)$$

By substituting the values of G_{00}, G_{10}, G_{01}, and G_{20} in $R_3 = 0$, we obtain the following two independent equations:

$$G_{11} - 4b_1 b_3 m = 0,$$
$$3G_{11} H''(0) - (G_{30} - mH^{(3)}(0))b_3 + 12m(7+m)b_1^3 b_3 - 6b_1\left(2G_{11} b_2 + mH''(0)b_3\right) - 3b_1^2\left(G_{11}(9+m) - 12mb_2 b_3\right) = 0, \qquad (28)$$

which further give:

$$G_{11} = 4b_1b_3m, \quad G_{30} = m(H^{(3)}(0) + 6b_1H''(0) - 24mb_1^3 - 12mb_1^2b_2). \tag{29}$$

Now, in order to obtain eighth-order convergence of the proposed scheme (7), the coefficients of e_n^4, R_1, R_2, R_3 defined in (24) must be equal to zero. Therefore, using the value of $G_{00} = m$ and substituting the values of R_i^s (i = 1, 2, 3) from Relations (25)–(29) in (7), one gets the following error equation:

$$e_{n+1} = \frac{1}{48m^8 b_1^5 b_3^2} \Big[c_1(4b_2c_1^2 + b_1((9+m)c_1^2 - 2mc_2)) \Big[-24G_{21}b_1b_2b_3c_1^4 + (-G_{40} + mH^{(4)}(0))b_3^2c_1^4 - 24b_1^3b_2c_1^2$$
$$\Big((G_{02}(9+m) - m(23+3m)b_3^2 - 2m(9+m)b_3b4\Big)c_1^2 + 2m\Big(-G_{02} + 3mb_3^2 + 2mb_3b_4\Big)c_2\Big) - 6b_1^2\Big(4b_2^2\Big(2G_{02}$$
$$- 3mb_3^2 - 4mb_3b_4\Big)c_1^4 + G_{21}b_3c_1^2\Big((9+m)c_1^2 - 2mc_2\Big)\Big) + b_1^4\Big(\Big(-3G_{02}(9+m)^2 + 2m\Big(431 + 102m + 7m^2\Big)b_3^2 \tag{30}$$
$$+ 6m(9+m)^2b_3b_4\Big)c_1^4 - 12m\Big(-G_{02}(9+m) + 2m(17+2m)b_3^2 + 2m(9+m)b_3b_4\Big)c_1^2c_2 + 12m^2\Big(-G_{02}$$
$$+ 2mb_3^2 + 2mb_3b_4\Big)c_2^2 + 24m^3b_3^2c_1c_3\Big)\Big]e_n^8 + O(e_n^9).$$

The consequence of the above error analysis is that (7) acquires eighth-order convergence using only four functional evaluations (viz. $f(x_n)$, $f'(x_n)$, $f(y_n)$ and $f(z_n)$) per full iteration. This completes the proof. □

2.1. Some Special Cases of the Proposed Class

In this section, we discuss some interesting special cases of the new class (7) by assigning different forms of weight functions $H(t_n)$ and $G(t_n, s_n)$ employed in the second and third steps, respectively.

1. Let us consider the following optimal class of eighth-order methods for multiple roots, with the weight functions chosen directly from Theorem 1:

$$y_n = x_n - m\frac{f(x_n)}{f'(x_n)},$$
$$z_n = y_n - mu_n\frac{f(x_n)}{f'(x_n)}\left[1 + 2t_nb_1 + \frac{1}{2}t_n^2 H''(0) + \frac{1}{3!}t_n^3 H^{(3)}(0) + \frac{1}{4!}t_n^4 H^{(4)}(0) + \frac{1}{5!}t_n^5 H^{(5)}(0)\right],$$
$$x_{n+1} = z_n - u_nv_n\frac{f(x_n)}{f'(x_n)}\Big[m + 2mb_1t_n + mb_3s_n + \frac{1}{2!}\Big((H''(0)m + 2mb_1^2)t_n^2 + 8mb_1b_3t_ns_n + G_{02}s_n^2\Big) \tag{31}$$
$$+ \frac{1}{3!}\Big\{(H^{(3)}(0) + 6H''(0)b_1 - 24b_1^3 - 12b_1^2b_2)mt_n^3 + 3G_{21}t_n^2s_n + 3G_{12}t_ns_n^2 + G_{03}s_n^3\Big\}$$
$$+ \frac{1}{4!}\Big(G_{40}t_n^4 + 4G_{31}t_n^3s_n + 6G_{22}t_n^2s_n^2 + 4G_{13}t_ns_n^3 + G_{04}s_n^4\Big)\Big],$$

where b_i (i = 1, 2, 3, 4), $H''(0), H^{(3)}(0), H^{(4)}(0), H^{(5)}(0), G_{02}, G_{12}, G_{21}, G_{03}, G_{40}, G_{31}, G_{22}, G_{13}$ and G_{04} are free parameters.

Subcases of the given scheme (31):

(a) Let us consider $H''(0) = H^{(3)}(0) = H^{(4)}(0) = H^{(5)}(0) = G_{02} = G_{12} = G_{21} = G_{03} = G_{31} = G_{22} = G_{13} = G_{04} = 0$ in Equation (31). Then, we obtain:

$$y_n = x_n - m\frac{f(x_n)}{f'(x_n)},$$
$$z_n = y_n - mu_n\frac{f(x_n)}{f'(x_n)}[1 + 2t_nb_1], \tag{32}$$
$$x_{n+1} = z_n - u_nv_n\frac{f(x_n)}{f'(x_n)}\left[m + ms_nb_3 + 2t_nmb_1(1 + 2s_nb_3) - 4t_n^3mb_1^3 + t_n^2mb_1^2(1 - 2t_nb_2) + \frac{G_{40}t_n^4}{24}\right].$$

2. Considering $H''(0) = H^{(3)}(0) = H^{(4)}(0) = H^{(5)}(0) = G_{12} = G_{03} = G_{31} = G_{22} = G_{13} = G_{04} = 0$ and $G_{21} = 2m$ in Equation (31), one gets:

$$\begin{aligned} y_n &= x_n - m\frac{f(x_n)}{f'(x_n)}, \\ z_n &= y_n - mu_n \frac{f(x_n)}{f'(x_n)}[1 + 2t_n b_1], \\ x_{n+1} &= z_n - u_n v_n \frac{f(x_n)}{f'(x_n)}\left[ms_n t_n^2 + \frac{G_{02}}{2}s_n^2 + m\left(1 - 4b_1^3 t_n^3 + b_1^2\left(t_n^2 - 2b_2 t_n^3\right) + b_3 s_n + 2b_1(t_n + 2b_3 t_n s_n)\right)\right]. \end{aligned} \tag{33}$$

3. A combination of polynomial and rational functions produces another optimal eighth-order scheme as follows:

$$\begin{aligned} y_n &= x_n - m\frac{f(x_n)}{f'(x_n)}, \\ z_n &= y_n - mu_n \frac{f(x_n)}{f'(x_n)}[1 + 2t_n b_1], \\ x_{n+1} &= z_n - u_n v_n \frac{f(x_n)}{f'(x_n)}\left[k_1 t_n^2 + k_2 s_n + \frac{k_3 t_n^2 + k_4 t_n + k_5 s_n + k_6}{k_7 t_n + s_n + 1}\right], \end{aligned} \tag{34}$$

where:

$$\begin{cases} k_1 = \dfrac{m\left(-24b_1^3 + 6b_1^2(-2b_2 + k_7)\right)}{6k_7}, \\ k_2 = \dfrac{m\left(b_1(2 + 4b_3) + b_3 k_7\right)}{k_7}, \\ k_3 = \dfrac{m\left(24b_1^3 + 12b_1^2 b_2 + 12b_1 k_7^2\right)}{6k_7}, \\ k_4 = m(2b_1 + k_7), \\ k_5 = \dfrac{m(-2b_1(1 + 2b_3) + k_7)}{k_7}, \\ k_6 = m. \end{cases} \tag{35}$$

Remark 1. *It is important to note that the weight functions $H(t_n)$ and $G(t_n, s_n)$ play a significant role in the construction of eighth-order techniques. Therefore, it is usual to display different choices of weight functions, provided they satisfy all the conditions of Theorem 1. Hence, we discussed above some special cases (32), (33), and (35) having simple body structures along with optimal eight-order convergence so that they can be easily implemented in numerical experiments.*

Remark 2. *The family (5) proposed by Behl et al. [19] can be obtained as a special case of (7) by selecting suitable values of free parameters as, namely, $b_1 = a_1$, $b_2 = a_2$, $b_3 = 1$, and $b_4 = 0$.*

3. Numerical Experiments

In this section, we analyze the computational aspects of the following cases: Equation (32) for $(b_1 = 1, b_2 = -2, b_3 = 1, b_4 = -2, G_{40} = 0)$ (MM1), Family (33) for $(b_1 = 1, b_2 = -2, b_3 = 1, b_4 = -2, G_{02} = 0)$ (MM2), and Equation (35) for $(b_1 = 1, b_2 = -2, b_3 = 1, b_4 = -2, k_7 = -\frac{3}{10})$ (MM3). Additionally, we compare the results with those of other techniques.

In this regard, we considered several test functions coming from real-life problems and linear algebra that represent Examples 1–7 in the follow-up. We compared them with the optimal eighth-order scheme (5) given by Behl et al. [19] for $Q(h_n) = m(1 + 2h_n + 3h_n^2)$ and $G(h_n, t_n) = m\left(\frac{1 + 2t_n + 3h_n^2 + h_n(2 + 6t_n + h_n)}{1 + t_n}\right)$ and the approach (6) of Zafar et al. [20] taking $H(u_n) = 6u_n^3 - u_n^2 + 2u_n + 1$, $P(v_n) = 1 + v_n$, and $G(w_n) = \frac{2w_n + 1}{A_2 P_0}$ for $(A_2 = P_0 = 1)$ denoted by (OM) and (ZM), respectively. Furthermore, we compared them with the family of two-point sixth-order methods proposed by Geum et al. in [17], choosing out of them Case 2A, given by:

$$y_n = x_n - m\frac{f(x_n)}{f'(x_n)}, \quad m > 1,$$
$$x_{n+1} = y_n - \left[\frac{m + b_1 u_n}{1 + a_1 u_n + a_2 s_n + a_3 s_n u_n}\right]\frac{f(y_n)}{f'(y_n)}, \tag{36}$$

where $a_1 = -\frac{2m(m-2)}{(m-1)}$, $b_1 = \frac{2m}{(m-1)}$, $a_2 = 2(m-1)$, and $a_3 = 3$.

Finally, we compared them with the non-optimal family of sixth-order methods based on the weight function approach presented by Geum et al. [18]; out of them, we considered Case 5YD, which is defined as follows:

$$y_n = x_n - m\frac{f(x_n)}{f'(x_n)}, \quad m \geq 1,$$
$$w_n = x_n - m\left[\frac{(u_n - 2)(2u_n - 1)}{(u_n - 1)(5u_n - 2)}\right]\frac{f(x_n)}{f'(x_n)}, \tag{37}$$
$$x_{n+1} = x_n - m\left[\frac{(u_n - 2)(2u_n - 1)}{(5u_n - 2)(u_n + v_n - 1)}\right]\frac{f(x_n)}{f'(x_n)}.$$

We denote Equations (36) and (37) by (GK1) and (GK2), respectively.

The numerical results listed in Tables 1–7, compare our techniques with the four ones described previously. Tables 1–7, include the number of iteration indices n, approximated zeros x_n, absolute residual error of the corresponding function $|f(x_n)|$, error in the consecutive iterations $|x_{n+1} - x_n|$, the computational order of convergence $\rho \approx \frac{\log|f(x_{n+1})/f(x_n)|}{\log|f(x_n)/f(x_{n-1})|}$ with $n \geq 2$ (the details of this formula can be seen in [22]), the ratio of two consecutive iterations based on the order of convergence $\left|\frac{x_{n+1} - x_n}{(x_n - x_{n-1})^p}\right|$ (where p is either six or eight, corresponding to the chosen iterative method), and the estimation of asymptotic error constant $\eta \approx \lim_{n \to \infty}\left|\frac{x_{n+1} - x_n}{(x_n - x_{n-1})^p}\right|$ at the last iteration. We considered 4096 significant digits of minimum precision to minimize the round off error.

We calculated the values of all the constants and functional residuals up to several significant digits, but we display the value of the approximated zero x_n up to 25 significant digits (although a minimum of 4096 significant digits were available). The absolute residual error in the function $|f(x_n)|$ and the error in two consecutive iterations $|x_{n+1} - x_n|$ are displayed up to two significant digits with exponent power. The computational order of convergence is reported with five significant digits, while $\left|\frac{x_{n+1} - x_n}{(x_n - x_{n-1})^p}\right|$ and η are displayed up to 10 significant digits. From Tables 1–7, it can be observed that a smaller asymptotic error constant implies that the corresponding method converged faster than the other ones. Nonetheless, it may happen in some cases that the method not only had smaller residual errors and smaller error differences between two consecutive iterations, but also larger asymptotic error.

All computations in the numerical experiments were carried out with the *Mathematica 10.4* programming package using multiple precision arithmetic. Furthermore, the notation $a(\pm b)$ means $a \times 10^{(\pm b)}$.

We observed that all methods converged only if the initial guess was chosen sufficiently close to the desired root. Therefore, going a step further, we decided to investigate the dynamical behavior of the test functions $f_i(x)$, $i = 1, 2, \ldots, 7$, in the complex plane. In other words, we numerically approximated the domain of attraction of the zeros as a qualitative measure of how the methods depend on the choice of the initial approximation of the root. To answer this important question on the behavior of the algorithms, we discussed the complex dynamics of the iterative maps (32), (33), and (35) and compared them with the schemes (36), (37), (5), and (6), respectively.

From the dynamical and graphical point of view [23,24], we took a 600 × 600 grid of the square $[-3, 3] \times [-3, 3] \in \mathbb{C}$ and assigned orange color to those points whose orbits converged to the multiple root. We represent a given point as black if the orbit converges to strange fixed points or diverges after at most 25 iterations using a tolerance of 10^{-3}. Note that the black color denotes lack of convergence of the algorithm to any of the roots. This happened, in particular, when the method converged to a strange fixed point (fixed points that were not roots of the nonlinear function), ended in a periodic cycle, or went to infinity.

Table 8 depicts the measures of convergence of different iterative methods in terms of the average number of iterations per point. The column I/P shows the average number of iterations per point until the algorithm decided that a root had been reached, otherwise it indicates that the point was non-convergent. The column $NC(\%)$ shows the percentage of non-convergent points, indicated as black zones in the fractal pictures represented in Figures 1–14. It is clear that the non-convergent points had a considerable influence on the values of I/P since these points contributed always with the maximum number of 25 allowed iterations. In contrast, convergent points were reached usually very quickly because we were working with multipoint iterative methods of higher order. Therefore, to minimize the effect of non-convergent points, we included the column I_C/C, which shows the average number of iterations per convergent point.

Example 1. *We considered the van der Waals equation of state [25]:*

$$\left(P + \frac{an^2}{V^2}\right)(V - nb) = nRT,$$

where a and b explain the behavior of a real gas by introducing in the ideal equations two parameters, a and b (known as van der Waals constants), specific for each case. The determination of the volume V of the gas in terms of the remaining parameters required the solution of a nonlinear equation in V:

$$PV^3 - (nbP + nRT)V^2 + an^2V - abn^2 = 0.$$

Given the constants a and b that characterize a particular gas, one can find values for n, P, and T, such that this equation has three roots. By using the particular values, we obtained the following nonlinear function (see [26] for more details)

$$f_1(x) = x^3 - 5.22x^2 + 9.0825x - 5.2675,$$

having three zeros, so that one is $x = 1.75$, of multiplicity of two, and the other $x = 1.72$. However, our desired root was $x = 1.75$.

The numerical results shown in Table 1 reveal that the new methods $MM1$, $MM2$, and $MM3$ had better performance than the others in terms of precision in the calculation of the multiple roots of $f_1(x) = 0$. On the other hand, the dynamical planes of different iterative methods for this problem are given in Figures 1 and 2. One can see that the new methods had a larger stable (area marked in orange) than the methods $GK1$, $GK2$, OM, and ZM. It can also be verified from Table 8 that the three new methods required a smaller average number of iterations per point (I/P) and a smaller percentage of non-convergent points ($NC(\%)$). Furthermore, we found that $MM1$ used the smallest number of iterations per point (I/P =5.95 on average), while $GK1$ required the highest number of iterations per point (I/P =14.82).

Table 1. Convergence behavior of seven different iterative methods on the test function $f_1(x)$.

Methods	n	x_n	$|f(x_n)|$	$|x_{n+1} - x_n|$	ρ	$\frac{x_{n+1}-x_n}{(x_n-x_{n-1})^8}$	η
GK1	0	1.8	2.0(−4)	4.9(−2)			
	1	1.75089525858009153564128O	2.5(−8)	9.0(−4)		6.385691220(+4)	
	2	1.75000000014299761415271	6.1(−24)	1.4(−11)		2.777396484(+7)	
	3	1.75000000000000000000000	2.7(−117)	3.0(−58)	5.9816	3.536522620(+7)	3.536522620(+7)
GK2	0	1.8	2.0(−4)	5.0(−2)			
	1	1.75038817279389159741273	4.6(−9)	3.9(−4)		2.603237303(+4)	
	2	1.75000000000001034322437	3.2(−30)	1.0(−14)		3.023468138(+6)	
	3	1.75000000000000000000000	4.6(−157)	3.9(−78)	5.9959	3.215020576(+6)	3.215020576(+6)
OM	0	1.8	2.0(−4)	4.9(−2)			
	1	1.75038817231982357536368O	9.9(−9)	5.7(−4)		1.599594295(+7)	
	2	1.75000000000001356336629	5.5(−32)	1.4(−15)		3.750857339(+11)	
	3	1.75000000000000000000000	8.4(−218)	1.7(−108)	7.9903	1.462834362(+11)	1.462834362(+11)
ZM	0	1.8	2.0(−4)	5.0(−2)			
	1	1.75038817231982357536368O	4.6(−9)	3.9(−4)		1.057651892(+7)	
	2	1.75000000000000051608567	8.0(−35)	5.2(−17)		1.001210273(+11)	
	3	1.75000000000000000000000	1.1(−240)	5.9(−120)	7.9928	1.178394347(+11)	1.178394347(+11)
MM1	0	1.8	2.0(−4)	5.0(−2)			
	1	1.75008304695029185333158T	2.1(−10)	8.3(−5)		2.154463519(+6)	
	2	1.75000000000000000000006	9.5(−49)	5.6(−24)		2.493663476(+9)	
	3	1.75000000000000000000000	2.0(−355)	2.6(−177)	7.9993	2.545224623(+9)	2.545224623(+9)
MM2	0	1.8	2.0(−4)	5.0(−2)			
	1	1.75007103801875080289624S	1.5(−10)	7.1(−5)		1.639376116(+6)	
	2	1.75000000000000000000001	3.7(−50)	1.1(−24)		1.712046103(+9)	
	3	1.75000000000000000000000	4.9(−367)	4.0(−183)	7.9994	1.741469479(+9)	1.741469479(+9)
MM3	0	1.8	2.0(−4)	4.9(−2)			
	1	1.75057007195078167220702	1.5(−8)	7.0(−4)		2.002134740(+7)	
	2	1.75000000000001356336629	4.6(−32)	1.2(−15)		2.174278591(+10)	
	3	1.75000000000000000000000	5.9(−220)	1.4(−109)	7.9904	2.569337277(+10)	2.569337277(+10)

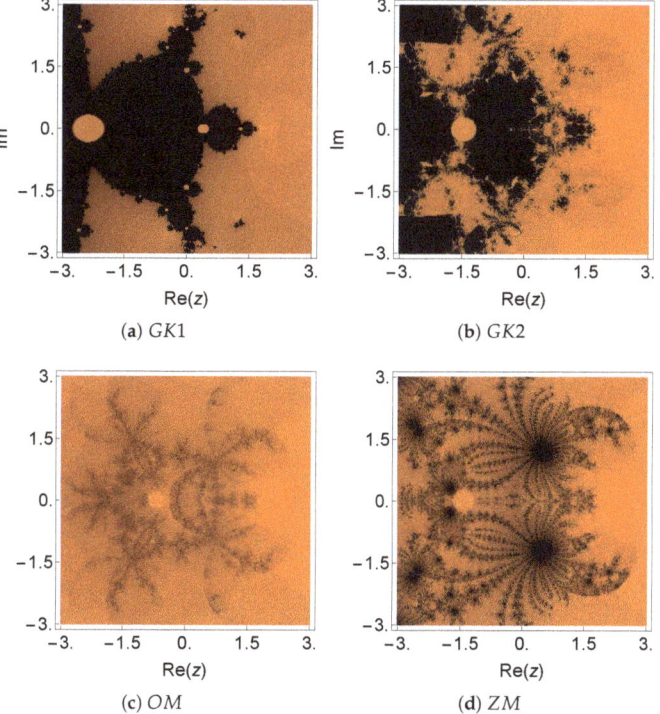

Figure 1. Dynamical plane of the methods GK1, GK2, OM, and ZM on $f_1(x)$.

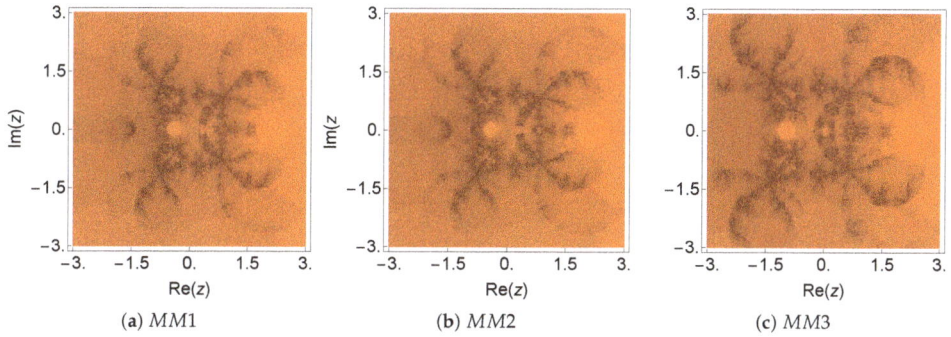

Figure 2. Dynamical plane of the methods MM1, MM2, and MM3 on $f_1(x)$.

Example 2. *Fractional conversion in a chemical reactor:*
Let us consider the following equation (see [27,28] for more details):

$$f_2(x) = \frac{x}{1-x} - 5\log\left[\frac{0.4(1-x)}{0.4 - 0.5x}\right] + 4.45977. \tag{38}$$

In the above equation, x represents the fractional conversion of Species A in a chemical reactor. There is no physical meaning of Equation (38) if x is less than zero or greater than one, since x is bounded in the region $0 \leq x \leq 1$. The required zero (that is simple) to this problem is $x \approx 0.75739624625375387945964129792$. Nonetheless, the above equation is undefined in the region $0.8 \leq x \leq 1$, which is very close to the desired zero.

Furthermore, there are some other properties of this function that make the solution more difficult to obtain. In fact, the derivative of the above equation is very close to zero in the region $0 \leq x \leq 0.5$, and there is an infeasible solution for $x = 1.098$.

From Figures 3 and 4, we verified that the basin of attraction of the searched root (orange color) was very small, or does not exist in most of the methods. The number of non-convergent points that corresponds to the black area was very large for all the considered methods (see Table 8). Moreover, an almost symmetric orange-colored area appeared in some cases, which corresponds to the solution without physical sense, that is an attracting fixed point. Except for *GK1* (not applicable for simple roots), all other methods converged to the multiple roost only if the initial guess was chosen sufficiently close to the required root, although the basins of attraction were quite small in all cases. The numerical results presented in Table 2 are compatible with the dynamical results in Figures 3 and 4. We can see that the new methods revealed smaller residual error and a smaller difference between the consecutive approximations in comparison to the existing ones. Moreover, the numerical estimation of the order of convergence coincided with the theoretical one in all cases. In Table 2, the symbol ∗ means that the corresponding method does not converge to the desired root.

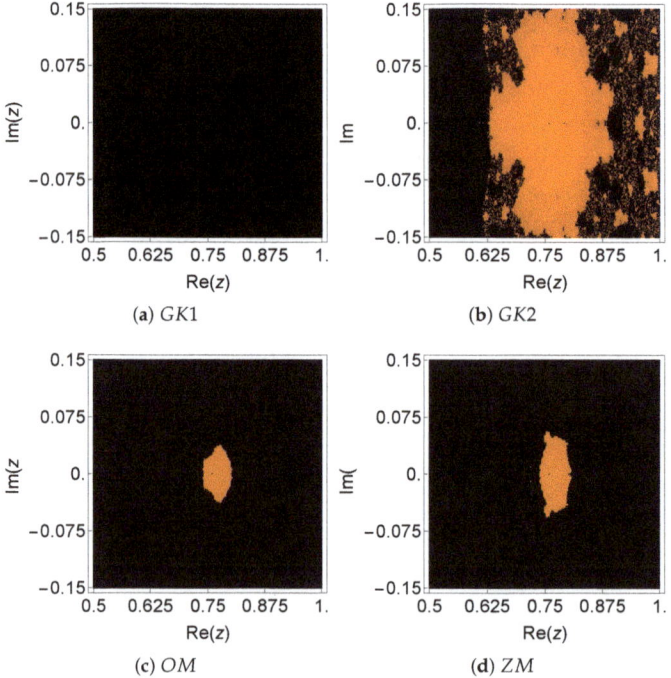

Figure 3. Dynamical plane of the methods *GK1*, *GK2*, *OM*, and *ZM* on $f_2(x)$.

Table 2. Convergence behavior of seven different iterative methods on the test function $f_2(x)$.

Methods	n	x_n	$|f(x_n)|$	$|x_{n+1} - x_n|$	ρ	$\frac{x_{n+1}-x_n}{(x_n-x_{n-1})^8}$	η
GK1	0	0.76					
	1		*	*		*	*
	2		*	*		*	
	3		*	*		*	
GK2	0	0.76	2.2(−1)	2.6(−3)			
	1	0.75739624607533362218998	1.4(−8)	1.8(−10)		5.725910242(+5)	5.257130496(+5)
	2	0.7573962462537538794596413	1.4(−51)	1.7(−53)		5.257130467(+5)	
	3	0.7573962462537538794596413	1.0(−309)	1.3(−311)	6.0000	5.257130496(+5)	
OM	0	0.76	2.2(−1)	2.6(−3)			
	1	0.7573962463137703385994168	4.8(−9)	6.0(−11)		2.840999693(+10)	3.013467463(+10)
	2	0.7573962462537538794596413	4.0(−70)	5.1(−72)		3.013467461(+10)	
	3	0.7573962462537538794596413	1.1(−558)	1.3(−560)	8.0000	3.013467463(+10)	
ZM	0	0.76	2.2(−1)	2.6(−3)			
	1	0.7573962463048948508621891	4.1(−9)	5.1(−11)		2.420860580(+10)	3.421344786(+10)
	2	0.7573962462537538794596413	1.3(−70)	1.6(−72)		3.421344762(+10)	
	3	0.7573962462537538794596413	1.2(−562)	1.5(−564)	8.0000	3.421344786(+10)	
MM1	0	0.76	2.2(−1)	2.6(−3)			
	1	0.7573962462537411428461	2.8(−13)	3.5(−15)		1.671792904(+6)	1.186467025(+6)
	2	0.7573962462537538794596413	2.3(−108)	2.9(−110)		1.186467025(+6)	
	3	0.7573962462537538794596413	4.4(−869)	5.5(−871)	8.0000	1.186467025(+6)	
MM2	0	0.76	2.2(−1)	2.6(−3)			
	1	0.7573962462537553703375248	1.2(−13)	1.5(−15)		7.057368744(+5)	4.421886626(+5)
	2	0.7573962462537538794596413	8.6(−112)	1.1(−113)		4.421886626(+5)	
	3	0.7573962462537538794596413	6.5(−897)	8.1(−899)	8.0000	4.421886626(+5)	
MM3	0	0.76	2.2(−1)	2.6(−3)			
	1	0.7573962462537526002632867	1.0(−13)	1.3(−15)		6.055331876(+5)	5.153221799(+5)
	2	0.7573962462537538794596413	2.9(−112)	3.7(−114)		5.153221799(+5)	
	3	0.7573962462537538794596413	1.4(−900)	1.8(−902)	8.0000	5.153221799(+5)	

(* means: the corresponding method does not work.)

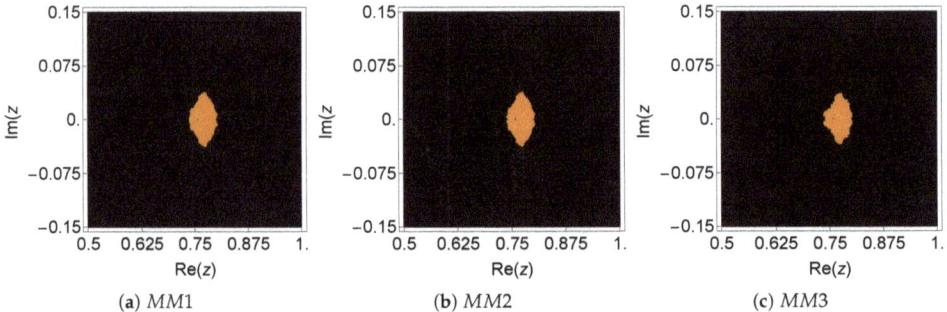

Figure 4. Dynamical plane of the methods MM1, MM2, and MM3 on $f_2(x)$.

Example 3. *Continuous stirred tank reactor (CSTR) [20,29]:*

In this third example, we considered the isothermal continuous stirred tank reactor (CSTR) problem. The following reaction scheme develops in the reactor (see [30] for more details):

$$\begin{aligned} A + R &\to B \\ B + R &\to C \\ C + R &\to D \\ C + R &\to E, \end{aligned} \qquad (39)$$

where components A and R are fed to the reactor at rates of Q and $q - Q$, respectively. The problem was analyzed in detail by Douglas [31] in order to design simple feedback control systems. In the modeling study, the following equation for the transfer function of the reactor was given:

$$K_C \frac{2.98(x + 2.25)}{(s + 1.45)(s + 2.85)^2(s + 4.35)} = -1, \qquad (40)$$

where K_C is the gain of the proportional controller. The control system is stable for values of K_C, which yields roots of the transfer function having a negative real part. If we choose $K_C = 0$, then we get the poles of the open-loop transfer function as roots of the nonlinear equation:

$$f_3(x) = x^4 + 11.50x^3 + 47.49x^2 + 83.06325x + 51.23266875 = 0 \qquad (41)$$

such as $x = -1.45, -2.85, -2.85, -4.35$. Therefore, we see that there is one root $x = -2.85$ with multiplicity two.

The numerical results for this example are listed in Table 3. The dynamical planes for this example are plotted in Figures 5 and 6. For methods GK1, GK2, and ZM, the black region of divergence was very large, which means that the methods would not converge if the initial point was located inside this region. This effect can also be observed from Table 8, where the average number of iterations per point and percentage of non-convergent points are high for methods GK1 ($I/P = 13.71$ on average), GK2 ($I/P = 12.18$ on average), and ZM ($I/P = 12.50$ on average), while the new methods have a comparatively smaller number of iterations per point. The results of method OM closely follow the new methods with an average number of 7.29 iterations per point.

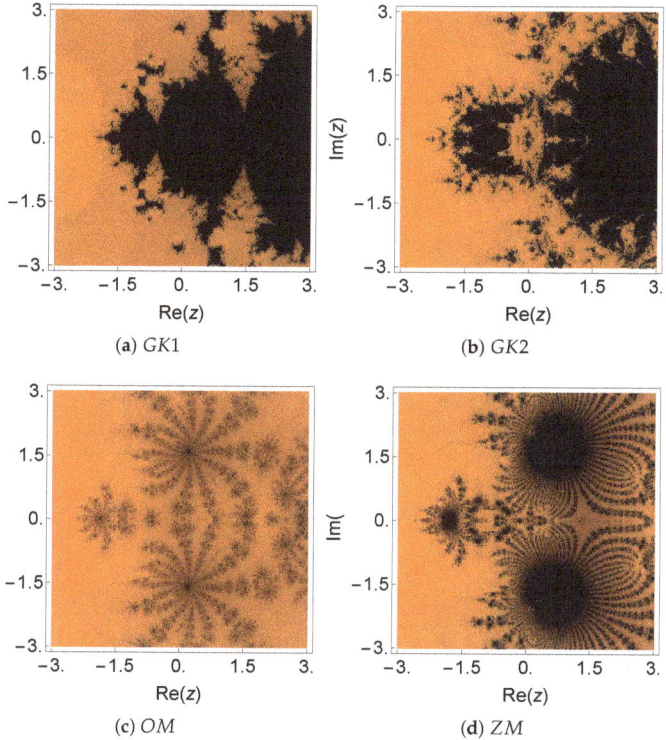

Figure 5. Dynamical plane of the methods *GK*1, *GK*2, *OM*, and *ZM* on $f_3(x)$.

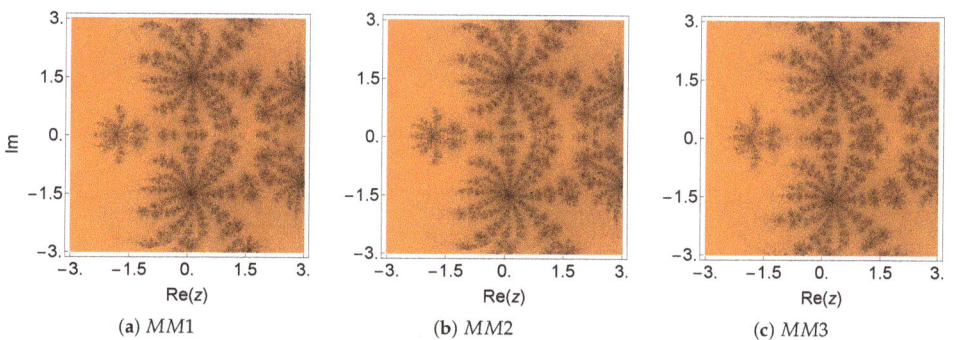

Figure 6. Dynamical plane of the methods *MM*1, *MM*2, and *MM*3 on $f_3(x)$.

Table 3. Convergence behavior of seven different iterative methods on the test function $f_3(x)$.

Methods	n	x_n	$|f(x_n)|$	$|x_{n+1} - x_n|$	ρ	$\dfrac{x_{n+1}-x_n}{(x_n-x_{n-1})^8}$	η
GK1	0	−3.0	4.7(−2)	1.5(−1)			
	1	−2.850032149435759899649078	2.2(−9)	3.2(−5)		2.826079363(+0)	4.198827967(−5)
	2	−2.850000000000000000000000	4.5(−63)	4.6(−32)		4.191188565(−5)	
	3	−2.850000000000000000000000	3.6(−385)	4.1(−193)	6.0000	4.198827967(−5)	
GK2	0	−3.0	4.7(−2)	1.5(−1)			
	1	−2.845530536829933778640841	4.2(−5)	4.5(−3)		3.291554609(+2)	1.360955722(−3)
	2	−2.850002074441970615144759	9.0(−12)	2.1(−6)		2.595135041(+8)	
	3	−2.850000000000000000000000	2.5(−74)	1.1(−37)	9.3846	1.360955722(−3)	
OM	0	−3.0	4.7(−2)	1.6(−1)			
	1	−2.844042602118935658056506	7.5(−5)	6.0(−3)		1.703635813(+4)	6.783289282(−4)
	2	−2.850005050121091781574571	5.4(−11)	5.1(−6)		3.161585672(+12)	
	3	−2.850000000000000000000000	1.7(−91)	2.9(−46)	13.102	6.783289282(−4)	
ZM	0	−3.0	4.7(−2)	1.6(−1)			
	1	−2.840827596075196247341513	1.8(−4)	9.2(−3)		2.230697732(+4)	3.402776481(−4)
	2	−2.850019022777759525868734	7.6(−10)	1.9(−5)		3.734311208(+11)	
	3	−2.850000000000000000000000	7.1(−83)	5.8(−42)	13.609	3.402776481(−4)	
MM1	0	−3.0	4.7(−2)	1.5(−1)			
	1	−2.847075767557386926817015	1.8(−5)	2.9(−3)		9.778827612(+3)	3.201998473(−4)
	2	−2.850000574904908612754099	6.9(−13)	5.7(−7)		1.073539173(+14)	
	3	−2.850000000000000000000000	3.1(−107)	3.8(−54)	12.729	3.201998473(−4)	
MM2	0	−3.0	4.7(−2)	1.5(−1)			
	1	2.847075846598888681386671	1.8(−5)	2.9(−3)		9.778603498(+3)	3.209704581(−4)
	2	−2.850000574872938822686310	6.9(−13)	5.7(−7)		1.073711858(+14)	
	3	−2.850000000000000000000000	3.1(−107)	3.8(−54)	12.728	3.209704581(−4)	
MM3	0	−3.0	4.7(−2)	9.4(−2)			
	1	−2.905607206926252789906690	6.5(−3)	5.5(−2)		8.756793722(+6)	6.122326772(−3)
	2	−2.850417788760620872669269	3.7(−7)	4.2(−4)		4.854116866(+6)	
	3	−2.850000000000000000000000	6.8(−59)	5.7(−30)	12.176	6.122326772(−3)	

Example 4. *Let us consider another nonlinear test function from [2], as follows:*

$$f_4(x) = ((x-1)^3 - 1)^{50}.$$

The above function has a multiple zero at $x = 2$ of multiplicity 50.

Table 4 shows the numerical results for this example. It can be observed that the results were very good for all the cases, the residuals being lower for the newly-proposed methods. Moreover, the asymptotic error constant (η) displayed in the last column of Table 4 was large for the methods OM and ZM in comparison to the other schemes. Based on the dynamical planes in Figures 7 and 8, it is observed that in all schemes, except $GK1$, the black region of divergence was very large. This is also justified from the observations of Table 8. We verified that ZM required the highest average number of iterations per point ($I/P = 17.74$), while $GK1$ required the smallest number of iterations per point ($I/P = 6.78$). All other methods required an average number of iterations per point in the range of 15.64–16.67. Furthermore, we observed that the percentage of non-convergent points $NC(\%)$ was very high for ZM (56.04%) followed by $GK2$ (45.88%). Furthermore, it can also be seen that the average number of iterations per convergent point (I_C/C) for the methods $GK1$, $GK2$, OM, and ZM was 6.55, 8.47, 9.58, and 8.52, respectively. On the other hand, the proposed methods $MM1$, $MM2$, and $MM3$ required 9.63, 9.67, and 10.13, respectively.

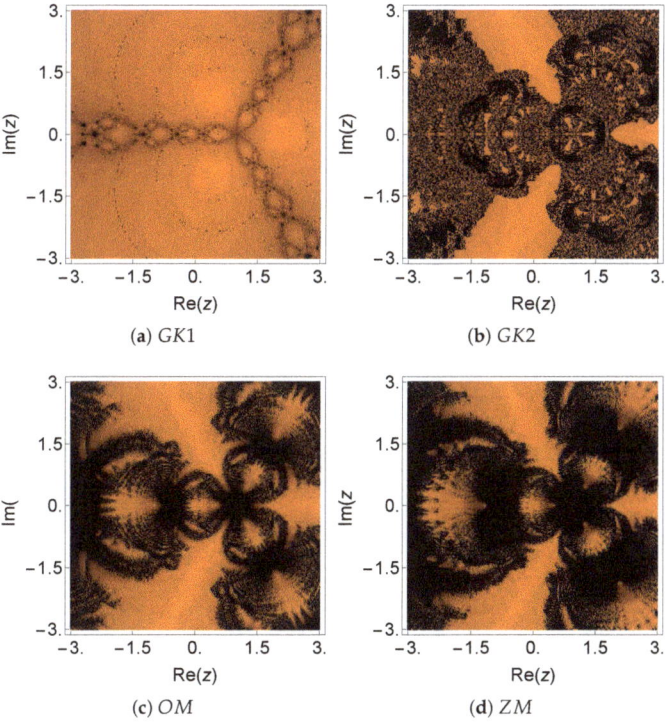

Figure 7. Dynamical plane of the methods $GK1$, $GK2$, OM, and ZM on $f_4(x)$.

Table 4. Convergence behavior of seven different iterative methods on the test function $f_4(x)$.

Methods	n	x_n	$\|f(x_n)\|$	$\|x_{n+1}-x_n\|$	ρ	$\frac{x_{n+1}-x_n}{(x_n-x_{n-1})^8}$	η
GK1	0	2.1	9.8(−25)	1.0(−1)			
	1	2.000002777374011867781357	1.1(−254)	2.8(−6)		2.777836885(+00)	5.504789671(+00)
	2	2.00000000000000000000000	9.6(−1607)	2.5(−33)		5.504677538(+00)	
	3	2.00000000000000000000000	4.5(−9719)	1.4(−195)	6.0000	5.504789671(+00)	
GK2	0	2.1	9.8(−25)	1.0(−1)			
	1	2.000000200989638086020762	1.0(−311)	2.0(−7)		2.009920619(−1)	2.777777778(−1)
	2	2.00000000000000000000000	9.8(−2014)	1.8(−41)		2.777775861(−1)	
	3	2.00000000000000000000000	7.3(−12226)	1.0(−245)	6.0000	2.777777778(−1)	
OM	0	2.1	9.8(−25)	1.0(−1)			
	1	2.000000785189010712446522	4.0(−282)	7.9(−7)		7.852383342(+1)	2.269259259(+2)
	2	2.00000000000000000000000	4.4(−2301)	3.3(−47)		2.269242109(+2)	
	3	2.00000000000000000000000	8.3(−18453)	3.0(−370)	8.0000	2.269259259(+2)	
ZM	0	2.1	9.8(−25)	1.0(−1)			
	1	2.000000477890417235498042	6.6(−293)	4.8(−7)		4.779086880(+1)	2.084074074(+2)
	2	2.00000000000000000000000	3.4(−2389)	5.7(−49)		2.084057463(+2)	
	3	2.00000000000000000000000	1.6(−19159)	2.2(−384)	8.0000	2.084074074(+2)	
MM1	0	2.1	9.8(−25)	1.0(−1)			
	1	2.000000073427782639970301	1.4(−383)	7.3(−9)		7.342782577(−1)	1.259259259(+00)
	2	2.00000000000000000000000	1.6(−3225)	1.1(−65)		1.259259209(+00)	
	3	2.00000000000000000000000	4.6(−25961)	2.1(−520)	8.0000	1.259259259(+00)	
MM2	0	2.1	9.8(−25)	1.0(−1)			
	1	2.000000049079860584101013	2.5(−392)	4.9(−9)		4.907982533(−1)	8.148148148(−1)
	2	2.00000000000000000000000	5.9(−3305)	2.7(−67)		8.148147946(−1)	
	3	2.00000000000000000000000	5.3(−26606)	2.6(−533)	8.0000	8.148148148(−1)	
MM3	0	2.1	9.8(−25)	1.0(−1)			
	1	2.000000037492911964195190	3.6(−348)	3.7(−8)		3.749302442(+00)	6.651851852(+00)
	2	2.00000000000000000000000	3.8(−2906)	2.6(−59)		6.651850415(+00)	
	3	2.00000000000000000000000	6.6(−23370)	1.4(−468)	8.0000	6.651851852(+00)	

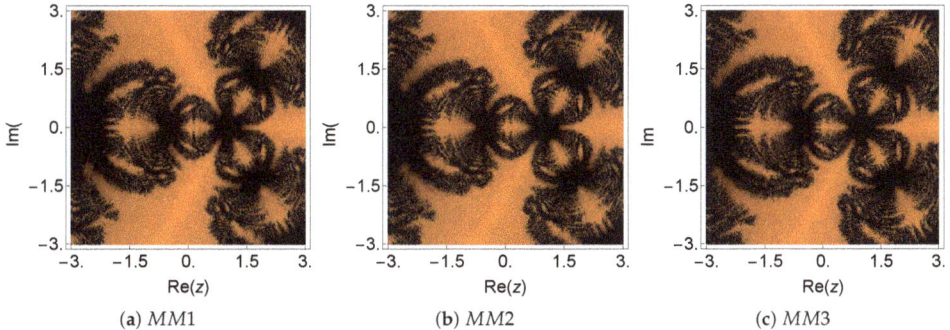

Figure 8. Dynamical plane of the methods MM1, MM2, and MM3 on $f_4(x)$.

Example 5. *Planck's radiation law problem* [32]:

We considered the following Planck's radiation law problem that calculates the energy density within an isothermal blackbody and is given by [33]:

$$\Psi(\lambda) = \frac{8\pi ch \lambda^{-5}}{e^{\frac{ch}{\lambda BT}} - 1},\qquad(42)$$

where λ represents the wavelength of the radiation, T stands for the absolute temperature of the blackbody, B is the Boltzmann constant, h denotes Planck's constant, and c is the speed of light. We were interested in determining the wavelength λ that corresponds to the maximum energy density $\Psi(\lambda)$.

The condition $\Psi'(\lambda) = 0$ implies that the maximum value of Ψ occurs when:

$$\frac{\frac{ch}{\lambda BT} e^{\frac{ch}{\lambda BT}}}{e^{\frac{ch}{\lambda BT}} - 1} = 5.\qquad(43)$$

If $x = \frac{ch}{\lambda BT}$, then (43) is satisfied when:

$$f_5(x) = e^{-x} + \frac{x}{5} - 1 = 0.\qquad(44)$$

Therefore, the solutions of $f_5(x) = 0$ give the maximum wavelength of radiation λ by means of the following formula:

$$\lambda \approx \frac{ch}{\alpha BT},\qquad(45)$$

where α is a solution of (44). The desired root is $x = 4.9651142317442$ with multiplicity $m = 1$.

The numerical results for the test equation $f_5(x) = 0$ are displayed in Table 5. It can be observed that MM1 and MM2 had small values of residual errors and asymptotic error constants (η), in comparison to the other methods, when the accuracy was tested in multi-precision arithmetic. Furthermore, the basins of attraction for all the methods are represented in Figures 9 and 10. One can see that the fractal plot of the method GK1 was completely black because the multiplicity of the desired root was unity in this case. On the other hand, method GK2 had the most reduced black area in Figure 9b, which is further justified in Table 8. The method GK2 had a minimum average number of iterations per point ($I/P = 2.54$) and the smallest percentage of non-convergent points (1.40%), while ZM had the highest percentage of non-convergent points (15.36%). For the other methods, the average number of iterations per point was in the range from 4.55–5.22 and the percentage of non-convergent points lies between 12.32 and 12.62.

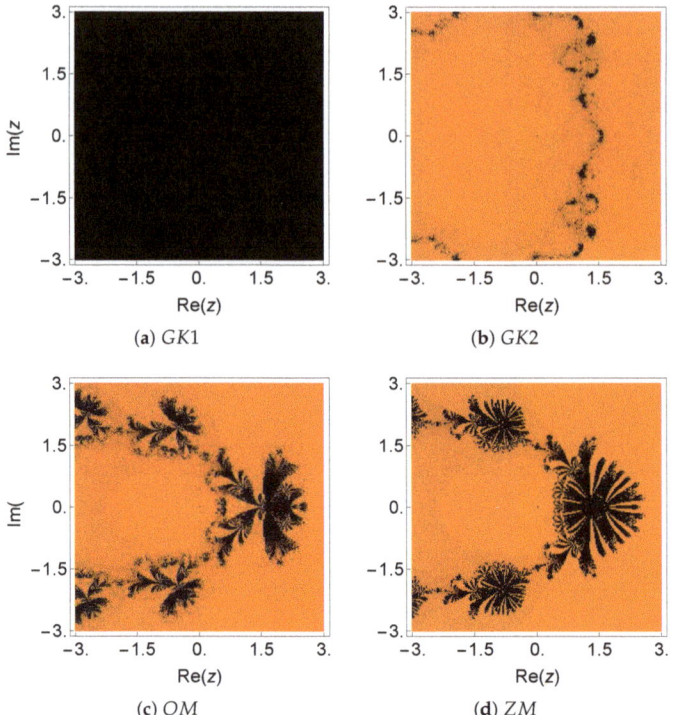

Figure 9. Dynamical plane of the methods $GK1$, $GK2$, OM, and ZM on $f_5(x)$.

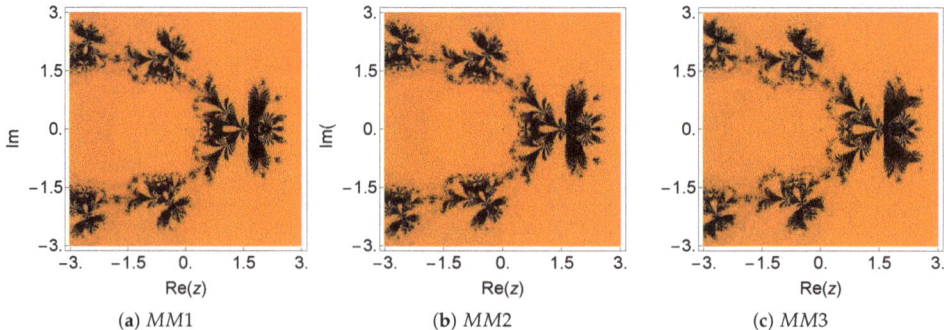

Figure 10. Dynamical plane of the methods $MM1$, $MM2$, and $MM3$ on $f_5(x)$.

Table 5. Convergence behavior of seven different iterative methods on the test function $f_5(x)$.

Methods	n	x_n	$|f(x_n)|$	$|x_{n+1}-x_n|$	ρ	$\frac{x_{n+1}-x_n}{(x_n-x_{n-1})^8}$	η
GK1	0	5.0	*	*		*	*
	1						
	2		*	*		*	
	3		*	*		*	
GK2	0	5.0	6.7(−3)	3.5(−2)			
	1	4.96511423174427756831718	2.4(−16)	1.3(−15)		7.015679382(−7)	7.468020979(−7)
	2	4.96511423174427630369875	5.9(−97)	3.1(−96)		7.468020979(−7)	
	3	4.96511423174427630369875	1.2(−580)	6.1(−580)	6.0000	7.468020979(−7)	
OM	0	5.0	6.7(−3)	3.5(−2)			
	1	4.96511423174427630374481	8.9(−21)	4.6(−20)		2.099233812(−8)	2.312146664(−8)
	2	4.96511423174427630369875	9.0(−164)	4.7(−163)		2.312146664(−8)	
	3	4.96511423174427630369875	1.0(−1307)	5.3(−1307)	8.0000	2.312146664(−8)	
ZM	0	5.0	6.7(−3)	3.5(−2)			
	1	4.96511423174427630372731	5.5(−21)	2.9(−20)		1.301869270(−8)	1.435568470(−8)
	2	4.96511423174427630369875	1.2(−165)	6.4(−165)		1.435568470(−8)	
	3	4.96511423174427630369875	7.4(−1323)	3.8(−1322)	8.0000	1.435568470(−8)	
MM1	0	5.0	6.7(−3)	3.5(−2)			
	1	4.96511423174427630369803	1.4(−22)	7.2(−22)		3.292330246(−10)	3.271194020(−10)
	2	4.96511423174427630369875	4.7(−180)	2.4(−179)		3.271194020(−10)	
	3	4.96511423174427630369875	7.5(−1440)	3.9(−1439)	8.0000	3.271194020(−10)	
MM2	0	5.0	6.7(−3)	3.5(−2)			
	1	4.96511423174427630369757	2.3(−22)	1.2(−21)		5.422796069(−10)	5.652515383(−10)
	2	4.96511423174427630369875	4.4(−178)	2.3(−177)		5.652515383(−10)	
	3	4.96511423174427630369875	7.6(−1424)	3.9(−1423)	8.0000	5.652515383(−10)	
MM3	0	5.0	6.7(−3)	3.5(−2)			
	1	4.96511423174427630388458	3.6(−20)	1.9(−19)		8.470476959(−8)	9.198872232(−8)
	2	4.96511423174427630369875	2.5(−158)	1.3(−157)		9.198872232(−8)	
	3	4.96511423174427630369875	1.5(−1263)	7.9(−1263)	8.0000	9.198872232(−8)	

(*means: the corresponding method does not work.)

Example 6. *Consider the following 5 × 5 matrix [29]:*

$$B = \begin{bmatrix} 29 & 14 & 2 & 6 & -9 \\ -47 & -22 & -1 & -11 & 13 \\ 19 & 10 & 5 & 4 & -8 \\ -19 & -10 & -3 & -2 & 8 \\ 7 & 4 & 3 & 1 & -3 \end{bmatrix}.$$

The corresponding characteristic polynomial of this matrix is as follows:

$$f_6(x) = (x-2)^4(x+1). \tag{46}$$

The characteristic equation has one root at $x = 2$ of multiplicity four.

Table 6 shows the numerical results for this example. It can be observed in Figures 11 and 12 that the orange areas dominate the plot. In fact, they correspond to the basin of attraction of the searched zero. This means that the method converges if the initial estimation was located inside this orange region. All schemes had a negligible black portion, as can be observed in Figures 11 and 12. Moreover, the numerical tests for this nonlinear function showed that $MM1$, $MM2$, and $MM3$ had the best results in terms of accuracy and estimation of the order of convergence. Consulting Table 8, we note that the average number of iterations per point (I/P) was almost identical for all methods (ranging from 3.25–3.52).

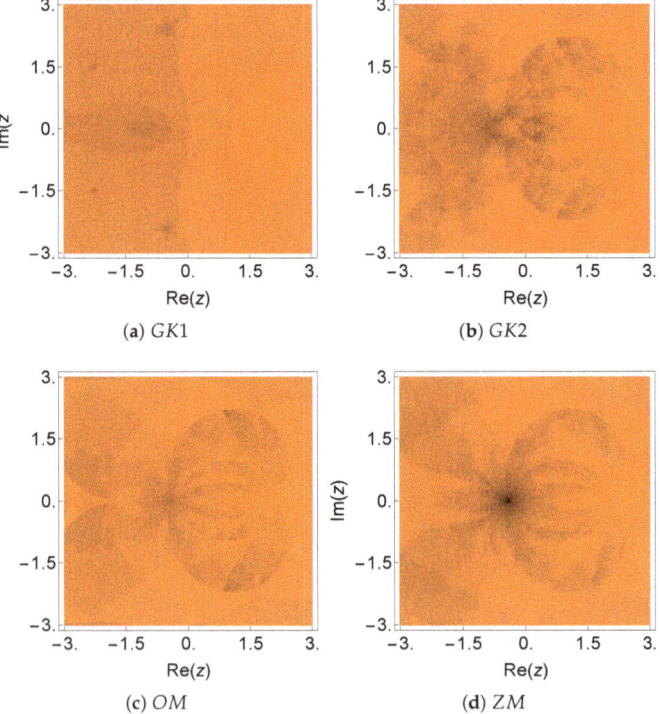

Figure 11. Dynamical plane of the methods $GK1$, $GK2$, OM, and ZM on $f_6(x)$.

Table 6. Convergence behavior of seven different iterative methods on the test function $f_6(x)$.

Methods	n	x_n	$\|f(x_n)\|$	$\|x_{n+1}-x_n\|$	ρ	$\dfrac{x_{n+1}-x_n}{(x_n-x_{n-1})^8}$	η
GK1	0	2.5	2.2(−1)	5.0(−1)			
	1	2.00000076296148293743725	1.0(−24)	7.6(−7)		4.882998197(−5)	1.120047678(−4)
	2	2.0000000000000000000000	7.1(−163)	2.2(−41)		1.120046114(−4)	
	3	2.0000000000000000000000	8.6(−992)	1.3(−248)	6.0000	1.120047678(−4)	
GK2	0	2.5	2.2(−1)	5.0(−1)			
	1	2.000000228864153793460042	8.2(−27)	2.3(−7)		1.464734607(−5)	2.813143004(−5)
	2	2.0000000000000000000000	8.0(−178)	4.0(−45)		2.813142103(−5)	
	3	2.0000000000000000000000	6.8(−1084)	1.2(−271)	6.0000	2.813143004(−5)	
OM	0	2.5	2.2(−1)	5.0(−1)			
	1	2.000000024064327301586022	1.0(−30)	2.4(−8)		6.160470161(−6)	2.026132759(−5)
	2	2.0000000000000000000000	8.1(−263)	2.3(−66)		2.026132634(−5)	
	3	2.0000000000000000000000	1.4(−2119)	1.5(−530)	8.0000	2.026132759(−5)	
ZM	0	2.5	2.2(−1)	5.0(−1)			
	1	2.000000015545259122950984	1.8(−31)	1.6(−8		3.979587325(−6)	1.501808114(−5)
	2	2.0000000000000000000000	2.1(−269)	5.1(−68)		1.501808045(−5)	
	3	2.0000000000000000000000	7.7(−2173)	7.1(−544)	8.0000	1.501808114(−5)	
MM1	0	2.5	2.2(−1)	5.0(−1)			
	1	2.000000008970643861208335	1.9(−36)	9.0(−10)		2.296484861(−7)	6.104911033(−7)
	2	2.0000000000000000000000	1.3(−314)	2.6(−79)		6.104911022(−7)	
	3	2.0000000000000000000000	4.8(−2540)	1.1(−635)	8.0000	6.104911033(−7)	
MM2	0	2.5	2.2(−1)	5.0(−1)			
	1	2.000000000657603174893603	5.6(−37)	6.6(−10)		1.683464145(−7)	4.360650738(−7)
	2	2.0000000000000000000000	1.6(−319)	1.5(−80)		4.360650732(−10)	
	3	2.0000000000000000000000	7.9(−2580)	1.3(−645)	8.0000	4.360650738(−7)	
MM3	0	2.5	2.2(−1)	5.0(−1)			
	1	2.000000013818852989402478	1.1(−31)	1.4(−8)		3.537627147(−6)	9.122481344(−6)
	2	2.0000000000000000000000	6.5(−272)	1.2(−68)		9.122481086(−6)	
	3	2.0000000000000000000000	1.0(−2193)	4.3(−549)	8.0000	9.122481344(−6)	

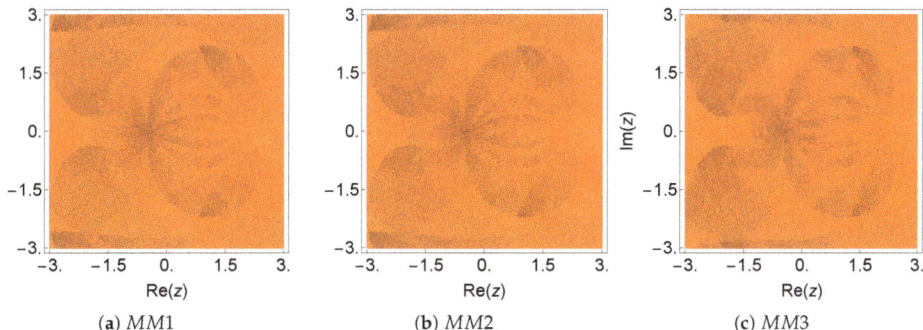

Figure 12. Dynamical plane of the methods MM1, MM2, and MM3 on $f_6(x)$.

Example 7. *Global CO_2 model by McHugh et al. [34] in ocean chemistry:*

In this example, we discuss the global CO_2 model by McHugh et al. [34] in ocean chemistry (please see [35] for more details). This problem leads to the numerical solution of a nonlinear fourth-order polynomial in the calculation of pH of the ocean. The effect of atmospheric CO_2 is very complex and varies with the location. Therefore, Babajee [35] considered a simplified approach based on the following assumptions:

1. Only the ocean upper layer is considered (not the deep layer),
2. Neglecting the spatial variations, an approximation of the ocean upper layer carbon distribution by perfect mixing is considered.

As the CO_2 dissolves in ocean water, it undergoes a series of chemical changes that ultimately lead to increased hydrogen ion concentration, denoted as $[H^+]$, and thus acidification. The problem was analyzed by Babajee [35] in order to find the solution of the following nonlinear function:

$$p([H^+]) = \sum_{n=0}^{4} \delta_n [H^+]^n, \tag{47}$$

so that:

$$\begin{cases} \delta_0 = 2N_0 N_1 N_2 P_t N_B, \\ \delta_1 = N_0 N_1 P_t N_B + 2N_0 N_1 N_2 P_t + N_W N_B, \\ \delta_2 = N_0 N_1 P_t + B N_B + N_W - A N_B, \\ \delta_3 = -N_B - A, \\ \delta_4 = -1. \end{cases} \tag{48}$$

where N_0, N_1, N_2, N_W, and N_B are equilibrium constants. The parameter A represents the alkalinity, which expresses the neutrality of the ocean water, and P_t is the gas phase CO_2 partial pressure. We assume the values of $A = 2.050$ and $B = 0.409$ taken by Sarmiento and Gruyber [36] and Bacastow and Keeling [37], respectively. Furthermore, choosing the values of N_0, N_1, N_2, N_W, N_B and P_t given by Babajee [35], we obtain the following nonlinear equation:

$$f_7(x) = x^4 - \frac{2309x^3}{250} - \frac{65226608163x^2}{500000} + \frac{425064009069x}{25000} - \frac{10954808368405209}{62500000} = 0. \tag{49}$$

The roots of $f_7(x) = 0$ are given by $x = -411.452, 11.286, 140.771, 268.332$. Hereafter, we pursue the root -411.452 having multiplicity $m = 1$.

The numerical experiments of this example are given in Table 7. The methods $MM1$, $MM2$, and $MM3$ had small residual errors and asymptotic error constants when compared to the other schemes. The computational order of convergence for all methods coincided with the theoretical ones in seven cases. Figures 13 and 14 show the dynamical planes of all the methods on test function $f_7(x)$. It can be observed that all methods showed stable behavior, except $GK1$, as can also be confirmed in Table 8.

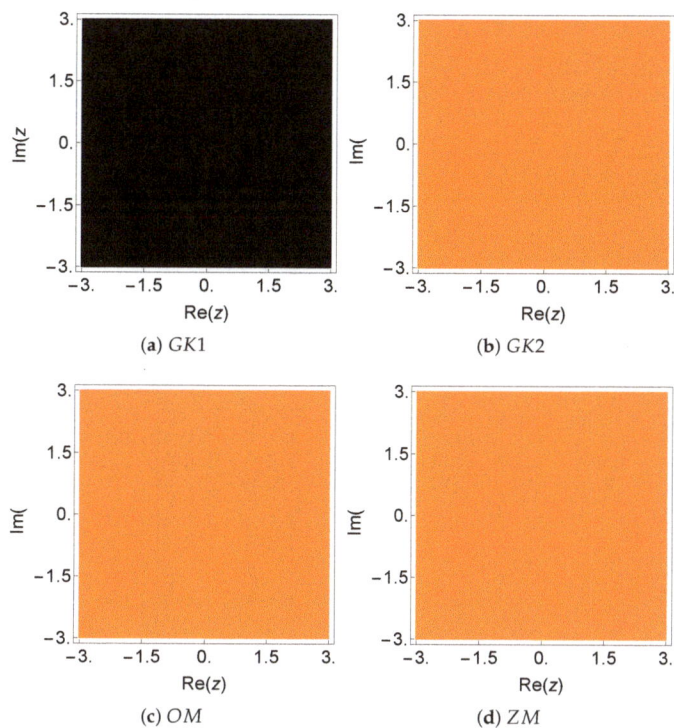

Figure 13. Dynamical plane of the methods $GK1$, $GK2$, OM, and ZM on $f_7(x)$.

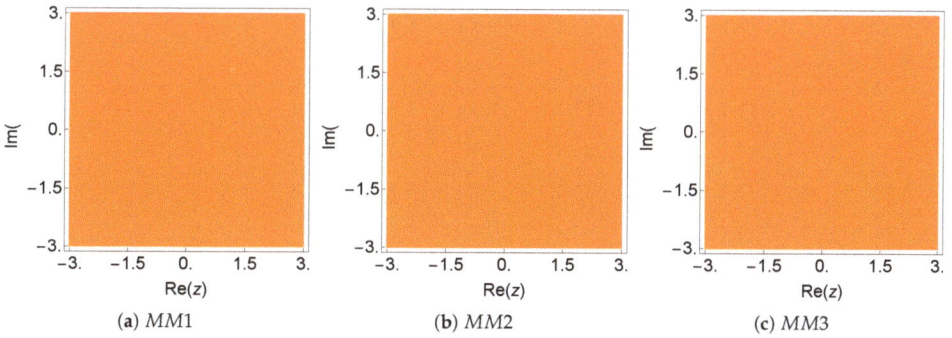

Figure 14. Dynamical plane of the methods $MM1$, $MM2$, and $MM3$ on $f_7(x)$.

Table 7. Convergence behavior of seven different iterative methods on the test function $f_7(x)$.

Cases	n	x_n	$|f(x_n)|$	$|x_{n+1} - x_n|$	ρ	$\frac{x_{n+1}-x_n}{(x_n-x_{n-1})^8}$	η
GK1	0	−412	*	*		*	*
	1		*	*			
	2		*	*	*		
GK2	0	−412	1.3(+8)	8.5(−1)			
	1	−411.15218696605467153730	9.6(−5)	6.1(−13)		1.636932403(−12)	1.668249119(−12)
	2	−411.152186966053959254939	1.3(−77)	8.4(−86)		1.668249119(−12)	
	3	−411.152186966053959254939	9.4(−515)	5.9(−523)	6.0000	1.668249119(−12)	
OM	0	−412	1.3(+8)	8.5(−1)			
	1	−411.152186966053969457729	1.7(−6)	1.1(−14)		4.005076808(−14)	4.198566741(−14)
	2	−411.152186966053959254939	1.1(−117)	7.2(−126)		4.198566741(−14)	
	3	−411.152186966053959254939	4.6(−1007)	2.9(−1015)	8.0000	4.198566741(−14)	
ZM	0	−412	1.3(+8)	8.5(−1)			
	1	−411.152186966053968748643	1.5(−6)	9.5(−15)		3.556599499(−14)	3.852323642(−14)
	2	−411.152186966053959254939	4.0(−118)	2.5(−126)		3.852323642(−14)	
	3	−411.152186966053959254939	1.1(−1010)	6.7(−1019)	8.0000	3.852323642(−14)	
MM1	0	−412	1.3(+8)	8.5(−1)			
	1	−411.152186966053959317572	9.9(−9)	6.3(−17)		2.346410356(−16)	6.104911033(−16)
	2	−411.152186966053959254939	9.0(−138)	5.7(−146)		2.411235469(−16)	
	3	−411.152186966053959254939	4.3(−1170)	2.7(−1178)	8.0000	2.411235469(−16)	
MM2	0	−412	1.3(+8)	8.5(−1)			
	1	−411.152186966053959295987	6.5(−9)	4.1(−17)		1.537771590(−16)	1.577812528(−16)
	2	−411.152186966053959254939	2.0(−139)	1.3(−147)		1.577812528(−16)	
	3	−411.152186966053959254939	1.7(−1183)	1.1(−1191)	8.0000	1.577812528(−16)	
MM3	0	−412	1.3(+8)	8.5(−1)			
	1	−411.152186966053959582284	5.2(−9)	3.3(−16)		1.226323165(−15)	1.263010316(−15)
	2	−411.152186966053959254939	2.6(−131)	1.7(−139)		1.263010316(−15)	
	3	−411.152186966053959254939	1.2(−1117)	7.5(−1126)	8.0000	1.263010316(−15)	

(* means: the corresponding method does not work.)

Table 8. Measures of convergence of the seven iterative methods for test functions $f_i(x) = 0$, $i = 1, \ldots, 7$.

$f(x)$	Methods	I/P	NC (%)	I_C/C	$f(x)$	Methods	I/P	NC (%)	I_C/C
$f_1(x)$	GK1	14.82	38.57	8.43	$f_5(x)$	GK1	0.99	99.99	0.00
	GK2	12.06	36.09	4.75		GK2	2.54	1.40	2.30
	OM	6.79	0.12	6.77		OM	4.86	12.32	2.84
	ZM	11.08	9.15	9.67		ZM	4.55	15.36	2.58
	MM1	5.95	0.04	5.95		MM1	5.17	12.41	2.92
	MM2	5.99	0.03	5.99		MM2	5.19	12.62	2.90
	MM3	6.60	0.04	6.59		MM3	5.22	12.50	2.98
$f_2(x)$	GK1	1.00	100	6.32	$f_6(x)$	GK1	3.48	0.00	3.48
	GK2	8.93	99.58	1.6		GK2	3.52	0.00	3.52
	OM	11.05	99.99	1.6		OM	3.25	0.00	3.25
	ZM	10.23	99.98	2.15		ZM	3.52	0.01	3.53
	MM1	11.72	99.99	1.6		MM1	3.32	0.00	3.32
	MM2	11.87	99.99	1.6		MM2	3.32	0.00	3.32
	MM3	11.82	99.99	1.6		MM3	3.32	0.00	3.32
$f_3(x)$	GK1	13.71	42.03	5.52	$f_7(x)$	GK1	1	100	**
	GK2	12.18	38.63	4.11		GK2	1	0	1
	OM	7.29	1.13	7.09		OM	1	0	1
	ZM	12.50	22.31	8.91		ZM	1	0	1
	MM1	6.97	0.71	6.84		MM1	1	0	1
	MM2	6.99	0.63	6.88		MM2	1	0	1
	MM3	6.84	0.62	6.73		MM3	1	0	1
$f_4(x)$	GK1	6.78	1.25	6.55					
	GK2	16.05	45.88	8.47					
	OM	15.77	40.18	9.58					
	ZM	17.74	56.04	8.52					
	MM1	15.64	39.12	9.63					
	MM2	15.67	39.16	9.67					
	MM3	16.41	42.22	10.13					

** stands for indeterminate.

4. Conclusions

This paper developed a wide general three-step class of methods for approximating multiple zeros of nonlinear functions numerically. Optimal iteration schemes having eighth order for multiple zeros have been seldom considered in the literature. Therefore, the presented methods may be regarded as a further step in this area. Weight functions based on function-to-function ratios and free parameters were employed in the second and third steps of the family. This strategy allowed us to achieve the desired convergence order of eight. In the numerical section, we considered a large variety of real-life problems. The seven examples confirmed the efficiency of the proposed technique in comparison to the existing robust methods. From the computational results, we found that the new methods showed superior performance in terms of precision, the average number of iterations per point, and the percentage of non-convergent points for the considered seven test functions. The straightforward structure and high convergence order of the proposed class make it relevant both from the theoretical and practical points of view.

Author Contributions: All the authors made equal contributions to this paper.

Funding: This work was supported by the Deanship of Scientific Research (DSR), King Abdulaziz University, Jeddah, under Grant No. D-247-248-1440. The authors, therefore, gratefully acknowledge the DSR for technical and financial support.

Acknowledgments: We would like to express our gratitude to the anonymous reviewers for their constructive suggestions, which improved the readability of the paper.

Conflicts of Interest: The authors declare no conflict of interest.

References

1. Traub, J.F. *Iterative Methods for the Solution of Equations*; Prentice-Hall: Englewood Cliffs, NJ, USA, 1964.

2. Petković, M.S.; Neta, B.; Petković, L.D.; Džunić, J. *Multipoint Methods for Solving Nonlinear Equations*; Academic Press: Cambridge, MA, USA, 2013.
3. Abad, M.F.; Cordero, A.; Torregrosa, J.R. Fourth-and fifth-order methods for solving nonlinear systems of equations: An application to the global positioning system. *Abstr. Appl. Anal.* **2013**, *2013*. [CrossRef]
4. Farhane, N.; Boumhidi, I.; Boumhidi, J. Smart Algorithms to Control a Variable Speed Wind Turbine. *Int. J. Interact. Multimed. Artif. Intell.* **2017**, *4*, 88–95. [CrossRef]
5. Abdel-Nasser, M.; Mahmoud, K.; Kashef, H. A Novel Smart Grid State Estimation Method Based on Neural Networks. *Int. J. Interact. Multimed. Artif. Intell.* **2018**, *5*, 92–100. [CrossRef]
6. Arora, S.; Singh, S. An Effective Hybrid Butterfly Optimization Algorithm with Artificial Bee Colony for Numerical Optimization. *Int. J. Interact. Multimed. Artif. Intell.* **2017**, *4*, 14–21. [CrossRef]
7. Ostrowski, A.M. *Solution of Equations and Systems of Equations*; Academic Press: New York, NY, USA, 1960.
8. Behl, R.; Cordero, A.; Motsa, S.S.; Torregrosa, J.R. On developing fourth-order optimal families of methods for multiple roots and their dynamics. *Appl. Math. Comput.* **2015**, *265*, 520–532. [CrossRef]
9. Behl, R.; Cordero, A.; Motsa, S.S.; Torregrosa, J.R.; Kanwar, V. An optimal fourth-order family of methods for multiple roots and its dynamics. *Numer. Algor.* **2016**, *71*, 775–796. [CrossRef]
10. Li, S.; Liao, X.; Cheng, L. A new fourth-order iterative method for finding multiple roots of nonlinear equations. *Appl. Math. Comput.* **2009**, *215*, 1288–1292.
11. Neta, B.; Chun, C.; Scott, M. On the development of iterative methods for multiple roots. *Appl. Math. Comput.* **2013**, *224*, 358–361. [CrossRef]
12. Sharma, J.R.; Sharma, R. Modified Jarratt method for computing multiple roots. *Appl. Math. Comput.* **2010**, *217*, 878–881. [CrossRef]
13. Zhou, X.; Chen, X.; Song, Y. Constructing higher-order methods for obtaining the multiple roots of nonlinear equations. *J. Comput. Appl. Math.* **2011**, *235*, 4199–4206. [CrossRef]
14. Li, S.; Cheng, L.; Neta, B. Some fourth-order nonlinear solvers with closed formulae for multiple roots. *Comput. Math. Appl.* **2010**, *59*, 126–135. [CrossRef]
15. Neta, B. Extension of Murakami's high-order non-linear solver to multiple roots. *Int. J. Comput. Math.* **2010**, *87*, 1023–1031. [CrossRef]
16. Thukral, R. Introduction to higher-order iterative methods for finding multiple roots of nonlinear equations. *J. Math.* **2013**, *2013*. [CrossRef]
17. Geum, Y.H.; Kim, Y.I.; Neta, B. A class of two-point sixth-order multiple-zero finders of modified double-Newton type and their dynamics. *Appl. Math. Comput.* **2015**, *270*, 387–400. [CrossRef]
18. Geum, Y.H.; Kim, Y.I.; Neta, B. A sixth-order family of three-point modified Newton-like multiple-root finders and the dynamics behind their extraneous fixed points. *Appl. Math. Comput.* **2016**, *283*, 120–140. [CrossRef]
19. Behl, R.; Cordero, A.; Motsa, S.S.; Torregrosa, J.R. An eighth-order family of optimal multiple root finders and its dynamics. *Numer. Algor.* **2018**, *77*, 1249–1272. [CrossRef]
20. Zafar, F.; Cordero, A.; Quratulain, R.; Torregrosa, J.R. Optimal iterative methods for finding multiple roots of nonlinear equations using free parameters. *J. Math. Chem.* **2017**. [CrossRef]
21. Artidiello, S.; Cordero, A.; Torregrosa, J.R.; Vassileva, M.P. Two weighted eight-order classes of iterative root-finding methods. *Int. J. Comput. Math.* **2015**, *92*, 1790–1805. [CrossRef]
22. Jay, L.O. A note on Q-order of convergence. *BIT Numer. Math.* **2001**, *41*, 422–429. [CrossRef]
23. Varona, J.L. Graphic and numerical comparison between iterative methods. *Math. Intell.* **2002**, *24*, 37–46. [CrossRef]
24. Matthies, G.; Salimi, M.; Varona, J.L.; Sharifi, S. An optimal three-point eighth-order iterative method without memory for solving nonlinear equations with its dynamics. *Jpn. J. Ind. Appl. Math.* **2016**, *33*, 751–766. [CrossRef]
25. Behl, R.; Alshomrani, A.S.; Motsa, S.S. An optimal scheme for multiple roots of nonlinear equations with eighth-order convergence. *J. Math. Chem.* **2018**, *56*, 2069–2084. [CrossRef]
26. Hueso, J.L.; Martínez, E.; Teruel, C. Determination of multiple roots of nonlinear equations and applications. *J. Math. Chem.* **2015**, *53*, 880–892. [CrossRef]
27. Magreñán, Á.A.; Argyros, I.K.; Rainer, J.J.; Sicilia, J.A. Ball convergence of a sixth-order Newton-like method based on means under weak conditions. *J. Math. Chem.* **2018**, *56*, 2117–2131. [CrossRef]

28. Shacham, M. Numerical solution of constrained nonlinear algebraic equations. *Int. J. Numer. Method Eng.* **1986**, *23*, 1455–1481. [CrossRef]
29. Kansal, M.; Behl, R.; Mahnashi, M.A.A.; Mallawi, F.O. Modified Optimal Class of Newton-Like Fourth-Order Methods for Multiple Roots. *Symmetry* **2019**, *11*, 526. [CrossRef]
30. Constantinides, A.; Mostoufi, N. *Numerical Methods for Chemical Engineers with MATLAB Applications*; Prentice Hall PTR: Englewood Cliffs, NJ, USA, 1999.
31. Douglas, J.M. *Process Dynamics and Control*; Prentice Hall: Englewood Cliffs, NJ, USA, 1972; Volume 2.
32. Maroju, P.; Magreñán, Á.A.; Motsa, S.S. Second derivative free sixth order continuation method for solving nonlinear equations with applications. *J. Math. Chem.* **2018**, *56*, 2099–2116. [CrossRef]
33. Jain, D. Families of Newton-like method with fourth-order convergence. *Int. J. Comput. Math.* **2013**, *90*, 1072–1082. [CrossRef]
34. McHugh, A.J.; Griffiths, G.W.; Schiesser, W.E. *An Introductory Global CO_2 Model: (with Companion Media Pack)*; World Scientific Pub Co Inc: Singapore, 2015.
35. Babajee, D.K.R. Analysis of Higher Order Variants of Newton's Method and Their Applications to Differential and Integral Equations and in Ocean Acidification. Ph.D. Thesis, University of Mauritius, Moka, Mauritius, 2010.
36. Sarmiento, J.L.; Gruber, N. *Ocean Biogeochemical Dynamics*; Princeton University Press: Princeton, NJ, USA, 2006.
37. Bacastow, R.; Keeling, C.D. Atmospheric carbon dioxide and radiocarbon in the natural carbon cycle: Changes from a.d. 1700 to 2070 as deduced from a geochemical model. In *Proceedings of the 24th Brookhaven Symposium in Biology*; Woodwell, G.W., Pecan, E.V., Eds.; The Technical Information Center, Office of Information Services, United State Atomic Energy Commission: Upton, NY, USA, 1972; pp. 86–133.

© 2019 by the authors. Licensee MDPI, Basel, Switzerland. This article is an open access article distributed under the terms and conditions of the Creative Commons Attribution (CC BY) license (http://creativecommons.org/licenses/by/4.0/).

Article

Development of Optimal Eighth Order Derivative-Free Methods for Multiple Roots of Nonlinear Equations

Janak Raj Sharma [1,*], **Sunil Kumar** [1] **and Ioannis K. Argyros** [2,*]

[1] Department of Mathematics, Sant Longowal Institute of Engineering and Technology, Longowal, Sangrur 148106, India; sfageria1988@gmail.com

[2] Department of Mathematical Sciences, Cameron University, Lawton, OK 73505, USA

* Correspondence: jrshira@yahoo.co.in (J.R.S.); iargyros@cameron.edu (I.K.A.)

Received: 27 April 2019; Accepted: 1 June 2019; Published: 5 June 2019

Abstract: A number of higher order iterative methods with derivative evaluations are developed in literature for computing multiple zeros. However, higher order methods without derivative for multiple zeros are difficult to obtain and hence such methods are rare in literature. Motivated by this fact, we present a family of eighth order derivative-free methods for computing multiple zeros. Per iteration the methods require only four function evaluations, therefore, these are optimal in the sense of Kung-Traub conjecture. Stability of the proposed class is demonstrated by means of using a graphical tool, namely, basins of attraction. Boundaries of the basins are fractal like shapes through which basins are symmetric. Applicability of the methods is demonstrated on different nonlinear functions which illustrates the efficient convergence behavior. Comparison of the numerical results shows that the new derivative-free methods are good competitors to the existing optimal eighth-order techniques which require derivative evaluations.

Keywords: nonlinear equations; multiple roots; derivative-free method; optimal convergence

MSC: 65H05; 41A25; 49M15

---

1. Introduction

Approximating a root (say, α) of a function is a very challenging task. It is also very important in many diverse areas such as Mathematical Biology, Physics, Chemistry, Economics and also Engineering to mention a few [1–4]. This is the case since problems from these areas are reduced to finding α. Researchers are utilizing iterative methods for approximating α since closed form solutions can not be obtained in general. In particular, we consider derivative-free methods to compute a multiple root (say, α) with multiplicity m, i.e., $f^{(j)}(\alpha) = 0, j = 0, 1, 2, ..., m-1$ and $f^{(m)}(\alpha) \neq 0$, of the equation $f(x) = 0$.

A number of higher order methods, either independent or based on the Newton's method ([5])

$$x_{k+1} = x_k - m\frac{f(x_k)}{f'(x_k)} \qquad (1)$$

have been proposed and analyzed in literature, see [6–24]. Such methods require the evaluation of derivative. However, higher order derivative-free methods to handle the case of multiple roots are yet to be investigated. Main reason of the non-availability of such methods is due to the difficulty in obtaining their order of convergence. The derivative-free methods are important in the situations where derivative of the function f is complicated to evaluate or is expensive to obtain. One such derivative-free method is the classical Traub-Steffensen method [1] which actually replaces the derivative f' in the classical Newton's method by a suitable approximation based on finite difference quotient,

$$f'(x_k) \simeq \frac{f(x_k + \beta f(x_k)) - f(x_k)}{\beta f(x_k)} = f[w_k, x_k]$$

where $w_k = x_k + \beta f(x_k)$ and $\beta \in \mathbb{R} - \{0\}$. In this way the modified Newton's method (1) becomes the modified Traub-Steffensen method

$$x_{k+1} = x_k - m \frac{f(x_k)}{f[w_k, x_k]}. \tag{2}$$

The modified Traub-Steffensen method (2) is a noticeable improvement of Newton's iteration, since it preserves the order of convergence without using any derivative.

Very recently, Sharma et al. in [25] have developed a family of three-point derivative free methods with seventh order convergence to compute the multiple zeros. The techniques of [25] require four function evaluations per iteration and, therefore, according to Kung-Traub hypothesis these do not possess optimal convergence [26]. According to this hypothesis multipoint methods without memory based on n function evaluations have optimal order 2^{n-1}. In this work, we introduce a family of eighth order derivative-free methods for computing multiple zeros that require the evaluations of four functions per iteration, and hence the family has optimal convergence of eighth order in the sense of Kung-Traub hypothesis. Such methods are usually known as optimal methods. The iterative scheme uses the modified Traub-Steffensen iteration (2) in the first step and Traub-Steffensen-like iterations in the second and third steps.

Rest of the paper is summarized as follows. In Section 2, optimal family of eighth order is developed and its local convergence is studied. In Section 3, the basins of attractors are analyzed to check the convergence region of new methods. In order to check the performance and to verify the theoretical results some numerical tests are performed in Section 4. A comparison with the existing methods of same order requiring derivatives is also shown in this section. Section 5 contains the concluding remarks.

2. Development of Method

Given a known multiplicity $m > 1$, we consider the following three-step scheme for multiple roots:

$$\begin{aligned} y_k &= x_k - m \frac{f(x_k)}{f[w_k, x_k]} \\ z_k &= y_k - mh(A_1 + A_2 h) \frac{f(x_k)}{f[w_k, x_k]} \\ x_{k+1} &= z_k - mutG(h, t) \frac{f(x_k)}{f[w_k, x_k]} \end{aligned} \tag{3}$$

where $h = \frac{u}{1+u}$, $u = \sqrt[m]{\frac{f(y_k)}{f(x_k)}}$, $t = \sqrt[m]{\frac{f(z_k)}{f(y_k)}}$ and $G : \mathbb{C}^2 \to \mathbb{C}$ is analytic in a neighborhood of $(0,0)$. Note that this is a three-step scheme with first step as the Traub-Steffensen iteration (2) and next two steps are Traub-Steffensen-like iterations. Notice also that third step is weighted by the factor $G(h, t)$, so this factor is called weight factor or weight function.

We shall find conditions under which the scheme (3) achieves eighth order convergence. In order to do this, let us prove the following theorem:

Theorem 1. *Let $f : \mathbb{C} \to \mathbb{C}$ be an analytic function in a domain enclosing a multiple zero (say, α) with multiplicity m. Suppose that initial guess x_0 is sufficiently close to α, then the local order of convergence of scheme (3) is at least 8, provided that $A_1 = 1$, $A_2 = 3$, $G_{00} = 1$, $G_{10} = 2$, $G_{01} = 1$, $G_{20} = -4$, $G_{11} = 4$, $G_{30} = -72$, $|G_{02}| < \infty$ and $|G_{21}| < \infty$, where $G_{ij} = \frac{\partial^{i+j}}{\partial h^i \partial t^j} G(h,t)|_{(0,0)}$, $i, j \in \{0, 1, 2, 3, 4\}$.*

Proof. Let $e_k = x_k - \alpha$, be the error in the k-th iteration. Taking into account that $f^{(j)}(\alpha) = 0, j = 0, 1, 2, \ldots, m-1$ and $f^m(\alpha) \neq 0$, the Taylor's expansion of $f(x_k)$ about α yields

$$f(x_k) = \frac{f^m(\alpha)}{m!} e_k^m \left(1 + C_1 e_k + C_2 e_k^2 + C_3 e_k^3 + C_4 e_k^4 + C_5 e_k^5 + C_6 e_k^6 + C_7 e_k^7 + + C_8 e_k^8 + \cdots \right), \quad (4)$$

where $C_n = \frac{m!}{(m+n)!} \frac{f^{(m+n)}(\alpha)}{f^{(m)}(\alpha)}$ for $n \in \mathbb{N}$.

Using (4) in $w_k = x_k + \beta f(x_k)$, we obtain that

$$\begin{aligned} w_k - \alpha &= x_k - \alpha + \beta f(x_k) \\ &= e_k + \frac{\beta f^{(m)}(\alpha)}{m!} e_k^m \left(1 + C_1 e_k + C_2 e_k^2 + C_3 e_k^3 + C_4 e_k^4 + C_5 e_k^5 + C_6 e_k^6 + C_7 e_k^7 + C_8 e_k^8 + \cdots \right). \end{aligned} \quad (5)$$

Expanding $f(w_k)$ about α

$$f(w_k) = \frac{f^m(\alpha)}{m!} e_{w_k}^m \left(1 + C_1 e_{w_k} + C_2 e_{w_k}^2 + C_3 e_{w_k}^3 + C_4 e_{w_k}^4 + \cdots \right), \quad (6)$$

where $e_{w_k} = w_k - \alpha$.

Then the first step of (3) yields

$$\begin{aligned} e_{y_k} =\;& y_k - \alpha \\ =\;& \frac{C_1}{m} e_k^2 + \frac{-(1+m)C_1^2 + 2mC_2}{m^2} e_k^3 + \frac{(1+m)^2 C_1^3 - m(4+3m)C_1 C_2 + 3m^2 C_3}{m^3} e_k^4 \\ & - \frac{(1+m)^3 C_1^4 - 2m(3+5m+2m^2)C_1^2 C_2 + 2m^2(2+m)C_2^2 + 2m^2(3+2m)C_1 C_3}{m^4} e_k^5 \\ & + \frac{1}{m^5}\big((1+m)^4 C_1^5 - m(1+m)^2(8+5m)C_1^3 C_2 + m^2(9+14m+5m^2)C_1^2 C_3 \\ & + m^2 C_1((12+16m+5m^2)C_2^2 - m^2 C_4) - m^3((12+5m)C_2 C_3 + m^2 C_5)\big) e_k^6 \\ & - P1 e_k^7 + P2 e_k^8 + O(e_k^9), \end{aligned} \quad (7)$$

where

$$\begin{aligned} P1 =\;& \frac{1}{m^6}\big((1+m)^5 C_1^6 - 2m(1+m)^3(5+3m)C_1^4 C_2 + 6m^2(1+m)^2(2+m)C_1^3 C_3 \\ & + m^2(1+m)C_1^2(3(8+10m+3m^2)C_2^2 - 2m^2 C_4) - m^3 C_1(4(9+11m+3m^2)C_2 C_3 \\ & + m^2(1+m)C_5) + m^3(-2(2+m)^2 C_2^3 + 2m^2 C_2 C_4 + m(3(3+m)C_3^2 + m^2 C_6))\big), \\ P2 =\;& \frac{1}{m^7}\big((1+m)^6 C_1^7 - m(1+m)^4(12+7m)C_1^5 C_2 + m^2(1+m)^3(15+7m)C_1^4 C_3 \\ & + m^2(1+m)^2 C_1^3(2(20+24m+7m^2)C_2^2 - 3m^2 C_4) - m^3(1+m)C_1^2(3(24+27m \\ & + 7m^2)C_2 C_3 + m^2(1+m)C_5) + m^3 C_1(-(2+m)^2(8+7m)C_2^3 + 2m^2(4+3m)C_2 C_4 \\ & + m((27+30m+7m^2)C_3^2 + m^2(1+m)C_6)) + m^4((36+32m+7m^2)C_2^2 C_3 \\ & + m^2(2+m)C_2 C_5 - m^2(3C_3 C_4 + mC_7))\big). \end{aligned}$$

Expanding $f(y_k)$ about α, we have that

$$f(y_k) = \frac{f^m(\alpha)}{m!} e_{y_k}^m \left(1 + C_1 e_{y_k} + C_2 e_{y_k}^2 + C_3 e_{y_k}^3 + C_4 e_{y_k}^4 + \cdots \right). \quad (8)$$

Also,

$$u = \frac{C_1}{m}e_k + \frac{(-(2+m)C_1^2 + 2mC_2)}{m^2}e_k^2 + \frac{(7+7m+2m^2)C_1^3 - 2m(7+3m)C_1C_2 + 6m^2C_3}{2m^3}e_k^3 \\
- \frac{1}{6m^4}\left((34+51m+29m^2+6m^3)C_1^4 - 6m(17+16m+4m^2)C_1^2C_2 + 12m^2(3+m)C_2^2\right. \\
\left. +12m^2(5+2m)C_1C_3\right)e_k^4 + \frac{1}{24m^5}\left((209+418m+355m^2+146m^3+24m^4)C_1^5 - 4m(209+306m\right. \\
+163m^2+30m^3)C_1^3C_2 + 12m^2(49+43m+10m^2)C_1^2C_3 + 12m^2C_1((53+47m+10m^2)C_2^2 \\
\left. -2m(1+m)C_4) - 24m^3((17+5m)C_2C_3 + m^2C_5)\right)e_k^5 + O(e_k^6) \quad (9)$$

and

$$h = \frac{C_1}{m}e_k + \frac{(-(3+m)C_1^2 + 2mC_2)}{m^2}e_k^2 + \frac{(17+11m+2m^2)C_1^3 - 2m(11+3m)C_1C_2 + 6m^2C_3}{2m^3}e_k^3 \\
- \frac{1}{6m^4}\left((142+135m+47m^2+6m^3)C_1^4 - 6m(45+26m+4m^2)C_1^2C_2 + 12m^2(5+m)C_2^2\right. \\
\left. +24m^2(4+m)C_1C_3\right)e_k^4 + \frac{1}{24m^5}\left((1573+1966m+995m^2+242m^3+24m^4)C_1^5 - 4m(983\right. \\
+864m+271m^2+30m^3)C_1^3C_2 + 12m^2(131+71m+10m^2)C_1^2C_3 + 12m^2C_1((157+79m \\
\left. +10m^2)C_2^2 - 2m(1+m)C_4) - 24m^3((29+5m)C_2C_3 + m^2C_5)\right)e_k^5 + O(e_k^6). \quad (10)$$

By inserting (4)–(10) in the second step of (3), we have

$$\begin{aligned} e_{z_k} &= z_k - \alpha \\ &= -\frac{(-1+A_1)C_1}{m}e_k^2 - \frac{(1+A_2+m-A_1(4+m))C_1^2 + 2(-1+A_1)mC_2}{m^2}e_k^3 \\ &\quad + \sum_{n=1}^{5}\delta_n e_k^{n+3} + O(e_k^9), \end{aligned} \quad (11)$$

where $\delta_n = \delta_n(A_1, A_2, m, C_1, C_2, C_3, \ldots, C_8)$, $n = 1, 2, 3, 4, 5$. Here, expressions of δ_n are not being produced explicitly since they are very lengthy.

In order to obtain fourth-order convergence, the coefficients of e_k^2 and e_k^3 should be equal to zero. This is possible only for the following values of A_1 and A_2, which can be calculated from the expression (11):

$$A_1 = 1 \text{ and } A_2 = 3. \quad (12)$$

Then, the error Equation (11) is given by

$$e_{z_k} = \frac{(19+m)C_1^3 - 2mC_1C_2}{2m^3}e_k^4 + \sum_{n=1}^{4}\phi_n e_k^{n+4} + O(e_k^9),$$

where $\phi_n = \phi_n(m, C_1, C_2, C_3, \ldots, C_8)$, $n = 1, 2, 3, 4$.

Expansion of $f(z_k)$ about α leads us to the expression

$$f(z_k) = \frac{f^m(\alpha)}{m!}e_{z_k}^m\left(1 + C_1 e_{z_k} + C_2 e_{z_k}^2 + C_3 e_{z_k}^3 + C_4 e_{z_k}^4 + \cdots\right).$$

and so $t = \sqrt[m]{\frac{f(z_k)}{f(y_k)}}$ yields

$$t = \frac{(19+m)C_1^2 - 2mC_2}{2m^2}e_k^2 - \frac{(163+57m+2m^2)C_1^3 - 6m(19+m)C_1C_2 + 6m^2C_3}{3m^3}e_k^3 \\
+ \frac{1}{24m^4}\left((5279+3558m+673m^2+18m^3)C_1^4 - 12m(593+187m+6m^2)C_1^2C_2\right. \\
\left. +24m^2(56+3m)C_1C_3 + 12m^2(3(25+m)C_2^2 + 2mC_4)\right)e_k^4 - \frac{1}{60m^5}\left((47457+46810m\right. \\
+16635m^2+2210m^3+48m^4)C_1^5 - 20m(4681+2898m+497m^2+12m^3)C_1^3C_2 + 60m^2(429 \\
+129m+4m^2)C_1^2C_3 + 60m^2C_1((537+147m+4m^2)C_2^2 - 2mC_4) - 60m^3(2(55+2m)C_2C_3 \\
\left. +m(1+m)C_5)\right)e_k^5 + O(e_k^6). \quad (13)$$

Expanding $G(h,t)$ in neighborhood of origin $(0,0)$ by Taylor's series, it follows that

$$
\begin{aligned}
G(h,t) \approx\ & G_{00} + G_{10}h + G_{01}t + \frac{1}{2!}(G_{20}h^2 + 2G_{11}ht + G_{02}t^2) \\
& + \frac{1}{3!}(G_{30}h^3 + 3G_{21}h^2t + 3G_{12}ht^2 + G_{03}t^3) \\
& + \frac{1}{4!}(G_{40}h^4 + 4G_{31}h^3t + 6G_{22}h^2t^2 + 4G_{13}ht^3 + G_{04}t^4),
\end{aligned}
\tag{14}
$$

where $G_{ij} = \frac{\partial^{i+j}}{\partial h^i \partial t^j} G(h,t)|_{(0,0)}$, $i,j \in \{0,1,2,3,4\}$.

By using (4), (6), (9), (10), (13) and (14) in third step of (3), we have

$$
e_{k+1} = -\frac{(G_{00}-1)C_1((19+m)C_1^2 - 2mC_2)}{2m^4}e_k^4 + \sum_{n=1}^{4}\psi_n e_k^{n+4} + O(e_k^9),
\tag{15}
$$

$\psi_n = \psi_n(m, G_{00}, G_{10}, G_{01}, G_{20}, G_{11}, G_{02}, G_{30}, G_{21}, C_1, C_2, \ldots, C_8)$, $n = 1,2,3,4$.

It is clear from the Equation (15) that we will obtain at least eighth order convergence if we choose $G_{00} = 1$, $\psi_1 = 0$, $\psi_2 = 0$ and $\psi_3 = 0$. We choose $G_{00} = 1$ in $\psi_1 = 0$. Then, we get

$$
G_{10} = 2.
\tag{16}
$$

By using G_{00} and (16) in $\psi_2 = 0$, we will obtain

$$
G_{01} = 1 \text{ and } G_{20} = -4.
\tag{17}
$$

Using G_{00}, (16) and (17) in $\psi_3 = 0$, we obtain that

$$
G_{11} = 4 \text{ and } G_{30} = -72.
\tag{18}
$$

Inserting G_{00} and (16)–(18) in (15), we will obtain the error equation

$$
\begin{aligned}
e_{k+1} =\ & -\frac{1}{48m^7}\Big(C_1((19+m)C_1^2 - 2mC_2)((3G_{02}(19+m)^2 + 2(-1121 \\
& -156m - 7m^2 + 3G_{21}(19+m)))C_1^4 - 12m(-52 + G_{21} - 4m \\
& + G_{02}(19+m))C_1^2C_2 + 12(-2 + G_{02})m^2C_2^2 - 24m^2C_1C_3)\Big)e_k^8 + O(e_k^9).
\end{aligned}
\tag{19}
$$

Thus, the eighth order convergence is established. □

Remark 1. *It is important to note that the weight function $G(h,t)$ plays a significant role in the attainment of desired convergence order of the proposed family of methods. However, only G_{02} and G_{21} are involved in the error Equation (19). On the other hand, $G_{12}, G_{03}, G_{40}, G_{31}, G_{22}, G_{13}$ and G_{04} do not affect the error Equation (19). So, we can assume them as dummy parameters.*

Remark 2. *The error Equation (19) shows that the proposed scheme (3) reaches at eighth-order convergence by using only four evaluations namely, $f(x_k), f(w_k), f(y_k)$ and $f(z_k)$ per iteration. Therefore, the scheme (3) is optimal according to Kung-Traub hypothesis [26] provided the conditions of Theorem 1 are satisfied.*

Remark 3. *Notice that the parameter β, which is used in the iteration w_k, does not appear in the expression (7) of e_{y_k} and also in later expressions. We have observed that this parameter has the appearance in the terms e_k^m and higher order. However, these terms are difficult to compute in general. Moreover, we do not need these in order to show the eighth convergence.*

Some Particular Forms of Proposed Family

(1) Let us consider the following function $G(h,t)$ which satisfies the conditions of Theorem 1

$$G(h,t) = 1 + 2h + t - 2h^2 + 4ht - 12h^3.$$

Then, the corresponding eighth-order iterative scheme is given by

$$x_{k+1} = z_k - mut\left[1 + 2h + t - 2h^2 + 4ht - 12h^3\right]\frac{f(x_k)}{f[w_k, x_k]}. \qquad (20)$$

(2) Next, consider the rational function

$$G(h,t) = \frac{1 + 2h + 2t - 2h^2 + 6ht - 12h^3}{1 + t}$$

Satisfying the conditions of Theorem 1. Then, corresponding eighth-order iterative scheme is given by

$$x_{k+1} = z_k - mut\left[\frac{1 + 2h + 2t - 2h^2 + 6ht - 12h^3}{1 + t}\right]\frac{f(x_k)}{f[w_k, x_k]}. \qquad (21)$$

(3) Consider another rational function satisfying the conditions of Theorem 1, which is given by

$$G(h,t) = \frac{1 + 3h + t + 5ht - 14h^3 - 12h^4}{1 + h}.$$

Then, corresponding eighth-order iterative scheme is given by

$$x_{k+1} = z_k - mut\left[\frac{1 + 3h + t + 5ht - 14h^3 - 12h^4}{1 + h}\right]\frac{f(x_k)}{f[w_k, x_k]}. \qquad (22)$$

(4) Next, we suggest another rational function satisfying the conditions of Theorem 1, which is given by

$$G(h,t) = \frac{1 + 3h + 2t + 8ht - 14h^3}{(1 + h)(1 + t)}.$$

Then, corresponding eighth-order iterative scheme is given by

$$x_{k+1} = z_k - mut\left[\frac{1 + 3h + 2t + 8ht - 14h^3}{(1 + h)(1 + t)}\right]\frac{f(x_k)}{f[w_k, x_k]}. \qquad (23)$$

(5) Lastly, we consider yet another function satisfying the conditions of Theorem 1

$$G(h,t) = \frac{1 + t - 2h(2 + t) - 2h^2(6 + 11t) + h^3(4 + 8t)}{2h^2 - 6h + 1}.$$

Then, the corresponding eighth-order method is given as

$$x_{k+1} = z_k - mut\left[\frac{1 + t - 2h(2 + t) - 2h^2(6 + 11t) + h^3(4 + 8t)}{2h^2 - 6h + 1}\right]\frac{f(x_k)}{f[w_k, x_k]}. \qquad (24)$$

In above each case $y_k = x_k - m\frac{f(x_k)}{f[w_k,x_k]}$ and $z_k = y_k - mh(1 + 3h)\frac{f(x_k)}{f[w_k,x_k]}$. For future reference the proposed methods (20), (21), (22), (23) and (24) are denoted by M-1, M-2, M-3, M-4 and M-5, respectively.

3. Complex Dynamics of Methods

Our aim here is to analyze the complex dynamics of new methods based on graphical tool the basins of attraction of the zeros of a polynomial $P(z)$ in complex plane. Analysis of the basins of attraction gives an important information about the stability and convergence of iterative methods. This idea was floated initially by Vrscay and Gilbert [27]. In recent times, many researchers used this concept in their work, see, for example [28–30] and references therein. To start with, let us recall some basic dynamical concepts of rational function associated to an iterative method. Let $\phi : \mathbb{R} \to \mathbb{R}$ be a rational function, the orbit of a point $x_0 \in \mathbb{R}$ is defined as the set

$$\{x_0, \phi(x_0), \ldots, \phi^m(x_0), \ldots\},$$

of successive images of x_0 by the rational function.

The dynamical behavior of the orbit of a point of $\mathbb{R}$ can be classified depending on its asymptotic behavior. In this way, a point $x_0 \in \mathbb{R}$ is a fixed point of $\phi(x_0)$ if it satisfies $\phi(x_0) = x_0$. Moreover, x_0 is called a periodic point of period $p > 1$ if it is a point such that $\phi^p(x_0) = x_0$ but $\phi^k(x_0) \neq x_0$, for each $k < p$. Also, a point x_0 is called pre-periodic if it is not periodic but there exists a $k > 0$ such that $\phi^k(x_0)$ is periodic. There exist different type of fixed points depending on the associated multiplier $|\phi'(x_0)|$. Taking the associated multiplier into account, a fixed point x_0 is called: (a) attractor if $|\phi'(x_0)| < 1$, (b) superattractor if $|\phi'(x_0)| = 0$, (c) repulsor if $|\phi'(x_0)| > 1$ and (d) parabolic if $|\phi'(x_0)| = 1$.

If α is an attracting fixed point of the rational function ϕ, its basin of attraction $\mathcal{A}(\alpha)$ is defined as the set of pre-images of any order such that

$$\mathcal{A}(\alpha) = \{x_0 \in \mathbb{R} : \phi^m(x_0) \to \alpha, m \to \infty\}.$$

The set of points whose orbits tend to an attracting fixed point α is defined as the Fatou set. Its complementary set, called Julia set, is the closure of the set consisting of repelling fixed points, and establishes the borders between the basins of attraction. That means the basin of attraction of any fixed point belongs to the Fatou set and the boundaries of these basins of attraction belong to the Julia set.

The initial point z_0 is taken in a rectangular region $R \in \mathbb{C}$ that contains all the zeros of a polynomial $P(z)$. The iterative method when starts from point z_0 in a rectangle either converges to the zero $P(z)$ or eventually diverges. The stopping criterion for convergence is considered as 10^{-3} up to a maximum of 25 iterations. If the required tolerance is not achieved in 25 iterations, we conclude that the method starting at point z_0 does not converge to any root. The strategy adopted is as follows: A color is allocated to each initial point z_0 in the basin of attraction of a zero. If the iteration initiating at z_0 converges, then it represents the attraction basin with that particular assigned color to it, otherwise if it fails to converge in 25 iterations, then it shows the black color.

To view complex geometry, we analyze the basins of attraction of the proposed methods M-I (I = 1, 2, ..., 5) on following polynomials:

Test problem 1. Consider the polynomial $P_1(z) = (z^2 - 1)^2$ having two zeros $\{-1, 1\}$ with multiplicities $m = 2$. The basin of attractors for this polynomial are shown in Figures 1–3, for different choices of $\beta = 0.01, 10^{-6}, 10^{-10}$. A color is assigned to each basin of attraction of a zero. In particular, to obtain the basin of attraction, the red and green colors have been assigned for the zeros -1 and 1, respectively. Looking at the behavior of the methods, we see that the method M-2 and M-4 possess less number of divergent points and therefore have better convergence than rest of the methods. Observe that there is a small difference among the basins for the remaining methods with the same value of β. Note also that the basins are becoming larger as the parameter β assumes smaller values.

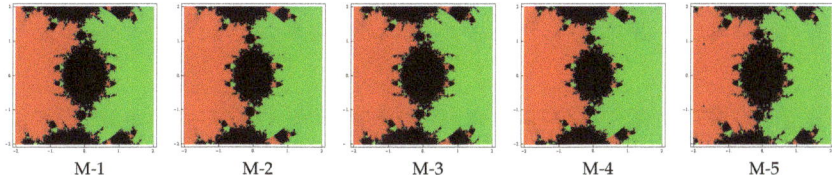

Figure 1. Basins of attraction for methods M-1 to M-5 ($\beta = 0.01$) in polynomial $P_1(z)$.

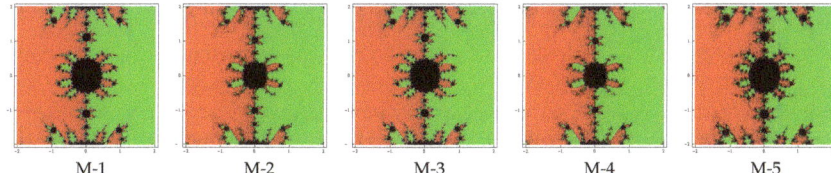

Figure 2. Basins of attraction for methods M-1 to M-5 ($\beta = 10^{-6}$) in polynomial $P_1(z)$.

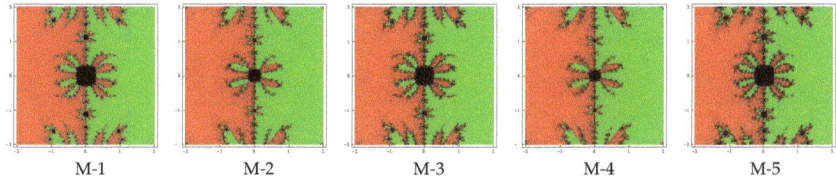

Figure 3. Basins of attraction for methods M-1 to M-5 ($\beta = 10^{-10}$) in polynomial $P_1(z)$.

Test problem 2. Let $P_2(z) = (z^3 + z)^2$ having three zeros $\{-i, 0, i\}$ with multiplicities $m = 2$. The basin of attractors for this polynomial are shown in Figures 4–6, for different choices of $\beta = 0.01, 10^{-6}, 10^{-10}$. A color is allocated to each basin of attraction of a zero. For example, we have assigned the colors: green, red and blue corresponding to the basins of the zeros $-i$, i and 0, From graphics, we see that the methods M-2 and M-4 have better convergence due to a lesser number of divergent points. Also observe that in each case, the basins are getting broader with the smaller values of β. The basins in methods M-1, M-3 are almost the same and method M-5 has more divergent points.

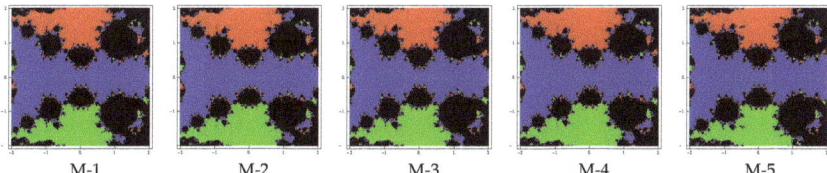

Figure 4. Basins of attraction for methods M-1 to M-5 ($\beta = 0.01$) in polynomial $P_2(z)$.

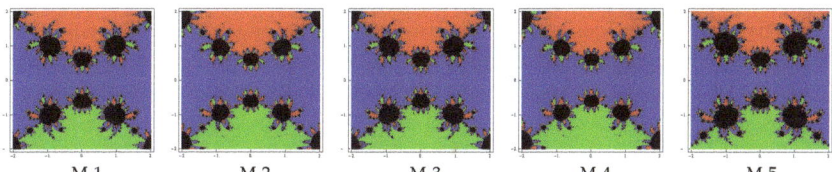

Figure 5. Basins of attraction for methods M-1 to M-5 ($\beta = 10^{-6}$) in polynomial $P_2(z)$.

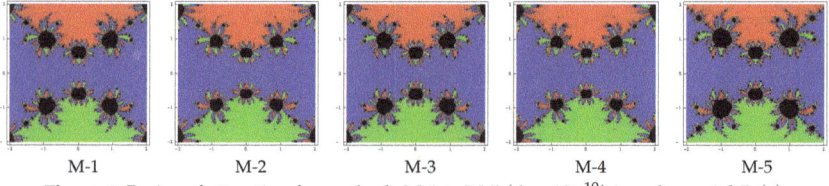

Figure 6. Basins of attraction for methods M-1 to M-5 ($\beta = 10^{-10}$) in polynomial $P_2(z)$.

Test problem 3. Let $P_3(z) = (z^2 - \frac{1}{4})(z^2 + \frac{9}{4})$ having four simple zeros $\{-\frac{1}{2}, \frac{1}{2}, -\frac{3}{2}i, \frac{3}{2}i, \}$. To see the dynamical view, we allocate the colors green, red, blue and yellow corresponding to basins of the zeros $-\frac{1}{2}, \frac{1}{2}, -\frac{3}{2}i$ and $\frac{3}{2}i$. The basin of attractors for this polynomial are shown in Figures 7–9, for different choices of $\beta = 0.01, 10^{-6}, 10^{-10}$. Looking at the graphics, we conclude that the methods M-2 and M-4 have better convergence behavior since they have lesser number of divergent points. The remaining methods have almost similar basins with the same value of β. Notice also that the basins are becoming larger with the smaller values of β.

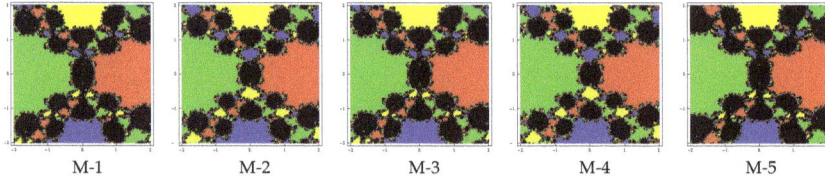

Figure 7. Basins of attraction for methods M-1 to M-5 ($\beta = 0.01$) in polynomial $P_3(z)$.

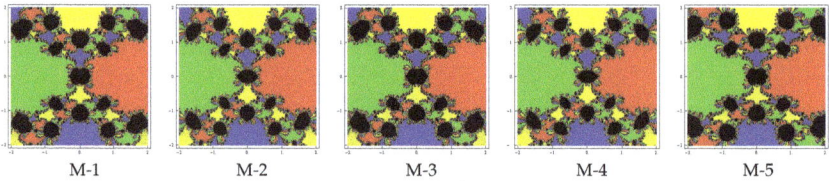

Figure 8. Basins of attraction for methods M-1 to M-5 ($\beta = 10^{-6}$) in polynomial $P_3(z)$.

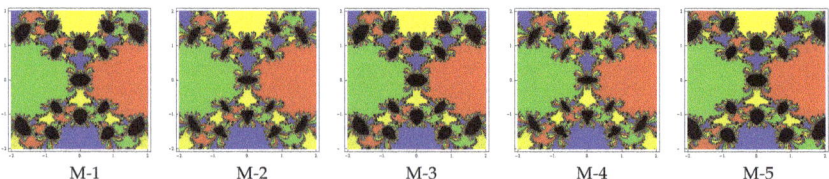

Figure 9. Basins of attraction for methods M-1 to M-5 ($\beta = 10^{-10}$) in polynomial $P_3(z)$.

From these graphics one can easily evaluate the behavior and stability of any method. If we choose an initial point z_0 in a zone where distinct basins of attraction touch each other, it is impractical to predict which root is going to be attained by the iterative method that starts in z_0. Hence, the choice of z_0 in such a zone is not a good one. Both the black zones and the regions with different colors are not suitable to take the initial guess z_0 when we want to acquire a unique root. The most adorable pictures appear when we have very tricky frontiers between basins of attraction and they correspond to the cases where the method is more demanding with respect to the initial point and its dynamic behavior is more unpredictable. We conclude this section with a remark that the convergence nature of proposed methods depends upon the value of parameter β. The smaller is the value of β, the better is the convergence of the method.

4. Numerical Results

In this section, we apply the methods M1–M5 of family (3) to solve few nonlinear equations, which not only depict the methods practically but also serve to verify the validity of theoretical results that we have derived. The theoretical order of convergence is verified by calculating the computational order of convergence (COC) using the formula (see [31])

$$\text{COC} = \frac{\ln|(x_{k+2} - \alpha)/(x_{k+1} - \alpha)|}{\ln|(x_{k+1} - \alpha)/(x_k - \alpha)|} \quad \text{for each } k = 1, 2, \ldots \tag{25}$$

Performance is compared with some existing eighth-order methods requiring derivative evaluations in their formulae. For example, we choose the methods by Zafar et al. [19] and Behl et al. [23,24]. These methods are expressed as follows:

Zafar et al. method (ZM-1):

$$y_k = x_k - m\frac{f(x_k)}{f'(x_k)}$$

$$z_k = y_k - mu_k \left(\frac{1 - 5u_k^2 + 8u_k^3}{-2u_k + 1}\right)\frac{f(x_k)}{f'(x_k)}$$

$$x_{k+1} = z_k - mu_k v_k (1 + 2u_k)(v_k + 1)(2w_k + 1)\frac{f(x_k)}{f'(x_k)}.$$

Zafar et al. method (ZM-2):

$$y_k = x_k - m\frac{f(x_k)}{f'(x_k)}$$

$$z_k = y_k - mu_k (6u_k^3 - u_k^2 + 2u_k + 1)\frac{f(x_k)}{f'(x_k)}$$

$$x_{k+1} = z_k - mu_k v_k e^{v_k} e^{2w_k}(1 + 2u_k)\frac{f(x_k)}{f'(x_k)}$$

where $u_k = \left(\frac{f(y_k)}{f(x_k)}\right)^{\frac{1}{m}}$, $v_k = \left(\frac{f(z_k)}{f(y_k)}\right)^{\frac{1}{m}}$ and $w_k = \left(\frac{f(z_k)}{f(x_k)}\right)^{\frac{1}{m}}$.

Behl et al. method (BM-1):

$$y_k = x_k - m\frac{f(x_k)}{f'(x_k)}$$

$$z_k = y_k - m\frac{f(x_k)}{f'(x_k)}u_k(1 + 2u_k - u_k^2)$$

$$x_{k+1} = z_k + m\frac{f(x_k)}{f'(x_k)}\frac{w_k u_k}{1 - w_k}\left[-1 - 2u_k + 6u_k^3 - \frac{1}{6}(85 + 21m + 2m^2)u_k^4 - 2v_k\right].$$

Behl et al. method (BM-2):

$$y_k = x_k - m\frac{f(x_k)}{f'(x_k)}$$

$$z_k = y_k - m\frac{f(x_k)}{f'(x_k)}u_k(1 + 2u_k)$$

$$x_{k+1} = z_k - m\frac{f(x_k)}{f'(x_k)}\frac{w_k u_k}{1 - w_k}\left[\frac{1 + 9u_k^2 + 2v_k + u_k(6 + 8v_k)}{1 + 4u_k}\right]$$

where $u_k = \left(\frac{f(y_k)}{f(x_k)}\right)^{\frac{1}{m}}$, $v_k = \left(\frac{f(z_k)}{f(x_k)}\right)^{\frac{1}{m}}$, $w_k = \left(\frac{f(z_k)}{f(y_k)}\right)^{\frac{1}{m}}$.

Behl et al. method (BM-3):

$$y_k = x_k - m \frac{f(x_k)}{f'(x_k)}$$

$$z_k = y_k - mu \frac{f(x_k)}{f'(x_k)} \frac{1+\gamma u}{1+(\gamma-2)u}$$

$$x_{k+1} = z_k - \frac{m}{2} uv \frac{f(x_k)}{f'(x_k)} \left[1 - \frac{(2v+1)(2u(2\gamma - \gamma^2 + 4\gamma u + u + 4) - 2\gamma + 5)}{2\gamma + 2(\gamma^2 - 6\gamma + 6)u - 5}\right]$$

where $u = \left(\frac{f(y_k)}{f(x_k)}\right)^{\frac{1}{m}}$, $v = \left(\frac{f(z_k)}{f(y_k)}\right)^{\frac{1}{m}}$ and $\gamma = \frac{1}{3}$.

All computations are performed in the programming package Mathematica [32] in PC with Intel(R) Pentium(R) CPU B960 @ 2.20 GHz, 2.20 GHz (32-bit Operating System) Microsoft Windows 7 Professional and 4 GB RAM using multiple-precision arithmetic. Performance of the new methods is tested by choosing value of the parameter $\beta = 0.01$. Choice of the initial approximation x_0 in the examples is obtained by using the procedure proposed in [33]. For example, the procedure when applied to the function of Example 2 in the interval [2, 3.5] using the statements

```
f[x_]=x^9-29x^8+349x^7-2261x^6+8455x^5-17663x^4+15927x^3+6993x^2-24732x+12960;
a=2; b=3.5; k=1; x0=0.5*(a+b+Sign[f[a]]*NIntegrate[Tanh[k*f[x]],{x,a,b}])
```

in programming package Mathematica yields a close initial approximation $x_0 = 3.20832$ to the root $\alpha = 3$.

Numerical results displayed in Tables 1–6 contain: (i) values of first three consecutive errors $|x_{k+1} - x_k|$, (ii) number of iterations (k) needed to converge to the required solution with the stopping criterion $|x_{k+1} - x_k| + |f(x_k)| < 10^{-100}$, (iii) computational order of convergence (COC) using (25) and (iv) the elapsed CPU-time (CPU-time) in seconds computed by the *Mathematica* command "TimeUsed[ ]". Further, the meaning of $a \times e \pm b$ is $a \times 10^{\pm b}$ in Tables 1–6.

The following examples are chosen for numerical tests:

Example 1. *We consider the Planck's radiation law problem [34]:*

$$\varphi(\lambda) = \frac{8\pi ch \lambda^{-5}}{e^{ch/\lambda kT} - 1} \tag{26}$$

which determines the energy density with in an isothermal black body. Here, c is the speed of light, λ is the wavelength of the radiation, k is Boltzmann's constant, T is the absolute temperature of the black body and h is the Planck's constant. Suppose, we would like to calculate wavelength λ which corresponds to maximum energy density $\varphi(\lambda)$. From (26), we get

$$\varphi'(\lambda) = \left(\frac{8\pi ch \lambda^{-6}}{e^{ch/\lambda kT} - 1}\right)\left(\frac{(ch/\lambda kT)e^{ch/\lambda kT}}{e^{ch/\lambda kT} - 1} - 5\right) = AB.$$

It can be seen that a maximum value for φ occurs when $B = 0$, that is, when

$$\left(\frac{(ch/\lambda kT)e^{ch/\lambda kT}}{e^{ch/\lambda kT} - 1}\right) = 5.$$

Then, setting $x = ch/\lambda kT$, the above equation becomes

$$1 - \frac{x}{5} = e^{-x}. \tag{27}$$

We consider this case for four times and obtained the required nonlinear function

$$f_1(x) = \left(e^{-x} - 1 + \frac{x}{5}\right)^4. \tag{28}$$

The aim is to find a multiple root of the equation $f_1(x) = 0$. Obviously, one of the multiple root $x = 0$ is not taken into account. As argued in [34], the left-hand side of (27) is zero for $x = 5$ and right-hand side is $e^{-5} \approx 6.74 \times 10^{-3}$. Hence, it is expected that another multiple root of the equation $f_1(x) = 0$ might exist near to $x = 5$. The calculated value of this multiple root is given by $\alpha \approx 4.96511423$ with $x_0 = 3.5$. As a result, the wavelength (λ) corresponding to which the energy density is maximum is approximately given as

$$\lambda \approx \frac{ch}{(kT)4.96511423}.$$

Numerical results are shown in Table 1.

Example 2. *Finding eigen values of a large sparse matrix is a challenging task in applied mathematics and engineering. Calculating even the roots of a characteristic equation of square matrix greater than 4 is another big challenge. So, we consider the following 9×9 matrix (see [23])*

$$M = \frac{1}{8}\begin{bmatrix} -12 & 0 & 0 & 19 & -19 & 76 & -19 & 18 & 437 \\ -64 & 24 & 0 & -24 & 24 & 64 & -8 & 32 & 376 \\ -16 & 0 & 24 & 4 & -4 & 16 & -4 & 8 & 92 \\ -40 & 0 & 0 & -10 & 50 & 40 & 2 & 20 & 242 \\ -4 & 0 & 0 & -1 & 41 & 4 & 1 & 2 & 25 \\ -40 & 0 & 0 & 18 & -18 & 104 & -18 & 20 & 462 \\ -84 & 0 & 0 & -29 & 29 & 84 & 21 & 42 & 501 \\ 16 & 0 & 0 & -4 & 4 & -16 & 4 & 16 & -92 \\ 0 & 0 & 0 & 0 & 0 & 0 & 0 & 0 & 24 \end{bmatrix}.$$

The characteristic polynomial of the matrix (M) is given as

$$f_2(x) = x^9 - 29x^8 + 349x^7 - 2261x^6 + 8455x^5 - 17663x^4 + 15927x^3 + 6993x^2 - 24732x + 12960.$$

This function has one multiple zero at $\alpha = 3$ of multiplicity 4. We find this zero with initial approximation $x_0 = 3.2$. Numerical results are shown in Table 2.

Example 3. *Consider an isentropic supersonic flow along a sharp expansion corner (see [2]). Then relationship between the Mach number before the corner (i.e., M_1) and after the corner (i.e., M_2) is given by*

$$\delta = b^{1/2}\left(\tan^{-1}\left(\frac{M_2^2-1}{b}\right)^{1/2} - \tan^{-1}\left(\frac{M_1^2-1}{b}\right)^{1/2}\right) - \left(\tan^{-1}(M_2^2-1)^{1/2} - \tan^{-1}(M_1^2-1)^{1/2}\right)$$

where $b = \frac{\gamma+1}{\gamma-1}$, γ is the specific heat ratio of the gas.

For a special case study, we solve the equation for M_2 given that $M_1 = 1.5$, $\gamma = 1.4$ and $\delta = 10^0$. In this case, we have

$$\tan^{-1}\left(\frac{\sqrt{5}}{2}\right) - \tan^{-1}(\sqrt{x^2-1}) + \sqrt{6}\left(\tan^{-1}(\sqrt{\frac{x^2-1}{6}}) - \tan^{-1}\left(\frac{1}{2}\sqrt{\frac{5}{6}}\right)\right) - \frac{11}{63} = 0,$$

where $x = M_2$.

We consider this case for ten times and obtained the required nonlinear function

$$f_3(x) = \left[\tan^{-1}\left(\frac{\sqrt{5}}{2}\right) - \tan^{-1}(\sqrt{x^2-1}) + \sqrt{6}\left(\tan^{-1}(\sqrt{\frac{x^2-1}{6}}) - \tan^{-1}\left(\frac{1}{2}\sqrt{\frac{5}{6}}\right)\right) - \frac{11}{63}\right]^{10}.$$

The above function has zero at $\alpha = 1.8411027704926161\ldots$ with multiplicity 10. This zero is calculated using initial approximation $x_0 = 2$. Numerical results are shown in Table 3.

Example 4. *The van der Waals equation-of-state*

$$\left(P + \frac{a_1 n^2}{V^2}\right)(V - na_2) = nRT,$$

explains the behavior of a real gas by introducing in the ideal gas equations two parameters, a_1 and a_2, specific for each gas. Determination of the volume V of the gas in terms of the remaining parameters requires the solution of a nonlinear equation in V.

$$PV^3 - (na_2 P + nRT)V^2 + a_1 n^2 V = a_1 a_2 n^3.$$

Given the parameters a_1 and a_2 of a particular gas, one can obtain values for n, P and T, such that this equation has three real zeros. By using the particular values (see [23]), we obtain the nonlinear equation

$$x^3 - 5.22 x^2 + 9.0825 x - 5.2675 = 0,$$

where $x = V$. This equation has a multiple root $\alpha = 1.75$ with multiplicity 2. We further increase the multiplicity of this root to 8 by considering this case for four times and so obtain the nonlinear function

$$f_4(x) = (x^3 - 5.22 x^2 + 9.0825 x - 5.2675)^4.$$

The initial guess chosen to obtain the solution 1.75 is $x_0 = 1.5$. Numerical results are shown in Table 4.

Example 5. *Next, we assume a standard nonlinear test function from Behl et al. [17] which is defined by*

$$f_5(x) = \left(-\sqrt{1 - x^2} + x + \cos\left(\frac{\pi x}{2}\right) + 1\right)^6.$$

The function f_5 has multiple zero at $\alpha = -0.728584046\ldots$ of multiplicity 6. We select initial approximation $x_0 = -0.76$ to obtain zero of this function. Numerical results are exhibited in Table 5.

Example 6. *Lastly, we consider another standard test function which is defined as*

$$f_6(x) = x(x^2 + 1)(2e^{x^2+1} + x^2 - 1)\cosh^2\left(\frac{\pi x}{2}\right).$$

This function has multiple zero $\alpha = i$ of multiplicity 4. Let us choose the initial approximation $x_0 = 1.5i$ to compute this zero. The computed results are displayed in Table 6.

Table 1. Performance of methods for example 1.

Methods	$|x_2 - x_1|$	$|x_3 - x_2|$	$|x_4 - x_3|$	k	COC	CPU-Time
ZM-1	2.13	4.82×10^{-8}	4.27×10^{-67}	4	8.000	0.608
ZM-2	6.43	5.30×10^{-7}	6.10×10^{-59}	4	8.000	0.671
BM-1	1.03×10^{-1}	3.34×10^{-6}	9.73×10^{-20}	5	3.000	0.687
BM-2	1.03×10^{-1}	3.35×10^{-6}	9.82×10^{-20}	5	3.000	0.702
BM-3	1.85	2.44×10^{-8}	1.15×10^{-69}	4	8.000	0.640
M-1	1.65	1.86×10^{-8}	3.08×10^{-70}	4	8.000	0.452
M-2	9.64×10^{-1}	1.86×10^{-9}	5.08×10^{-78}	4	8.000	0.453
M-3	1.64	1.81×10^{-8}	2.80×10^{-70}	4	8.000	0.468
M-4	9.55×10^{-1}	1.84×10^{-9}	5.09×10^{-78}	4	8.000	0.437
M-5	1.65	1.86×10^{-8}	3.29×10^{-70}	4	8.000	0.421

Table 2. Performance of methods for example 2.

Methods	$|x_2 - x_1|$	$|x_3 - x_2|$	$|x_4 - x_3|$	k	COC	CPU-Time
ZM-1	2.24×10^{-1}	3.06×10^{-8}	3.36×10^{-62}	4	8.000	0.140
ZM-2	6.45×10^{-1}	1.99×10^{-6}	5.85×10^{-48}	4	8.000	0.187
BM-1	9.85×10^{-3}	4.51×10^{-7}	4.14×10^{-20}	5	3.000	0.140
BM-2	9.86×10^{-3}	4.52×10^{-7}	4.18×10^{-20}	5	3.000	0.140
BM-3	1.97×10^{-1}	5.21×10^{-9}	4.23×10^{-69}	4	8.000	0.125
M-1	2.07×10^{-1}	6.58×10^{-8}	5.78×10^{-59}	4	8.000	0.125
M-2	1.21×10^{-1}	2.12×10^{-9}	1.01×10^{-70}	4	8.000	0.110
M-3	2.05×10^{-1}	6.68×10^{-8}	7.64×10^{-59}	4	8.000	0.125
M-4	1.20×10^{-1}	2.24×10^{-9}	1.79×10^{-70}	4	8.000	0.109
M-5	2.07×10^{-1}	8.86×10^{-8}	7.65×10^{-58}	4	8.000	0.093

Table 3. Performance of methods for example 3.

Methods	$|x_2 - x_1|$	$|x_3 - x_2|$	$|x_4 - x_3|$	k	COC	CPU-Time
ZM-1	3.19×10^{-2}	2.77×10^{-16}	0	3	7.995	2.355
ZM-2	7.25×10^{-2}	5.76×10^{-14}	0	3	7.986	2.371
BM-1	5.84×10^{-4}	1.78×10^{-11}	5.08×10^{-34}	4	3.000	2.683
BM-2	5.84×10^{-4}	1.78×10^{-11}	5.09×10^{-34}	4	3.000	2.777
BM-3	3.07×10^{-2}	4.39×10^{-17}	0	3	8.002	2.324
M-1	3.05×10^{-2}	4.52×10^{-16}	0	3	7.993	1.966
M-2	1.96×10^{-2}	2.65×10^{-17}	0	3	7.996	1.982
M-3	3.04×10^{-2}	5.46×10^{-16}	0	3	7.993	1.965
M-4	1.96×10^{-2}	3.05×10^{-17}	0	3	7.996	1.981
M-5	3.05×10^{-2}	5.43×10^{-16}	0	3	7.992	1.903

Table 4. Performance of methods for example 4.

Methods	$|x_2 - x_1|$	$|x_3 - x_2|$	$|x_4 - x_3|$	k	COC	CPU-Time
ZM-1	2.21×10^{-1}	1.83×10^{-1}	7.19×10^{-3}	6	8.000	0.124
ZM-2	Fails	–	–	–	–	–
BM-1	1.15	1.06	5.83×10^{-2}	7	3.000	0.109
BM-2	2.44×10^{-2}	4.15×10^{-3}	5.41×10^{-4}	7	3.000	0.110
BM-3	2.67×10^{-2}	3.06×10^{-3}	9.21×10^{-4}	5	7.988	0.109
M-1	3.55×10^{-2}	2.32×10^{-3}	1.42×10^{-10}	5	8.000	0.084
M-2	3.05×10^{-2}	7.06×10^{-3}	2.94×10^{-3}	6	8.000	0.093
M-3	3.30×10^{-2}	5.82×10^{-4}	4.26×10^{-5}	5	8.000	0.095
M-4	2.95×10^{-2}	1.22×10^{-2}	6.70×10^{-3}	6	8.000	0.094
M-5	5.01×10^{-2}	1.20×10^{-2}	5.06×10^{-6}	5	8.000	0.089

Table 5. Performance of methods for example 5.

Methods	$|x_2 - x_1|$	$|x_3 - x_2|$	$|x_4 - x_3|$	k	COC	CPU-Time
ZM-1	1.02×10^{-2}	1.56×10^{-14}	0	3	7.983	0.702
ZM-2	2.40×10^{-2}	5.32×10^{-14}	7.45×10^{-89}	4	8.000	0.873
BM-1	2.55×10^{-4}	7.84×10^{-11}	2.26×10^{-30}	5	3.000	0.920
BM-2	2.55×10^{-4}	7.84×10^{-11}	2.26×10^{-30}	5	3.000	0.795
BM-3	9.57×10^{-3}	2.50×10^{-15}	0	3	7.989	0.671
M-1	9.44×10^{-3}	2.07×10^{-14}	0	3	7.982	0.593
M-2	5.96×10^{-3}	1.02×10^{-15}	0	3	7.990	0.608
M-3	9.42×10^{-3}	2.48×10^{-14}	0	3	7.982	0.562
M-4	5.95×10^{-3}	1.18×10^{-15}	0	3	7.989	0.530
M-5	9.44×10^{-3}	2.62×10^{-14}	0	3	7.982	0.499

Table 6. Performance of methods for example 6.

Methods	$|x_2 - x_1|$	$|x_3 - x_2|$	$|x_4 - x_3|$	k	COC	CPU-Time
ZM-1	1.38×10^{-2}	5.09×10^{-4}	2.24×10^{-27}	4	8.000	1.217
ZM-2	3.13×10^{-2}	4.80×10^{-3}	7.00×10^{-20}	4	7.998	1.357
BM-1	4.76×10^{-5}	1.26×10^{-36}	0	3	8.000	0.874
BM-2	4.76×10^{-5}	2.57×10^{-36}	0	3	8.000	0.889
BM-3	1.37×10^{-2}	4.98×10^{-4}	3.38×10^{-28}	4	8.000	1.201
M-1	7.34×10^{-6}	1.14×10^{-41}	0	3	8.000	0.448
M-2	8.25×10^{-6}	4.84×10^{-41}	0	3	8.000	0.452
M-3	7.71×10^{-6}	2.09×10^{-41}	0	3	8.000	0.460
M-4	8.68×10^{-6}	8.58×10^{-41}	0	3	8.000	0.468
M-5	8.32×10^{-6}	4.03×10^{-41}	0	3	8.000	0.436

From the numerical values of errors we examine that the accuracy in the values of successive approximations increases as the iteration proceed. This explains the stable nature of methods. Also, like the existing methods the new methods show consistent convergence nature. At the stage when stopping criterion $|x_{k+1} - x_k| + |f(x_k)| < 10^{-100}$ has been satisfied we display the value '0' of $|x_{k+1} - x_k|$. From the calculation of computational order of convergence shown in the penultimate column in each table, we verify the theoretical eighth order of convergence. However, this is not true for the existing eighth-order methods BM-1 and BM-2, since the eighth order convergence is not maintained. The efficient nature of proposed methods can be observed by the fact that the amount of CPU time consumed by the methods is less than that of the time taken by existing methods. In addition, the new methods are more accurate because error becomes much smaller with increasing n as compare to the error of existing techniques. The main purpose of implementing the new derivative-free methods for solving different type of nonlinear equations is purely to illustrate the exactness of the approximate solution and the stability of the convergence to the required solution. Similar numerical experiments, performed for many other different problems, have confirmed this conclusion to a good extent.

5. Conclusions

In the foregoing study, we have proposed the first ever, as far as we know, class of optimal eighth order derivative-free iterative methods for solving nonlinear equations with multiple roots. Analysis of the local convergence has been carried out, which proves the order eight under standard assumptions of the function whose zeros we are looking for. Some special cases of the class are presented. These are implemented to solve nonlinear equations including those arising in practical problems. The methods are compared with existing techniques of same order. Testing of the numerical results shows that the presented derivative-free methods are good competitors to the existing optimal eighth-order techniques that require derivative evaluations in their algorithm. We conclude the work with a remark that derivative-free techniques are good options to Newton-type iterations in the cases when derivatives are expensive to compute or difficult to evaluate.

Author Contributions: Methodology, J.R.S.; Writing—review & editing, J.R.S.; Investigation, S.K.; Data Curation, S.K.; Conceptualization, I.K.A.; Formal analysis, I.K.A.

Funding: This research received no external funding.

Conflicts of Interest: The authors declare no conflict of interest.

References

1. Traub, J.F. *Iterative Methods for the Solution of Equations*; Chelsea Publishing Company: New York, NY, USA, 1982.
2. Hoffman, J.D. *Numerical Methods for Engineers and Scientists*; McGraw-Hill Book Company: New York, NY, USA, 1992.

3. Argyros, I.K. *Convergence and Applications of Newton-Type Iterations*; Springer-Verlag: New York, NY, USA, 2008.
4. Argyros, I.K.; Magreñán, Á.A. *Iterative Methods and Their Dynamics with Applications*; CRC Press: New York, NY, USA, 2017.
5. Schröder, E. Über unendlich viele Algorithmen zur Auflösung der Gleichungen. *Math. Ann.* **1870**, *2*, 317–365. [CrossRef]
6. Hansen, E.; Patrick, M. A family of root finding methods. *Numer. Math.* **1977**, *27*, 257-269. [CrossRef]
7. Victory, H.D.; Neta, B. A higher order method for multiple zeros of nonlinear functions. *Int. J. Comput. Math.* **1983**, *12*, 329–335. [CrossRef]
8. Dong, C. A family of multipoint iterative functions for finding multiple roots of equations. *Int. J. Comput. Math.* **1987**, *21*, 363–367. [CrossRef]
9. Osada, N. An optimal multiple root-finding method of order three. *J. Comput. Appl. Math.* **1994**, *51*, 131–133. [CrossRef]
10. Neta, B. New third order nonlinear solvers for multiple roots. *Appl. Math. Comput.* **2008**, *202*, 162–170. [CrossRef]
11. Li, S.; Liao, X.; Cheng, L. A new fourth-order iterative method for finding multiple roots of nonlinear equations. *Appl. Math. Comput.* **2009**, *215*, 1288–1292.
12. Li, S.G.; Cheng, L.Z.; Neta, B. Some fourth-order nonlinear solvers with closed formulae for multiple roots. *Comput. Math. Appl.* **2010**, *59*, 126–135. [CrossRef]
13. Sharma, J.R.; Sharma, R. Modified Jarratt method for computing multiple roots. *Appl. Math. Comput.* **2010**, *217*, 878–881. [CrossRef]
14. Zhou, X.; Chen, X.; Song, Y. Constructing higher-order methods for obtaining the multiple roots of nonlinear equations. *J. Comput. Math. Appl.* **2011**, *235*, 4199–4206. [CrossRef]
15. Kansal, M.; Kanwar, V.; Bhatia, S. On some optimal multiple root-finding methods and their dynamics. *Appl. Appl. Math.* **2015**, *10*, 349–367.
16. Geum, Y.H.; Kim, Y.I.; Neta, B. A class of two-point sixth-order multiplezero finders of modified double-Newton type and their dynamics. *Appl. Math. Comput.* **2015**, *270*, 387–400.
17. Behl, R.; Cordero, A.; Motsa, S.S.; Torregrosa, J.R.; Kanwar, V. An optimal fourth-order family of methods for multiple roots and its dynamics. *Numer. Algor.* **2016**, *71*, 775–796. [CrossRef]
18. Geum, Y.H.; Kim, Y.I.; Neta, B. Constructing a family of optimal eighth-order modified Newton-type multiple-zero finders along with the dynamics behind their purely imaginary extraneous fixed points. *J. Comp. Appl. Math.* **2017**. [CrossRef]
19. Zafar, F.; Cordero, A.; Quratulain, R.; Torregrosa, J.R. Optimal iterative methods for finding multiple roots of nonlinear equations using free parameters. *J. Math. Chem.* **2017**. [CrossRef]
20. Zafar, F.; Cordero, A.; Sultana, S.; Torregrosa, J.R. Optimal iterative methods for finding multiple roots of nonlinear equations using weight functions and dynamics. *J. Comp. Appl. Math.* **2018**, *342*, 352–374. [CrossRef]
21. Zafar, F.; Cordero, A.; Torregrosa, J.R. An efficient family of optimal eighth-order multiple root finders. *Mathematics* **2018**, *6*, 310. [CrossRef]
22. Behl, R.; Cordero, A.; Motsa, S.S.; Torregrosa, J.R. An eighth-order family of optimal multiple root finders and its dynamics. *Numer. Algor.* **2018**, *77*, 1249-1272. [CrossRef]
23. Behl, R.; Zafar, F.; Alshormani, A.S.; Junjua, M.U.D.; Yasmin, N. An optimal eighth-order scheme for multiple zeros of unvariate functions. *Int. J. Comput. Meth.* **2018**. [CrossRef]
24. Behl, R.; Alshomrani, A.S.; Motsa, S.S. An optimal scheme for multiple roots of nonlinear equations with eighth-order convergence. *J. Math. Chem.* **2018**. [CrossRef]
25. Sharma, J.R.; Kumar, D.; Argyros, I.K. An efficient class of Traub-Steffensen-like seventh order multiple-root solvers with applications. *Symmetry* **2019**, *11*, 518. [CrossRef]
26. Kung, H.T.; Traub, J.F. Optimal order of one-point and multipoint iteration. *J. Assoc. Comput. Mach.* **1974**, *21*, 643–651. [CrossRef]
27. Vrscay, E.R.; Gilbert, W.J. Extraneous fixed points, basin boundaries and chaotic dynamics for Schröder and König rational iteration functions. *Numer. Math.* **1988**, *52*, 1–16. [CrossRef]
28. Varona, J.L. Graphic and numerical comparison between iterative methods. *Math. Intell.* **2002**, *24*, 37–46. [CrossRef]

29. Scott, M.; Neta, B.; Chun, C. Basin attractors for various methods. *Appl. Math. Comput.* **2011**, *218*, 2584–2599. [CrossRef]
30. Lotfi, T.; Sharifi, S.; Salimi, M.; Siegmund, S. A new class of three-point methods with optimal convergence order eight and its dynamics. *Numer. Algorithms* **2015**, *68*, 261–288. [CrossRef]
31. Weerakoon, S.; Fernando, T.G.I. A variant of Newton's method with accelerated third-order convergence. *Appl. Math. Lett.* **2000**, *13*, 87–93. [CrossRef]
32. Wolfram, S. *The Mathematica Book*, 5th ed.; Wolfram Media: Champaign, IL, USA, 2003.
33. Yun, B.I. A non-iterative method for solving non-linear equations. *Appl. Math. Comput.* **2008**, *198*, 691–699. [CrossRef]
34. Bradie, B. *A Friendly Introduction to Numerical Analysis*; Pearson Education Inc.: New Delhi, India, 2006.

© 2019 by the authors. Licensee MDPI, Basel, Switzerland. This article is an open access article distributed under the terms and conditions of the Creative Commons Attribution (CC BY) license (http://creativecommons.org/licenses/by/4.0/).

Article

Strong Convergence of a System of Generalized Mixed Equilibrium Problem, Split Variational Inclusion Problem and Fixed Point Problem in Banach Spaces

Mujahid Abbas [1,2], Yusuf Ibrahim [3], Abdul Rahim Khan [4] and Manuel de la Sen [5,*]

[1] Department of Mathematics, Government College University, Katchery Road, Lahore 54000, Pakistan; abbas.mujahid@gmail.com
[2] Department of Mathematics and Applied Mathematics University of Pretoria, Pretoria 0002, South Africa
[3] Department of Mathematics, Sa'adatu Rimi College of Education, Kumbotso Kano, P.M.B. 3218 Kano, Nigeria; danustazz@gmail.com
[4] Department of Mathematics and Statistics, King Fahad University of Petroleum and Minerals, Dhahran 31261, Saudi Arabia; arahim@kfupm.edu.sa
[5] Institute of Research and Development of Processes, University of The Basque Country, Campus of Leioa (Bizkaia), 48080 Leioa, Spain
* Correspondence: manuel.delasen@ehu.eus

Received: 2 April 2019; Accepted: 21 May 2019; Published: 27 May 2019

Abstract: The purpose of this paper is to introduce a new algorithm to approximate a common solution for a system of generalized mixed equilibrium problems, split variational inclusion problems of a countable family of multivalued maximal monotone operators, and fixed-point problems of a countable family of left Bregman, strongly asymptotically non-expansive mappings in uniformly convex and uniformly smooth Banach spaces. A strong convergence theorem for the above problems are established. As an application, we solve a generalized mixed equilibrium problem, split Hammerstein integral equations, and a fixed-point problem, and provide a numerical example to support better findings of our result.

Keywords: split variational inclusion problem; generalized mixed equilibrium problem; fixed point problem; maximal monotone operator; left Bregman asymptotically nonexpansive mapping; uniformly convex and uniformly smooth Banach space

1. Introduction and Preliminaries

Let E be a real normed space with dual E^*. A map $B : E \to E^*$ is called:

(i) monotone if, for each $x, y \in E$, $\langle \eta - \nu, x - y \rangle \geq 0$, $\forall\ \eta \in Bx$, $\nu \in By$, where $\langle \cdot, \cdot \rangle$ denotes duality pairing,

(ii) ϵ-inverse strongly monotone if there exists $\epsilon > 0$, such that $\langle Bx - By, x - y \rangle \geq \epsilon ||Bx - By||^2$,

(iii) maximal monotone if B is monotone and the graph of B is not properly contained in the graph of any other monotone operator. We note that B is maximal monotone if, and only if it is monotone, and $R(J + tB) = E^*$ for all $t > 0$, J is the normalized duality map on E and $R(J + tB)$ is the range of $(J + tB)$ (cf. [1]).

Let H_1 and H_2 be Hilbert spaces. For the maximal monotone operators $B_1 : H_1 \to 2^{H_1}$ and $B_2 : H_2 \to 2^{H_2}$, Moudafi [2] introduced the following split monotone variational inclusion:

$$\text{find } x^* \in H_1 \text{ such that } 0 \in f(x^*) + B_1(x^*),$$
$$y^* = Ax^* \in H_2 \text{ solves } 0 \in g(y^*) + B_2(y^*),$$

where $A: H_1 \to H_2$ is a bounded linear operator, $f: H_1 \to H_1$ and $g: H_2 \to H_2$ are given operators. In 2000, Moudafi [3] proposed the viscosity approximation method, which is formulated by considering the approximate well-posed problem and combining the non-expansive mapping S with a contraction mapping f on a non-empty, closed, and convex subset C of H_1. That is, given an arbitrary x_1 in C, a sequence $\{x_n\}$ defined by

$$x_{n+1} = \alpha_n f(x_n) + (1 - \alpha_n) S x_n,$$

converges strongly to a point of $F(S)$, the set of fixed point of S, whenever $\{\alpha_n\} \subset (0,1)$ such that $\alpha_n \to 0$ as $n \to \infty$.

In [4,5], the viscosity approximation method for split variational inclusion and the fixed point problem in a Hilbert space was presented as follows:

$$\begin{aligned} u_n &= J_\lambda^{B_1}(x_n + \gamma_n A^*(J_\lambda^{B_2} - I) A x_n); \\ x_{n+1} &= \alpha_n f(x_n) + (1 - \alpha_n) T^n(u_n), \forall n \geq 1, \end{aligned} \quad (1)$$

where B_1 and B_2 are maximal monotone operators, $J_\lambda^{B_1}$ and $J_\lambda^{B_2}$ are resolvent mappings of B_1 and B_2, respectively, f is the Meir Keeler function, T a non-expansive mapping, and A^* is the adjoint of A, $\gamma_n, \alpha_n \in (0,1)$ and $\lambda > 0$.

The algorithm introduced by Schopfer et al. [6] involves computations in terms of Bregman distance in the setting of p-uniformly convex and uniformly smooth real Banach spaces. Their iterative algorithm given below converges weakly under some suitable conditions:

$$x_{n+1} = \Pi_C J^{-1}(Jx_n + \gamma A^* J(P_Q - I) A x_n), \ n \geq 0, \quad (2)$$

where Π_C denotes the Bregman projection and P_C denotes metric projection onto C. However, strong convergence is more useful than the weak convergence in some applications. Recently, strong convergence theorems for the split feasibility problem (SFP) have been established in the setting of p-uniformly convex and uniformly smooth real Banach spaces [7–10].

Suppose that

$$F(x,y) = f(x,y) + g(x,y)$$

where $f, g: C \times C \longrightarrow \mathbb{R}$ are bifunctions on a closed and convex subset C of a Banach space, which satisfy the following special properties $(A_1) - (A_4), (B_1) - (B_3)$ and (C):

$$\begin{cases} (A_1) \ f(x,y) = 0, \forall x \in C; \\ (A_2) \ f \text{ is maximal monotone;} \\ (A_3) \ \forall x,y,z \in C \text{ and } t \in [0,1] \text{ we have } \limsup_{n \to 0^+} (f(tz + (1-t)x, y) \leq f(x,y)); \\ (A_4) \ \forall x \in C, \text{the function } y \mapsto f(x,y) \text{ is convex and weakly lower semi-continuous;} \\ (B_1) \ g(x,x) = 0 \ \forall \ x \in C; \\ (B_2) \ g \text{ is maximal monotone, and weakly upper semi-continuous in the first variable;} \\ (B_3) \ g \text{ is convex in the second variable;} \\ (C) \text{ for fixed } \lambda > 0 \text{ and } x \in C, \text{there exists a bounded set } K \subset C \\ \text{and } a \ \in K \text{ such that } f(a,z) + g(z,a) + \frac{1}{\lambda}(a-z,z-x) < 0 \ \forall x \in C \backslash K. \end{cases} \quad (3)$$

The well-known, generalized mixed equilibrium problem (GMEP) is to find an $x \in C$, such that

$$F(x,y) + \langle Bx, y - x \rangle \geq 0 \quad \forall \ y \in C,$$

where B is nonlinear mapping.

In 2016, Payvand and Jahedi [11] introduced a new iterative algorithm for finding a common element of the set of solutions of a system of generalized mixed equilibrium problems, the set of common fixed points of a finite family of pseudo contraction mappings, and the set of solutions of the variational inequality for inverse strongly monotone mapping in a real Hilbert space. Their sequence is defined as follows:

$$\begin{cases} g_i(u_{n,i}, y) + \langle C_i u_{n,i} + S_{n,i} x_n, y - u_{n,i} \rangle + \theta_i(y) - \theta_i(u_{n,i}) \\ + \frac{1}{r_{n,i}} \langle y - u_{n,i}, u_{n,i} - x_n \rangle \geq 0 \, \forall y \in K, \, \forall i \in I, \\ y_n = \alpha_n v_n + (1 - \alpha_n (I - f)) P_K(\sum_{i=0}^{\infty} \delta_{n,i} u_{n,i} - \lambda_n A \sum_{i=0}^{\infty} \delta_{n,i} u_{n,i}), \\ x_{n+1} = \beta_n x_n + (1 + \beta_n)(\gamma_0 + \sum_{j=1}^{\infty} \gamma_j T_j) P_K(y_n - \lambda_n A y_n) n \geq 1, \end{cases} \quad (4)$$

where g_i are bifunctions, S_i are $\epsilon-$ inverse strongly monotone mappings, C_i are monotone and Lipschitz continuous mappings, θ_i are convex and lower semicontinuous functions, A is a $\Phi-$ inverse strongly monotone mapping, and f is an $\iota-$contraction mapping and $\alpha_n, \delta_n, \beta_n, \lambda_n, \gamma_0 \in (0, 1)$.

In this paper, inspired by the above cited works, we use a modified version of (1), (2) and (4) to approximate a solution of the problem proposed here. Both the iterative methods and the underlying space used here are improvements and extensions of those employed in [2,6,7,9–11] and the references therein.

Let $p, q \in (1, \infty)$ be conjugate exponents, that is, $\frac{1}{p} + \frac{1}{q} = 1$. For each $p > 1$, let $g(t) = t^{p-1}$ be a gauge function where $g : \mathbb{R}^+ \longrightarrow \mathbb{R}^+$ with $g(0) = 0$ and $\lim_{t \to \infty} g(t) = \infty$. We define the generalized duality map $J_p : E \longrightarrow 2^{E^*}$ by

$$J_{g(t)} = J_p(x) = \{x^* \in E^*; \langle x, x^* \rangle = \|x\| \|x^*\|, \|x^*\| = g(\|x\|) = \|x\|^{p-1}\}.$$

In the sequel, $a \vee b$ denotes $\max\{a, b\}$.

Lemma 1 ([12]). *In a smooth Banach space E, the Bregman distance $\triangle_p$ of x to y, with respect to the convex continuous function $f : E \to R$, such that $f(x) = \frac{1}{p} \|x\|^p$, is defined by*

$$\triangle_p(x, y) = \frac{1}{q} \|x\|^p - \langle J^p(x), y \rangle + \frac{1}{p} \|y\|^p,$$

for all $x, y \in E$ and $p > 1$.

A Banach space E is said to be uniformly convex if, for $x, y \in E$, $0 < \delta_E(\epsilon) \leq 1$, where $\delta_E(\epsilon) = \inf\{1 - \|\frac{1}{2}(x+y)\|; \|x\| = \|y\| = 1, \|x - y\| \geq \epsilon$, where $0 \leq \epsilon \leq 2\}$.

Definition 1. *A Banach space E is said to be uniformly smooth, if for $x, y \in E$, $\lim_{r \to 0}(\frac{\rho_E(r)}{r}) = 0$ where $\rho_E(r) = \frac{1}{2} \sup\{\|x + y\| + \|x - y\| - 2 : \|x\| = 1, \|y\| \leq r; 0 \leq r < \infty$ and $0 \leq \rho_E(r) < \infty\}$.*
It is shown in [12] that:

1. ρ_E is continuous, convex, and nondecreasing with $\rho_E(0) = 0$ and $\rho_E(r) \leq r$
2. The function $r \mapsto \frac{\rho_E(r)}{r}$ is nondecreasing and fulfils $\frac{\rho_E(r)}{r} > 0$ for all $r > 0$.

Definition 2 ([13]). *Let E be a smooth Banach space. Let $\triangle_p$ be the Bregman distance. A mapping $T : E \longrightarrow E$ is said to be a strongly non-expansive left Bregman with respect to the non-empty fixed point set of T, $F(T)$, if $\triangle_p(T(x), v) \leq \triangle_p(x, v) \, \forall \, x \in E$ and $v \in F(T)$.*

Furthermore, if $\{x_n\} \subset C$ is bounded and $\lim_{n \to \infty}(\triangle_p(x_n, v) - \triangle_p(Tx_n, v)) = 0$, then it follows that $\lim_{n \to \infty} \triangle_p(x_n, Tx_n) = 0$.

Definition 3. Let E be a smooth Banach space. Let $\triangle_p$ be the Bregman distance. A mapping $T : E \longrightarrow E$ is said to be a strongly asymptotically non-expansive left Bregman with $\{k_n\} \subset [1,\infty)$ if there exists non-negative real sequences $\{k_n\}$ with $\lim_{n\to\infty} k_n = 1$, such that $\triangle_p(T^n(x), T^n(v)) \leq k_n \triangle_p(x,v)$, $\forall (x,v) \in E \times F(T)$.

Lemma 2 ([14]). *Let E be a real uniformly convex Banach space, K a non-empty closed subset of E, and $T : K \to K$ an asymptotically non-expansive mapping. Then, $I - T$ is demi-closed at zero, if $\{x_n\} \subset K$ converges weakly to a point $p \in K$ and $\lim_{n\to\infty} \|Tx_n - x_n\| = 0$, then $p = Tp$.*

Lemma 3 ([12]). *In a smooth Banach space E, let $x_n \in E$. Consider the following assertions:*

1. $\lim_{n\to\infty} \|x_n - x\| = 0$
2. $\lim_{n\to\infty} \|x_n\| = \|x\|$ and $\lim_{n\to\infty} \langle J^p(x_n), x \rangle = \langle J^p(x), x \rangle$
3. $\lim_{n\to\infty} \triangle_p(x_n, x) = 0$.

The implication (1) $\Longrightarrow$ (2) $\Longrightarrow$ (3) are valid. If E is also uniformly convex, then the assertions are equivalent.

Lemma 4. *Let E be a smooth Banach space. Let $\triangle_p$ and V_p be the mappings defined by $\triangle_p(x,y) = \frac{1}{q}\|x\|^p - \langle J_E^p x, y \rangle + \frac{1}{p}\|y\|^p$ for all $(x,y) \in E \times E$ and $V_p(x^*, x) = \frac{1}{q}\|x^*\|^q - \langle x^*, x \rangle + \frac{1}{p}\|x\|^p$ for all $(x, x^*) \in E \times E^*$. Then, $\triangle_p(x,y) = V_p(x^*, y)$ for all $x, y \in E$.*

Lemma 5 ([12]). *Let E be a reflexive, strictly convex, and smooth Banach space, and J^p be a duality mapping of E. Then, for every closed and convex subset $C \subset E$ and $x \in E$, there exists a unique element $\Pi_C^p(x) \in C$, such that $\triangle_p(x, \Pi_C^p(x)) = \min_{y \in C} \triangle_p(x, y)$; here, $\Pi_C^p(x)$ denotes the Bregman projection of x onto C, with respect to the function $f(x) = \frac{1}{p}\|x\|^p$. Moreover, $x_0 \in C$ is the Bregman projection of x onto C if*

$$\langle J^p(x_0 - x), y - x_0 \rangle \geq 0$$

or equivalently

$$\triangle_p(x_0, y) \leq \triangle_p(x, y) - \triangle_p(x, x_0) \text{ for every } y \in C.$$

Lemma 6 ([15]). *In the case of a uniformly convex space, E, with the duality map J^q of E^*, $\forall x^*, y^* \in E^*$ we have*

$$\|x^* - y^*\|^q \leq \|x^*\|^q - q\langle J^q(x^*), y^* \rangle + \bar{\sigma}_q(x^*, y^*), \text{ where}$$

$$\bar{\sigma}_q(x^*, y^*) = qG_q \int_0^1 \frac{(\|x^* - ty^*\| \vee \|x^*\|)^q}{t} \rho_{E^*}\left(\frac{t\|y^*\|}{2(\|x^* - ty^*\| \vee \|x^*\|)}\right) dt \quad (5)$$

and $G_q = 8 \vee 64cK_q^{-1}$ with $c, K_q > 0$.

Lemma 7 ([12]). *Let E be a reflexive, strictly convex, and smooth Banach space. If we write $\triangle_q^*(x,y) = \frac{1}{p}\|x^*\|^q - \langle J_{E^*}^q x^*, y^* \rangle + \frac{1}{q}\|y^*\|^q$ for all $(x^*, y^*) \in E^* \times E^*$ for the Bregman distance on the dual space E^* with respect to the function $f_q^*(x^*) = \frac{1}{q}\|x^*\|^q$, then we have $\triangle_p(x,y) = \triangle_q^*(x^*, y^*)$.*

Lemma 8 ([16]). *Let $\{\alpha_n\}$ be a sequence of non-negative real numbers, such that $\alpha_{n+1} \leq (1 - \beta_n)\alpha_n + \delta_n$, $n \geq 0$, where $\{\beta_n\}$ is a sequence in $(0,1)$ and $\{\delta_n\}$ is a sequence in R, such that*

1. $\lim_{n\to\infty} \beta_n = 0, \sum_{n=1}^{\infty} \beta_n = \infty$;
2. $\limsup_{n\to\infty} \frac{\delta_n}{\beta_n} \leq 0$ or $\sum_{n=1}^{\infty} |\delta_n| < \infty$.

Then, $\lim_{n\to\infty} \alpha_n = 0$.

Lemma 9. Let E be reflexive, smooth, and strictly convex Banach space. Then, for all $x, y, z \in E$ and $x^*, z^* \in E^*$ the following facts hold:

1. $\triangle_p(x, y) \geq 0$ and $\triangle_p(x, y) = 0$ iff $x = y$;
2. $\triangle_p(x, y) = \triangle_p(x, z) + \triangle_p(z, y) + \langle x^* - z^*, z - y \rangle$.

Lemma 10 ([17]). Let E be a real uniformly convex Banach space. For arbitrary $r > 1$, let $B_r(0) = \{x \in E : \|x\| \leq r\}$. Then, there exists a continuous strictly increasing convex function

$$g : [0, \infty) \longrightarrow [0, \infty), g(0) = 0$$

such that for every $x, y \in B_r(0), f_x \in J_p(x), f_y \in J_p(y)$ and $\lambda \in [0, 1]$, the following inequalities hold:

$$\|\lambda x + (1-\lambda)y\|^p \leq \lambda \|x\|^p + (1-\lambda)\|y\|^p - (\lambda^p(1-\lambda) + (1-\lambda)^p\lambda)g(\|x-y\|)$$

and

$$\langle x - y, f_x - f_y \rangle \geq g(\|x - y\|).$$

Lemma 11 ([18]). Suppose that $\sum_{n=1}^{\infty} \sup\{\|T_{n+1}z - T_nz\| : z \in C\} < \infty$. Then, for each $y \in C$, $\{T_ny\}$ converges strongly to some point of C. Moreover, let T be a mapping of C onto itself, defined by $Ty = \lim_{n \to \infty} T_ny$ for all $y \in C$. Then, $\lim_{n \to \infty} \sup\{\|Tz - T_nz\| : z \in C\} = 0$. Consequently, by Lemma 3, $\lim_{n \to \infty} \sup\{\triangle_p(Tz, T_nz) : z \in C\} = 0$.

Lemma 12 ([19]). Let E be a reflexive, strictly convex, and smooth Banach space, and C be a non-empty, closed convex subset of E. If $f, g : C \times C \longrightarrow \mathbb{R}$ be two bifunctions which satisfy the conditions $(A_1) - (A_4), (B_1) - (B_3)$ and (C), in (3), then for every $x \in E$ and $r > 0$, there exists a unique point $z \in C$ such that $f(z, y) + g(z, y) + \frac{1}{r}\langle y - z, jz - jx \rangle \geq 0 \ \forall \ y \in C$.

For $f(x) = \frac{1}{p}\|x\|^p$, Reich and Sabach [20] obtained the following technical result:

Lemma 13. Let E be a reflexive, strictly convex, and smooth Banach space, and C be a non-empty, closed, and convex subset of E. Let $f, g : C \times C \longrightarrow \mathbb{R}$ be two bifunctions which satisfy the conditions $(A_1) - (A_4), (B_1) - (B_3)$ and (C), in (3). Then, for every $x \in E$ and $r > 0$, we define a mapping $S_r : E \longrightarrow C$ as follows;

$$S_r(x) = \{z \in C : f(z, y) + g(z, y) + \frac{1}{r}\langle y - z, J_E^p z - J_E^p x \rangle \geq 0 \forall y \in C\}. \quad (6)$$

Then, the following conditions hold:

1. S_r is single-valued;
2. S_r is a Bregman firmly non-expansive-type mapping, that is,

$$\forall x, y \in E \langle S_r x - S_r y, J_E^p S_r x - J_E^p S_r y \rangle \leq \langle S_r x - S_r y, J_E^p x - J_E^p y \rangle$$

or equivalently

$$\triangle_p(S_r x, S_r y) + \triangle_p(S_r y, S_r x) + \triangle_p(S_r x, x) + \triangle_p(S_r y, y) \leq \triangle_p(S_r x, y) + \triangle_p(S_r y, x);$$

3. $F(S_r) = MEP(f, g)$, here MEP stands for mixed equilibrium problem;
4. $MEP(f, g)$ is closed and convex;
5. for all $x \in E$ and for all $v \in F(S_r)$, $\triangle_p(v, S_r x) + \triangle_p(S_r x, x) \leq \triangle_p(v, x)$.

2. Main Results

Let E_1 and E_2 be uniformly convex and uniformly smooth Banach spaces and E_1^* and E_2^* be their duals, respectively. For $i \in I$, let $U_i : E_1 \to 2^{E_1^*}$ and $T_i : E_2 \to 2^{E_2^*}, i \in I$ be multi-valued maximal monotone operators. For $i \in I$, $\delta > 0$, $p,q \in (1, \infty)$ and $K \subset E_1$ closed and convex, let $\Phi_i : K \times K \to \mathbb{R}, i \in I$, be bifunctions satisfying $(A1) - (A4)$ in (3), let $B_\delta^{U_i} : E_1 \to E_1$ be resolvent operators defined by $B_\delta^{U_i} = (J_{E_1}^p + \delta U_i)^{-1} J_{E_1}^p$ and $B_\delta^{T_i} : E_2 \to E_2$ be resolvent operators defined by $B_\delta^{T_i} = (J_{E_2}^p + \delta T_i)^{-1} J_{E_2}^p$. Let $A : E_1 \to E_2$ be a bounded and linear operator, A^* denotes the adjoint of A and AK be closed and convex. For each $i \in I$, let $S_i : E_1 \to E_1$ be a uniformly continuous Bregman asymptotically non-expansive operator with the sequences $\{k_{n,i}\} \subset [1, \infty)$ satisfying $\lim_{n \to \infty} k_{n,i} = 1$. Denote by $Y : E_1^* \to E_1^*$ a firmly non-expansive mapping. Suppose that, for $i \in I$, $\theta_i : K \to \mathbb{R}$ are convex and lower semicontinuous functions, $G_i : K \to E_1$ are ε-inverse strongly monotone mappings and $C_i : K \to E_1$, are monotone and Lipschitz continuous mappings. Let $f : E_1 \to E_1$ be a ζ-contraction mapping, where $\zeta \in (0,1)$. Suppose that $\Pi_{AK}^p : E_2 \to AK$ is a generalized Bregman projection onto AK. Let $\Omega = \{x^* \in \cap_{i=1}^\infty SOLVIP(U_i); Ax^* \in \cap_{i=1}^\infty SOLVIP(T_i)\}$ be the set of solution of the split variational inclusion problem, $\omega = \{x^* \in \cap_{i=1}^\infty GMEP(G_i, C_i, \theta_i, g_i)\}$ be the solution set of a system of generalized mixed equilibrium problems, and $\Im = \{x^* \in \cap_{i=1}^\infty F(S_i)\}$ be the common fixed-point set of S_i for each $i \in I$. Let the sequence $\{x_n\}$ be defined as follows:

$$\begin{cases} \Phi_i(u_{n,i}, y) + \langle J_{E_1}^p G_{n,i} x_n, y - u_{n,i} \rangle + \frac{1}{r_{n,i}} \langle y - u_{n,i}, J_{E_1}^p u_{n,i} - J_{E_1}^p x_n \rangle \geq 0 \forall y \in K, \\ \forall i \in I, \\ x_{n+1} = J_{E_1^*}^q \left(\sum_{i=0}^\infty \alpha_{n,i} B_{\delta_n}^{U_i} \left(J_{E_1}^p x_n - \sum_{i=0}^\infty \beta_{n,i} \lambda_n A^* J_{E_2}^p (I - \Pi_{AK}^p B_{\delta_n}^{T_i}) A u_{n,i} \right) \right), \end{cases} \quad (7)$$

where $\Phi_i(x,y) = g_i(x,y) + \langle J_{E_1}^p C_i x, y - x \rangle + \theta_i(y) - \theta_i(x)$.
We shall strictly employ the above terminology in the sequel.

Lemma 14. *Suppose that $\bar{\sigma}_q$ is the function (5) in Lemma 6 for the characteristic inequality of the uniformly smooth dual E_1^*. For the sequence $\{x_n\} \subset E_1$ defined by (7), let $0 \neq x_n \in E_1$, $0 \neq A$, $0 \neq J_{E_1}^p G_{n,i} x_n \in E_1^*$ and $0 \neq \sum_{i=0}^\infty \beta_{n,i} J_{E_2}^p (I - \Pi_{AK}^p B_{\delta_n}^{T_i}) A u_{n,i} \in E_2^*, i \in I$. Let, for $\lambda_{n,i} > 0$ and $r_{n,i} > 0$, $i \in I$ be defined by*

$$\lambda_{n,i} = \frac{1}{\|A\|} \frac{1}{\|\sum_{i=0}^\infty \beta_{n,i} J_{E_2}^p (I - \Pi_{AK}^p B_{\delta_n}^{T_i}) A u_{n,i}\|}, \text{ and} \quad (8)$$

$$r_{n,i} = \frac{1}{\|J_{E_1}^p G_{n,i} x_n\|}, \text{ respectively.} \quad (9)$$

Then for $\mu_{n,i} = \frac{1}{\|x_n\|^{p-1}}$,

$$2^q G_q \|J_{E_1}^p x_n\|^p \rho_{E_1^*}(\mu_{n,i}) \geq \begin{cases} \frac{1}{q} \bar{\sigma}_q (J_{E_1}^p x_n, r_{n,i} J_{E_1}^p G_{n,i} x_n) \\ \frac{1}{q} \bar{\sigma}_q \left(J_{E_1}^p x_n, \sum_{i=0}^\infty \beta_{n,i} \lambda_n A^* \sum_{i=0}^\infty \beta_{n,i} J_{E_2}^p (I - \Pi_{AK}^p B_{\delta_n}^{T_i}) A u_{n,i} \right), \end{cases} \quad (10)$$

where G_q is the constant defined in Lemma 6 and $\rho_{E_1^*}$ is the modulus of smoothness of E_1^*.

Proof. By Lemma 12, (6) in Lemma 13 and (7), for each $i \in I$, we have that $u_{n,i} = J_{E_1^*}^q(Y_{r_{n,i}}(J_{E_1}^p x_n - r_{n,i} J_{E_1}^p G_{n,i} x_n))$. By Lemma 6, we get

$$\frac{1}{q} \bar{\sigma}_q(J_{E_1}^p x_n, r_{n,i} J_{E_1}^p G_{n,i} x_n) = G_q \int_0^1 \frac{(\|J_{E_1}^p x_n - t r_{n,i} J_{E_1}^p G_{n,i} x_n\| \vee \|J_{E_1}^p x_n\|)^q}{t} \times$$

$$\rho_{E^*} \left(\frac{t \|r_{n,i} J_{E_1}^p G_{n,i} x_n\|}{(\|J_{E_1}^p x_n - t r_{n,i} J_{E_1}^p G_{n,i} x_n\| \vee \|J_{E_1}^p x_n\|)} \right) dt, \quad (11)$$

for every $t \in [0,1]$.

However, by (9) and Definition 1(2), we have

$$\rho_{E_1^*}\left(\frac{t\|r_{n,i}J_{E_1}^p G_{n,i}x_n\|}{(\|J_{E_1}^p x_n - tr_{n,i}J_{E_1}^p G_{n,i}x_n\| \vee \|J_{E_1}^p x_n\|)}\right) \leq \rho_{E_1^*}\left(\frac{t\|r_{n,i}J_{E_1}^p G_{n,i}x_n\|}{\|x_n\|^{p-1}}\right)$$
$$= \rho_{E_1^*}(t\mu_{n,i}). \quad (12)$$

Substituting (12) into (11), and using the nondecreasing of function $\rho_{E_1^*}$, we have

$$\frac{1}{q}\bar{\sigma}_q(J_{E_1}^p x_n, r_{n,i}J_{E_1}^p G_{n,i}x_n) \leq 2^q G_q \|x_n\|^p \rho_{E_1^*}(\mu_{n,i}). \quad (13)$$

In addition, by Lemma 6, we have

$$\frac{1}{q}\bar{\sigma}_q\left(J_{E_1}^p x_n, \sum_{i=0}^{\infty} \beta_{n,i}\lambda_n A^* J_{E_2}^p (I - \Pi_{AK}^p B_{\delta_n}^{T_i})Au_{n,i}\right)$$
$$= G_q \int_0^1 \frac{\left(\left\|J_{E_1}^p x_n - \sum_{i=0}^{\infty} \beta_{n,i}\lambda_n A^* J_{E_2}^p (I - \Pi_{AK}^p B_{\delta_n}^{T_i})Au_{n,i}\right\| \vee \|J_{E_1}^p x_n\|\right)^q}{t} \times$$
$$\rho_{E^*}\left(\frac{t\|\sum_{i=0}^{\infty} \beta_{n,i}\lambda_n A^* J_{E_2}^p (I - \Pi_{AK}^p B_{\delta_n}^{T_i})Au_{n,i}\|}{\left(\left\|J_{E_1}^p x_n - \sum_{i=0}^{\infty} \beta_{n,i}\lambda_n A^* J_{E_2}^p (I - \Pi_{AK}^p B_{\delta_n}^{T_i})Au_{n,i}\right\| \vee \|J_{E_1}^p x_n\|\right)}\right) dt, \quad (14)$$

for every $t \in [0,1]$.

However, by (8) and Definition 1(2), we have

$$\rho_{E_1^*}\left(\frac{t\left\|\sum_{i=0}^{\infty} \beta_{n,i}\lambda_n A^* J_{E_2}^p (I - \Pi_{AK}^p B_{\delta_n}^{T_i})Au_{n,i}\right\|}{\left(\left\|J_{E_1}^p x_n - t\sum_{i=0}^{\infty} \beta_{n,i}\lambda_n A^* J_{E_2}^p (I - \Pi_{AK}^p B_{\delta_n}^{T_i})Au_{n,i}\right\| \vee \|J_{E_1}^p x_n\|\right)}\right)$$
$$\leq \rho_{E_1^*}\left(\frac{t\left\|\sum_{i=0}^{\infty} \beta_{n,i}\lambda_{n,i} A^* J_{E_2}^p (I - \Pi_{AK}^p B_{\delta_n}^{T_i})Au_{n,i}\right\|}{\|x_n\|^{p-1}}\right) = \rho_{E_1^*}(t\mu_{n,i}). \quad (15)$$

Substituting (15) into (14), and using the nondecreasing of function $\rho_{E_1^*}$, we get

$$\frac{1}{q}\bar{\sigma}_q\left(J_{E_1}^p x_n, \sum_{i=0}^{\infty} \beta_{n,i}\lambda_n A^* J_{E_2}^p (I - \Pi_{AK}^p B_{\delta_n}^{T_i})Au_{n,i}\right)$$
$$\leq 2^q G_q \|x_n\|^p \rho_{E_1^*}(\mu_{n,i}). \quad (16)$$

By (13) and (16), the result follows. □

Lemma 15. *For the sequence $\{x_n\} \subset E_1$, defined by (7), $i \in I$, let $0 \neq \sum_{i=0}^{\infty} \beta_{n,i} J_{E_2}^p (I - \Pi_{AK}^p B_{\delta_n}^{T_i})Au_{n,i} \in E_2^*$, $0 \neq J_{E_1}^p G_{n,i} x_n \in E_1^*$, and $\lambda_n > 0$ and $r_{n,i} > 0$, $i \in I$, be defined by*

$$\lambda_n = \frac{1}{\|A\|} \frac{1}{\|\sum_{i=0}^{\infty} \beta_{n,i} J_{E_2}^p (I - \Pi_{AK}^p B_{\delta_n}^{T_i})Au_{n,i}\|} \quad (17)$$

and

$$r_{n,i} = \frac{1}{\|J_{E_1}^p G_{n,i} x_n\|}, \quad (18)$$

where $\iota, \gamma \in (0,1)$ and $\mu_{n,i} = \frac{1}{\|x_n\|^{p-1}}$ are chosen such that

$$\rho_{E_1^*}(\mu_{n,i}) = \frac{\iota}{2^q G_q \|A\|} \times \frac{\|\sum_{i=0}^{\infty} \beta_{n,i} J_{E_2}^p (I - \Pi_{AK}^p B_{\delta_n}^{T_i}) A u_{n,i}\|^p}{\|x_n\|^p \|\sum_{i=0}^{\infty} \beta_{n,i} J_{E_2}^p (I - \Pi_{AK}^p B_{\delta_n}^{T_i}) A u_{n,i}\|^{p-1}}, \tag{19}$$

and

$$\rho_{E_1^*}(\mu_{n,i}) = \frac{\gamma \langle J_{E_1}^p G_{n,i} x_n, x_n - v \rangle}{2^q G_q \|x_n\|^p \|J_{E_1}^p G_{n,i} x_n\|}. \tag{20}$$

Then, for all $v \in \Gamma$, we get

$$\triangle_p(x_{n+1}, v) \leq \triangle_p(x_n, v)$$
$$- [1-\iota] \times \frac{\langle \sum_{i=0}^{\infty} \beta_{n,i} J_{E_2}^p (I - \Pi_{AK}^p B_{\delta_n}^{T_i}) A u_{n,i}, \sum_{i=0}^{\infty} \beta_{n,i} (I - \Pi_{AK}^p B_{\delta_n}^{T_i}) A u_{n,i} \rangle}{\|A\| \|\sum_{i=0}^{\infty} \beta_{n,i} J_{E_2}^p (I - \Pi_{AK}^p B_{\delta_n}^{T_i}) A u_{n,i}\|} \tag{21}$$

and

$$\triangle_p(u_n, v) \leq \triangle_p(x_n, v) - [1 - \gamma] \times \frac{\langle J_{E_1}^p G_{n,i} x_n, x_n - v \rangle}{\|J_{E_1}^p G_{n,i} x_n\|}, \text{ respectively.} \tag{22}$$

Proof. By Lemmas 13, 4 and 6, for each $i \in I$, we get that $u_{n,i} = J_{E_1^*}^q (\Upsilon_{r_{n,i}} (J_{E_1}^p x_n - r_{n,i} J_{E_1}^p G_{n,i} x_n))$, and hence it follows that

$$\triangle_p(u_{n,i}, v) \leq V_p(J_{E_1}^p x_n - r_{n,i} J_{E_1}^p G_{n,i} x_n, v)$$
$$= -\langle J_{E_1}^p x_n, v \rangle + r_{n,i} \langle J_{E_1}^p G_{n,i} x_n, v \rangle$$
$$+ \frac{1}{q} \|J_{E_1}^p x_n - r_{n,i} J_{E_1}^p G_{n,i} x_n\|^q + \frac{1}{p} \|v\|^p. \tag{23}$$

By Lemmas 6 and 14, we have

$$\frac{1}{q} \|J_{E_1}^p x_n - r_{n,i} J_{E_1}^p G_{n,i} x_n\|^q$$
$$\leq \frac{1}{q} \|J_{E_1}^p x_n\|^q - r_{n,i} \langle J_{E_1}^p G_{n,i} x_n, x_n \rangle + 2^q G_q \|J_{E_1}^p x_n\|^p \rho_{E_1^*}(\mu_{n,i}). \tag{24}$$

Substituting (24) into (23), we have, by Lemma 4

$$\triangle_p(u_{n,i}, v) \leq \triangle_p(x_n, v) + 2^q G_q \|J_{E_1}^p x_n\|^p \rho_{E_1^*}(\mu_{n,i})$$
$$- r_{n,i} \langle J_{E_1}^p G_{n,i} x_n, x_n - v \rangle \tag{25}$$

Substituting (18) and (20) into (25), we have

$$\triangle_p(u_{n,i}, v) \leq \triangle_p(x_n, v) + \frac{\gamma \langle J_{E_1}^p G_{n,i} x_n, x_n - v \rangle}{\|J_{E_1}^p G_{n,i} x_n\|} - \frac{\langle J_{E_1}^p G_{n,i} x_n, x_n - v \rangle}{\|J_{E_1}^p G_{n,i} x_n\|}$$
$$= \triangle_p(x_n, v) - [1 - \gamma] \times \frac{\langle J_{E_1}^p G_{n,i} x_n, x_n - v \rangle}{\|J_{E_1}^p G_{n,i} x_n\|}.$$

Thus, (22) holds.

Now, for each $i \in I$, let $v = B_\gamma^{U_i} v$ and $Av = B_\gamma^{T_i} Av$. By Lemma 4, we have

$$\triangle_p(y_n, v) \leq \frac{1}{q}\left\| J_{E_1}^p u_{n,i} - \sum_{i=0}^\infty \beta_{n,i}\lambda_n A^* J_{E_2}^p (I - \Pi_{AK}^p B_{\delta_n}^{T_i}) Au_{n,i}\right\|^q + \frac{1}{p}\|v\|^p$$
$$- \langle J_{E_1}^p u_{n,i}, v\rangle + \left\langle \sum_{i=0}^\infty \beta_{n,i}\lambda_n A^* J_{E_2}^p (I - \Pi_{AK}^p B_{\delta_n}^{T_i}) Au_{n,i}, v\right\rangle, \tag{26}$$

where,

$$\left\langle \sum_{i=0}^\infty \beta_{n,i}\lambda_n A^* J_{E_2}^p (I - \Pi_{AK}^p B_{\delta_n}^{T_i}) Au_{n,i}, v\right\rangle$$
$$= -\left\langle \sum_{i=0}^\infty \beta_{n,i}\lambda_n J_{E_2}^p (\Pi_{AK}^p B_{\delta_n}^{T_i} - I) Au_{n,i}, (Av - \sum_{i=0}^\infty \beta_{n,i} Au_{n,i}) - \sum_{i=0}^\infty \beta_{n,i}(\Pi_{AK}^p B_{\delta_n}^{T_i} - I) Au_{n,i}\right\rangle$$
$$- \left\langle \sum_{i=0}^\infty \beta_{n,i}\lambda_n J_{E_2}^p (I - \Pi_{AK}^p B_{\delta_n}^{T_i}) Au_{n,i}, \sum_{i=0}^\infty \beta_{n,i}(I - \Pi_{AK}^p B_{\delta_n}^{T_i}) Au_{n,i}\right\rangle$$
$$+ \left\langle \sum_{i=0}^\infty \beta_{n,i}\lambda_n J_{E_2}^p (I - \Pi_{AK}^p B_{\delta_n}^{T_i}) Au_{n,i}, Au_{n,i}\right\rangle.$$

As AK is closed and convex, by Lemma 5 and the variational inequality for the Bregman projection of zero onto $AK - \sum_{i=0}^\infty \beta_{n,i} Au_{n,i}$, we arrive at

$$\left\langle \sum_{i=0}^\infty \beta_{n,i}\lambda_n J_{E_2}^p (\Pi_{AK}^p B_{\delta_n}^{T_i} - I) Au_{n,i}, (Av - \sum_{i=0}^\infty \beta_{n,i} Au_{n,i}) - \sum_{i=0}^\infty \beta_{n,i}(\Pi_{AK}^p B_{\delta_n}^{T_i} - I) Au_{n,i}\right\rangle \geq 0$$

and therefore,

$$\left\langle \sum_{i=0}^\infty \beta_{n,i}\lambda_n A^* J_{E_2}^p (I - \Pi_{AK}^p B_{\delta_n}^{T_i}) Au_{n,i}, v\right\rangle$$
$$\leq -\left\langle \sum_{i=0}^\infty \beta_{n,i}\lambda_n J_{E_2}^p (I - \Pi_{AK}^p B_{\delta_n}^{T_i}) Au_{n,i}, \sum_{i=0}^\infty \beta_{n,i}(I - \Pi_{AK}^p B_{\delta_n}^{T_i}) Au_{n,i}\right\rangle$$
$$+ \left\langle \sum_{i=0}^\infty \beta_{n,i}\lambda_n J_{E_2}^p (I - \Pi_\Gamma^p B_{\delta_n}^{T_i}) Au_{n,i}, Au_{n,i}\right\rangle. \tag{27}$$

By Lemma 6, 14 and (27), we get

$$\triangle_p(y_n, v) \leq \triangle_p(u_{n,i}, v) + 2^p G_p \|J_{E_1}^p u_{n,i}\|^p \rho_{E_1^*}(\tau_{n,i})$$
$$- \left\langle \sum_{i=0}^\infty \beta_{n,i}\lambda_n J_{E_2}^p (I - \Pi_{AK}^p B_{\delta_n}^{T_i}) Au_{n,i}, \sum_{i=0}^\infty \beta_{n,i}(I - \Pi_{AK}^p B_{\delta_n}^{T_i}) Au_{n,i}\right\rangle. \tag{28}$$

Substituting (17) and (19) into (28), we have

$$\triangle_p(y_n, v) \leq \triangle_p(u_{n,i}, v) - [1 - \iota]$$
$$\times \frac{\left\langle \sum_{i=0}^\infty \beta_{n,i} J_{E_2}^p (I - \Pi_{AK}^p B_{\delta_n}^{T_i}) Au_{n,i}, \sum_{i=0}^\infty \beta_{n,i}(I - \Pi_{AK}^p B_{\delta_n}^{T_i}) Au_{n,i}\right\rangle}{\|A\|\|\sum_{i=0}^\infty \beta_{n,i} J_{E_2}^p (I - \Pi_{AK}^p B_{\delta_n}^{T_i}) Au_{n,i}\|}.$$

Thus, (21) holds as desired. □

We now prove our main result.

Theorem 1. *Let $g_i : K \times K \to R$, $i \in I$, be bifunctions satisfying $(A1) - (A4)$ in (3). For $\delta > 0$ and $p, q \in (1, \infty)$, let $(I - \Pi^p_{AK} B^{T_i}_{\delta})$, $i \in I$, be demi-closed at zero. Let $x_1 \in E_1$ be chosen arbitrarily and the sequence $\{x_n\}$ be defined as follows;*

$$\begin{cases} g_i(u_{n,i}, y) + \langle J^p_{E_1} C_i u_{n,i} + J^p_{E_1} G_{n,i} x_n, y - u_{n,i} \rangle + \theta_i(y) - \theta_i(u_{n,i}) \\ + \frac{1}{r_{n,i}} \langle y - u_{n,i}, J^p_{E_1} u_{n,i} - J^p_{E_1} x_n \rangle \geq 0 \; \forall y \in K, \; \forall i \in I, \\ y_n = J^q_{E_1^*} \left(\sum_{i=0}^{\infty} \alpha_{n,i} B^{U_i}_{\delta_n} \left(J^p_{E_1} u_{n,i} - \sum_{i=0}^{\infty} \beta_{n,i} \lambda_n A^* J^p_{E_2} (I - \Pi^p_{AK} B^{T_i}_{\delta_n}) A u_{n,i} \right) \right), \\ x_{n+1} = J^q_{E_1^*} \left(\eta_{n,0} J^p_{E_1} (f(x_n)) + \sum_{i=1}^{\infty} \eta_{n,i} J^p_{E_1} (S_{n,i}(y_n)) \right) \; n \geq 1, \end{cases} \quad (29)$$

where $r_{n,i} = \frac{1}{\|J^p_{E_1} G_{n,i} x_n\|}$, $\mu_{n,i} = \frac{1}{\|x_n\|^{p-1}}$ and $\gamma \in (0,1)$ such that $\rho_{E_1^*}(\mu_{n,i}) = \frac{\gamma \langle J^p_{E_1} G_{n,i} x_n, x_n - v \rangle}{2^q G_q \|x_n\|^p \|J^p_{E_1} G_{n,i} x_n\|}$,

$$\lambda_n = \begin{cases} \frac{1}{\|A\|} \frac{1}{\|\sum_{i=0}^{\infty} \beta_{n,i} J^p_{E_2} (I - \Pi^p_{AK} B^{T_i}_{\delta_n}) A u_{n,i}\|}, & u_{n,i} \neq 0 \\ \frac{1}{\|A\|^p} \frac{\|\sum_{i=0}^{\infty} \beta_{n,i} J^p_{E_2} (I - \Pi^p_{AK} B^{T_i}_{\delta_n}) A u_{n,i}\|^{p(p-1)}}{\|\sum_{i=0}^{\infty} \beta_{n,i} J^p_{E_2} (I - \Pi^p_{AK} B^{T_i}_{\delta_n}) A u_{n,i}\|^p}, & u_{n,i} = 0, \end{cases} \quad (30)$$

$\iota \in (0,1)$ and $\tau_{n,i} = \frac{1}{\|u_{n,i}\|^{p-1}}$ are chosen such that

$$\rho_{E_1^*}(\tau_{n,i}) = \frac{\iota}{2^q G_q \|A\|} \times \frac{\|\sum_{i=0}^{\infty} \beta_{n,i} J^p_{E_2} (I - \Pi^p_{AK} B^{T_i}_{\delta_n}) A u_{n,i}\|^p}{\|u_{n,i}\|^p \|\sum_{i=0}^{\infty} \beta_{n,i} J^p_{E_2} (I - \Pi^p_{AK} B^{T_i}_{\delta_n}) A u_{n,i}\|^{p-1}}, \quad (31)$$

with, $\lim_{n \to \infty} \eta_{n,0} = 0$, $\eta_{n,0} \leq \sum_{i=1}^{\infty} \eta_{n,i}$, for $M \geq 0$, $\eta_{n-1,0} \leq \sum_{i=1}^{\infty} \eta_{n-1,i} \leq \sum_{n=1}^{\infty} \sum_{i=1}^{\infty} \eta_{n-1,i} M < \infty$, $\sum_{i=0}^{\infty} \eta_{n,i} = \sum_{i=0}^{\infty} \alpha_{n,i} = \sum_{i=0}^{\infty} \beta_{n,i} = 1$ and $k_n = \max_{i \in I}\{k_{n,i}\}$. If $\Gamma = \Omega \cap \omega \cap \Im \neq \emptyset$, then $\{x_n\}$ converges strongly to $x^* \in \Gamma$, where $\sum_{i=0}^{\infty} \beta_{n,i} \Pi^p_{AK} B^{T_i}_{\delta_n}(x^*) = \sum_{i=0}^{\infty} \beta_{n,i} B^{T_i}_{\delta_n}(x^*)$, for each $i \in I$.

Proof. For $x, y \in K$ and $i \in I$, let $\Phi_i(x, y) = g_i(x, y) + \langle J^p_{E_1} C_i x, y - x \rangle + \theta_i(y) - \theta_i(x)$. Since g_i are bi-functions satisfying $(A1) - (A4)$ in (3) and C_i are monotone and Lipschitz continuous mappings, and θ_i are convex and lower semicontinuous functions, therefore $\Phi_i (i \in I)$ satisfy the conditions $(A1) - (A4)$ in (3), and hence the algorithm (29) can be written as follows:

$$\begin{cases} \Phi_i(u_{n,i}, y) + \langle J^p_{E_1} G_{n,i} x_n, y - u_{n,i} \rangle + \frac{1}{r_{n,i}} \langle y - u_{n,i}, J^p_{E_1} u_{n,i} - J^p_{E_1} x_n \rangle \geq 0 \\ \forall y \in K, \; \forall i \in I, \\ y_n = J^q_{E_1^*} \left(\sum_{i=0}^{\infty} \alpha_{n,i} B^{U_i}_{\delta_n} \left(J^p_{E_1} u_{n,i} - \sum_{i=0}^{\infty} \beta_{n,i} \lambda_n A^* J^p_{E_2} (I - \Pi^p_{AK} B^{T_i}_{\delta_n}) A u_{n,i} \right) \right), \\ x_{n+1} = J^q_{E_1^*} \left(\eta_{n,0} J^p_{E_1} (f(x_n)) + \sum_{i=1}^{\infty} \eta_{n,i} J^p_{E_1} (S_{n,i}(y_n)) \right) \; n \geq 1. \end{cases} \quad (32)$$

We will divide the proof into four steps.
Step One: We show that $\{x_n\}$ is a bounded sequence.
Assume that $\|\sum_{i=0}^{\infty} \beta_{n,i} J^p_{E_2} (I - \Pi^p_{AK} B^{T_i}_{\delta_n}) A u_{n,i}\| = 0$ and $\|J^p_{E_1} G_{n,i} x_n\| = 0$. Then, by (32), we have

$$\Phi_i(u_{n,i}, y) + \frac{1}{r_{n,i}} \langle y - u_{n,i}, J^p_{E_1} u_{n,i} - J^p_{E_1} x_n \rangle \geq 0 \; \forall y \in K, \; \forall i \in I. \quad (33)$$

By (33) and Lemma 13, for each $i \in I$, we have that $u_{n,i} = J^q_{E_1^*}(Y_{r_{n,i}}(J^p_{E_1} x_n))$. By Lemma 4 and for $v \in \Gamma$ and $v = Y_{r_{n,i}} v$, we have

$$\triangle_p(u_{n,i}, v) = V_p(Y_{r_{n,i}}(J^p_{E_1} x_n), v) \leq V_p(J^p_{E_1} x_n, v) = \triangle_p(x_n, v). \quad (34)$$

In addition, for each $i \in I$, let $v = B_\gamma^{U_i} v$. By Lemma 4 and for $v \in \Gamma$, we have

$$\triangle_p(y_n, v) = V_p\left(\sum_{i=0}^\infty \alpha_{n,i} B_{\delta_n}^{U_i} J_{E_1}^p u_{n,i}, v\right) \leq \triangle_p(u_{n,i}, v). \tag{35}$$

Now assume that $\|\sum_{i=0}^\infty \beta_{n,i} J_{E_2}^p (I - \Pi_{AK}^p B_{\delta_n}^{T_i}) A u_{n,i}\| \neq 0$ and $\|J_{E_1}^p G_{n,i} x_n\| \neq 0$. Then by (32), we have that

$$\Phi_i(u_{n,i}, y) + \frac{1}{r_{n,i}} \left\langle y - u_{n,i}, J_{E_1}^p u_{n,i} - (J_{E_1}^p x_n - r_{n,i} J_{E_1}^p G_{n,i} x_n) \right\rangle \geq 0 \ \forall y \in K, \ \forall i \in I. \tag{36}$$

By (36) and Lemma 13, for each $i \in I$, we have $u_{n,i} = J_{E_1^*}^q (Y_{r_{n,i}} (J_{E_1}^p x_n - r_{n,i} J_{E_1}^p G_{n,i} x_n))$. For $v \in \Gamma$, by (22) in Lemma 15, we get

$$\triangle_p(u_{n,i}, v) \leq \triangle_p(x_n, v). \tag{37}$$

In addition, for each $i \in I$, $v \in \Gamma$, (21) in Lemma 15 gives

$$\triangle_p(y_n, v) \leq \triangle_p(u_{n,i}, v). \tag{38}$$

Let $u_{n,i} = 0$. By Lemma 1, we have

$$\triangle_p(u_{n,i}, v) = \frac{1}{p} \|v\|^p \tag{39}$$

and by (27), (39), Lemmas 4 and 15, we have

$$\triangle_p(y_n, v) \leq \frac{1}{q} \left\| \sum_{i=0}^\infty \beta_{n,i} \lambda_n A^* J_{E_2}^p (I - \Pi_{AK}^p B_{\delta_n}^{T_i}) A u_{n,i} \right\|^p$$
$$+ \triangle_p(u_{n,i}, v) + \lambda_n \left\langle \sum_{i=0}^\infty \beta_{n,i} J_{E_2}^p (I - \Pi_{AK}^p B_{\delta_n}^{T_i}) A u_{n,i}, A u_{n,i} \right\rangle$$
$$- \lambda_n \left\langle \sum_{i=0}^\infty \beta_{n,i} J_{E_2}^p (I - \Pi_{AK}^p B_{\delta_n}^{T_i}) A u_{n,i}, \sum_{i=0}^\infty \beta_{n,i} (I - \Pi_{AK}^p B_{\delta_n}^{T_i}) A u_{n,i} \right\rangle. \tag{40}$$

However, by (30) and (40), we have

$$\triangle_p(y_n, v)$$
$$\leq \frac{1}{q} \frac{1}{\|A\|^p} \frac{\left\langle \sum_{i=0}^\infty \beta_{n,i} J_{E_2}^p (I - \Pi_{AK}^p B_{\delta_n}^{T_i}) A u_{n,i}, \sum_{i=0}^\infty \beta_{n,i} (I - \Pi_{AK}^p B_{\delta_n}^{T_i}) A u_{n,i} \right\rangle^p}{\left\| \sum_{i=0}^\infty \beta_{n,i} J_{E_2}^p (I - \Pi_{AK}^p B_{\delta_n}^{T_i}) A u_{n,i} \right\|^p}$$
$$+ \triangle_p(u_{n,i}, v) + \lambda_n \langle \sum_{i=0}^\infty \beta_{n,i} J_{E_2}^p (I - \Pi_{AK}^p B_{\delta_n}^{T_i}) A u_{n,i}, A u_{n,i} \rangle$$
$$- \lambda_n \langle \sum_{i=0}^\infty \beta_{n,i} J_{E_2}^p (I - \Pi_{AK}^p B_{\delta_n}^{T_i}) A u_{n,i}, \sum_{i=0}^\infty \beta_{n,i} (I - \Pi_{AK}^p B_{\delta_n}^{T_i}) A u_{n,i} \rangle$$
$$\leq \triangle_p(u_{n,i}, v)$$
$$- \frac{1}{\|A\|^p} \frac{\langle \sum_{i=0}^\infty \beta_{n,i} J_{E_2}^p (I - \Pi_{AK}^p B_{\delta_n}^{T_i}) A u_{n,i}, \sum_{i=0}^\infty \beta_{n,i} (I - \Pi_{AK}^p B_{\delta_n}^{T_i}) A u_{n,i} \rangle^p}{\| \sum_{i=0}^\infty \beta_{n,i} J_{E_2}^p (I - \Pi_{AK}^p B_{\delta_n}^{T_i}) A u_{n,i} \|^p}. \tag{41}$$

This implies that

$$\triangle_p(y_n, v) \leq \triangle_p(u_{n,i}, v). \tag{42}$$

By (42) and (37), we get

$$\triangle_p(y_n, v) \leq \triangle_p(x_n, v). \tag{43}$$

In addition, it follows from the assumption $\eta_{n,0} \leq \sum_{i=1}^{\infty} \eta_{n,i}$, (43), Definition 3, Lemmas 9 and 4

$$\begin{aligned}
&\triangle_p(x_{n+1}, v) \\
&= \triangle_p\left(J_{E_1^*}^q\left(\eta_{n,0}J_{E_1}^p(f(x_n)) + \sum_{i=1}^{\infty} \eta_{n,i}J_{E_1}^p(S_{n,i}(y_n))\right), v\right) \\
&= V_p\left(\eta_{n,0}J_{E_1}^p(f(x_n)) + \sum_{i=1}^{\infty} \eta_{n,i}J_{E_1}^p(S_{n,i}(y_n)), v\right) \\
&\leq \eta_{n,0} V_p\left(J_{E_1}^p(f(x_n)), v\right) + \sum_{i=1}^{\infty} \eta_{n,i} V_p\left(J_{E_1}^p(S_{n,i}(y_n)), v\right) \\
&\leq \eta_{n,0} \zeta \triangle_p(x_n, v) + \eta_{n,0}(\triangle_p(f(v), v) \\
&\quad + \langle J_{E_1}^p x_n - J_{E_1}^p f(v), f(v) - v\rangle) + \sum_{i=1}^{\infty} \eta_{n,i} k_{n,i} \triangle_p(y_n, v) \\
&\leq \eta_{n,0}\left(\triangle_p(f(v), v) + \langle J_{E_1}^p x_n - J_{E_1}^p f(v), f(v) - v\rangle\right) \\
&\quad + \left(\eta_{n,0}\zeta + \sum_{i=1}^{\infty} \eta_{n,i} k_{n,i}\right) \triangle_p(x_n, v) \\
&\leq \eta_{n,0}\left(\triangle_p(f(v), v) + \langle J_{E_1}^p x_n - J_{E_1}^p f(v), f(v) - v\rangle\right) \\
&\quad + \left(\sum_{i=1}^{\infty} \eta_{n,i}(\zeta + k_{n,i})\right) \triangle_p(x_n, v) \\
&\leq \max\left\{\frac{\left(\triangle_p(f(v), v) + \langle J_{E_1}^p x_1 - J_{E_1}^p f(v), f(v) - v\rangle\right)}{\zeta + k_{1,i}}, \triangle_p(x_1, v)\right\}. \tag{44}
\end{aligned}$$

By (44), we conclude that $\{x_n\}$ is bounded, and hence, from (42), (34), (35), (44), (38), and (37), $\{y_n\}$ and $\{u_{n,i}\}$ are also bounded.

Step Two: We show that $\lim_{m \to \infty} \triangle_p(x_{n+1}, x_n) = 0$. By Lemmas 1, 4, 10, and 7, we have, by the convexity of $\triangle_p$ in the first argument and for $\eta_{n-1,0} \leq \sum_{i=1}^{\infty} \eta_{n-1,i}$,

$$\triangle_p(x_{n+1}, x_n) = \triangle_p(J_{E_1^*}^q \left(\eta_{n,0} J_{E_1}^p(f(x_n)) + \sum_{i=1}^{\infty} \eta_{n,i} J_{E_1}^p(S_{n,i}(y_n)) \right),$$

$$J_{E_1^*}^q \left(\eta_{n-1,0} J_{E_1}^p(f(x_{n-1})) + \sum_{i=1}^{\infty} \eta_{n-1,i} J_{E_1}^p(S_{n-1,i}(y_{n-1})) \right))$$

$$\leq \eta_{n,0} \triangle_q^* (J_{E_1}^p(f(x_n)), \eta_{n-1,0} J_{E_1}^p(f(x_{n-1})) + \sum_{i=1}^{\infty} \eta_{n-1,i} J_{E_1}^p(S_{n-1,i}(y_{n-1})))$$

$$+ \sum_{i=1}^{\infty} \eta_{n,i} \triangle_q^* (J_{E_1}^p(S_{n,i}(y_n)), \eta_{n-1,0} J_{E_1}^p(f(x_{n-1})) + \sum_{i=1}^{\infty} \eta_{n-1,i} J_{E_1}^p(S_{n-1,i}(y_{n-1})))$$

$$\leq \eta_{n,0} \left(\triangle_q^*(J_{E_1}^p(f(x_n)), J_{E_1}^p(f(x_{n-1}))) \right)$$

$$+ \sum_{i=1}^{\infty} \eta_{n-1,i} \left(\sum_{i=1}^{\infty} \eta_{n,i} \frac{1}{p} \|S_{n-1,i}(y_{n-1})\|^p + \eta_{n,0} \|f(x_n)\| \left\| J_{E_1}^p(S_{n-1,i}(y_{n-1})) \right\| \right)$$

$$+ \eta_{n-1,0} \left(\eta_{n,0} \frac{1}{p} \|f(x_{n-1})\|^p + \sum_{i=1}^{\infty} \eta_{n,i} \|S_{n,i}(y_n)\| \left\| J_{E_1}^p(f(x_{n-1})) \right\| \right)$$

$$+ \sum_{i=1}^{\infty} \eta_{n,i} \triangle_q^* \left((J_{E_1}^p S_{n,i}(y_n), J_{E_1}^p S_{n-1,i}(y_{n-1})) \right)$$

$$\leq (1 - \eta_{n,0}(1 - \zeta)) \triangle_p(x_n, x_{n-1}) + \sum_{i=1}^{\infty} \eta_{n,i} \sup_{n,n-1 \geq 1} \{\triangle_p(S_{n,i}(y_n), S_{n-1,i}(y_{n-1}))\}$$

$$+ \sum_{i=1}^{\infty} \eta_{n-1,i} M, \tag{45}$$

where

$$M = \max \{\max\{\|f(x_n)\|, \|S_{n-1,i}(y_{n-1})\|\}, \max\{\|f(x_{n-1})\|, \|S_{n,i}(y_n)\|\}\}.$$

In view of the assumption $\sum_{n=1}^{\infty} \sum_{i=1}^{\infty} \eta_{n-1,i} M < \infty$ and (45), Lemmas 11 and 8 imply

$$\lim_{n \to \infty} \triangle_p(x_{n+1}, x_n) = 0. \tag{46}$$

Step Three: We show that $\lim_{n \to \infty} \triangle_p(S_{n,i} y_n, y_n) = 0$.
For each $i \in I$, we have

$$\triangle_p(S_i(y_n), v) \leq \triangle_p(y_n, v).$$

Then,

$$0 \leq \triangle_p(y_n, v) - \triangle_p(S_i(y_n), v)$$

$$= \triangle_p(y_n, v) - \triangle_p(x_{n+1}, v) + \triangle_p(x_{n+1}, v) - \triangle_p(S_i(y_n), v)$$

$$\leq \triangle_p(x_n, v) - \triangle_p(x_{n+1}, v) + \triangle_p(x_{n+1}, v) - \triangle_p(S_i(y_n), v)$$

$$= \triangle_p(x_n, v) - \triangle_p(x_{n+1}, v) + \triangle_p \left(J_{E_1^*}^q \left(\eta_{n,0} J_{E_1}^p(f(x_n)) + \sum_{i=1}^{\infty} \eta_{n,i} J_{E_1}^p(S_i(y_n)) \right), v \right)$$

$$- \triangle_p(S_i(y_n), v)$$

$$\leq \triangle_p(x_n, v) - \triangle_p(x_{n+1}, v) + \eta_{n,0} \triangle_p(f(x_n), v) - \eta_{n,0} \triangle_p(S_i(y_n), v)$$

$$\longrightarrow 0 \text{ as } n \to \infty. \tag{47}$$

By (47) and Definition 2, we get

$$\lim_{n\to\infty} \triangle_p (S_i y_n, y_n) = 0. \tag{48}$$

By uniform continuity of S, we have

$$\lim_{n\to\infty} \triangle_p (S_{n,i} y_n, y_n) = 0. \tag{49}$$

Step Four: We show that $x_n \to x^* \in \Gamma$.
Note that,

$$\triangle_p (x_{n+1}, y_n) = \triangle_p(J_{E_1^*}^q \left(\eta_{n,0} J_{E_1}^p (f(x_n)) + \sum_{i=1}^{\infty} \eta_{n,i} J_{E_1}^p (S_{n,i}(y_n)) \right), y_n)$$

$$\leq \eta_{n,0} \triangle_p (f(x_n), y_n) + \sum_{i=1}^{\infty} \eta_{n,i} \triangle_p (S_{n,i}(y_n), y_n)$$

$$\leq \eta_{n,0}(\zeta \triangle_p (x_n, y_n) + \triangle_p(f(y_n), y_n) + \langle f(x_n) - f(y_n), J_{E_1}^p f(y_n) - J_{E_1}^p y_n \rangle)$$

$$+ \sum_{i=1}^{\infty} \eta_{n,i} \triangle_p (S_{n,i}(y_n), y_n)$$

$$\leq (1 - \eta_{n,0}(1 - \zeta)) \triangle_p (x_n, y_n)$$

$$+ \eta_{n,0}(\triangle_p(f(y_n), y_n) + \langle f(x_n) - f(y_n), J_{E_1}^p f(y_n) - J_{E_1}^p y_n \rangle)$$

$$+ \sum_{i=1}^{\infty} \eta_{n,i} \triangle_p (S_{n,i}(y_n), y_n). \tag{50}$$

By (49), (50), and Lemma 8, we have

$$\lim_{n\to\infty} \triangle_p (x_n, y_n) = 0. \tag{51}$$

Therefore, by (51) and the boundedness of $\{y_n\}$, and since by (46), $\{x_n\}$ is Cauchy, we can assume without loss of generality that $y_n \rightharpoonup x^*$ for some $x^* \in E_1$. It follows from Lemmas 2, 3, and (48) that $x^* = S_i x^*$, for each $i \in I$. This means that $x^* \in \Im$.

In addition, by (31) and the fact that $u_{n,i} \to x^*$ as $n \to \infty$, we arrive at

$$\frac{(J_{E_1}^p u_{n,i} - J_{E_1}^p y_n) - \sum_{i=0}^{\infty} \beta_{n,i} \lambda_n A^* J_{E_2}^p (I - \Pi_{AK}^p B_{\delta_n}^{T_i}) A u_{n,i}}{\delta_n} \in \sum_{i=0}^{\infty} \alpha_{n,i} U_i(y_n). \tag{52}$$

By (21), we have

$$\| \sum_{i=0}^{\infty} \beta_{n,i} (I - \Pi_{AK}^p B_{\delta_n}^{T_i}) A u_{n,i} \| \leq \left[\frac{\triangle_p(u_{n,i}, v) - \triangle_p(y_n, v)}{\|A\|^{-1}[1 - \iota]} \right] \longrightarrow 0 \text{ as } n \to \infty, \tag{53}$$

and by (41), we have

$$\| \sum_{i=0}^{\infty} \beta_{n,i} (I - \Pi_{AK}^p B_{\delta_n}^{T_i}) A u_{n,i} \| \leq \left[\frac{\triangle_p(u_{n,i}, v) - \triangle_p(y_n, v)}{(p\|A\|)^{-1}} \right]^{\frac{1}{p}} \longrightarrow 0 \text{ as } n \to \infty. \tag{54}$$

From (53), (54), and (52), by passing n to infinity in (52), we have that $0 \in \sum_{i=0}^{\infty} \alpha_{n,i} U_i(x^*)$. This implies that $x^* \in SOLVIP(U_i)$. In addition, by (48), we have $Ay_n \rightharpoonup Ax^*$. Thus, by (53), (54) and an application of the demi-closeness of $\sum_{i=0}^{\infty} \beta_{n,i} (I - \Pi_{AK}^p B_{\delta_n}^{T_i})$ at zero, we have that $0 \in \sum_{i=0}^{\infty} \beta_{n,i} T_i(Ax^*)$. Therefore, $Ax \in SOLVIP(T_i)$ as $\sum_{i=0}^{\infty} \beta_{n,i} \Pi_{AK}^p B_{\delta}^{T_i}(Ax^*) = \sum_{i=0}^{\infty} \beta_{n,i} B_{\delta}^{T_i}(Ax^*)$. This means that $x^* \in \Omega$.

Now, we show that $x^* \in (\cap_{i=1}^{\infty} GMEP(\theta_i, C_i, G_i, g_i))$. By (32), we have

$$\Phi_i(u_{n,i}, y) + \langle J_{E_1}^p G_{n,i} x_n, y - u_{n,i} \rangle + \frac{1}{r_{n,i}} \langle y - u_{n,i}, J_{E_1}^p u_{n,i} - J_{E_1}^p x_n \rangle \geq 0$$

$$\forall y \in K, \forall i \in I,$$

Since Φ_i, for each $i \in I$, are monotone, that is, for all $y \in K$,

$$\Phi_i(u_{n,i}, y) + \Phi_i(y, u_{n,i}) \leq 0$$
$$\Rightarrow \frac{1}{r_{n,i}} \langle y - u_{n,i}, J_{E_1}^p u_{n,i} - J_{E_1}^p x_n \rangle$$
$$\geq \Phi_i(y, u_{n,i}) + \langle J_{E_1}^p G_{n,i} x_n, y - u_{n,i} \rangle,$$

therefore,

$$\frac{1}{r_{n,i}} \langle y - u_{n,i}, J_{E_1}^p u_{n,i} - J_{E_1}^p x_n \rangle \geq \Phi_i(y, u_{n,i}) + \langle J_{E_1}^p G_{n,i} x_n, y - u_{n,i} \rangle.$$

By the lower semicontinuity of Φ_i, for each $i \in I$, the weak upper semicontinuity of G, and the facts that, for each $i \in I$, $u_{n,i} \to x^*$ as $n \to \infty$ and J^p is $norm - to - weak^*$ uniformly continuous on a bounded subset of E_1, we have

$$0 \geq \Phi_i(y, x^*) + \langle J_{E_1}^p G_{n,i} x^*, y - x^* \rangle. \tag{55}$$

Now, we set $y_t = ty + (1-t)x^* \in K$. From (55), we get

$$0 \geq \Phi_i(y_t, x^*) + \langle J_{E_1}^p G_{n,i} x^*, y_t - x^* \rangle. \tag{56}$$

From (56), and by the convexity of Φ_i, for each $i \in I$, in the second variable, we arrive at

$$0 = \Phi_i(y_t, y_t) \leq t\Phi_i(y_t, y) + (1-t)\Phi_i(y_t, x^*)$$
$$\leq t\Phi_i(y_t, y) + (1-t)\langle J_{E_1}^p G_{n,i} x^*, y_t - x^* \rangle$$
$$\leq t\Phi_i(y_t, y) + (1-t)t\langle J_{E_1}^p G_{n,i} x^*, y - x^* \rangle,$$

which implies that

$$\Phi_i(y_t, y) + (1-t)\langle J_{E_1}^p G_{n,i} x^*, y - x^* \rangle \geq 0. \tag{57}$$

From (57), by the lower semicontinuity of Φ_i, for each $i \in I$, we have for $y_t \to x^*$ as $t \to 0$

$$\Phi_i(x^*, y) + \langle J_{E_1}^p G_{n,i} x^*, y - x^* \rangle \geq 0. \tag{58}$$

Therefore, by (58) we can conclude that $x^* \in (\cap_{i=1}^{\infty} GMEP(\theta_i, C_i, G_i, g_i))$. This means that $x^* \in \omega$. Hence, $x^* \in \Gamma$.

Finally, we show that $x_n \to x^*$, as $n \to \infty$. By Definition 3, we have

$$\triangle_p(x_{n+1}, x^*)$$
$$= \triangle_p(J_{E_1^*}^q\left(\eta_{n,0} J_{E_1}^p(f(x_n)) + \sum_{i=1}^{\infty} \eta_{n,i} J_{E_1}^p(G_{n,i}(y_n))\right), x^*)$$
$$\leq \eta_{n,0} \triangle_q^*(J_{E_1}^p(f(u_n)), J_{E_1}^p x^*) + \sum_{i=1}^{\infty} \eta_{n,i} \triangle_q^*(J_{E_1}^p(G_{n,i}(y_n)), J_{E_1}^p x^*)$$
$$\leq \eta_{n,0} \zeta \triangle_p(x_n, x^*) + \eta_{n,0}(\triangle_p(f(x^*), x^*))$$
$$+ \langle J_{E_1}^p x_n - J_{E_1}^p f(x^*), f(x^*) - x^* \rangle) + \sum_{i=1}^{\infty} \eta_{n,i} k_n \triangle_p(y_n, x^*)$$
$$\leq \eta_{n,0}\left(\triangle_p(f(x^*), x^*) + \langle J_{E_1}^p x_n - J_{E_1}^p f(x^*), f(x^*) - x^* \rangle\right)$$
$$+ \left(1 - \sum_{i=1}^{\infty} \eta_{n,i}(1 - k_n)\right) \triangle_p(x_n, x^*). \tag{59}$$

By (59) and Lemma 8, we have that

$$\lim_{n \to \infty} \triangle_p(x_n, x^*) = 0.$$

The proof is completed. □

In Theorem 1, $i = 0$ leads to the following new result.

Corollary 1. *Let $g : K \times K \to R$ be bifunctions satisfying $(A1) - (A4)$ in (3). Let $(I - \Pi_{AK}^p B_\delta^T)$ be demiclosed at zero. Suppose that $x_1 \in E_1$ is chosen arbitrarily and the sequence $\{x_n\}$ is defined as follows:*

$$\begin{cases} g(u_n, y) + \langle J_{E_1}^p C u_n + J_{E_1}^p G_n x_n, y - u_n \rangle + \theta(y) - \theta(u_n) \\ + \frac{1}{r_n} \langle y - u_n, J_{E_1}^p u_n - J_{E_1}^p x_n \rangle \geq 0 \, \forall y \in K, \\ y_n = J_{E_1^*}^q\left(B_{\delta_n}^U\left(J_{E_1}^p u_n - \lambda_n A^* J_{E_2}^p (I - \Pi_{AK}^p B_{\delta_n}^T) A u_n\right)\right), \\ x_{n+1} = J_{E_1^*}^q\left(\eta_n J_{E_1}^p(f(x_n)) + (1 - \eta_n) J_{E_1}^p(S_n(y_n))\right) n \geq 1, \end{cases} \tag{60}$$

where $r_n = \frac{1}{\|J_{E_1}^p G_n x_n\|}$, $\mu_n = \frac{1}{\|x_n\|^{p-1}}$ and $\gamma \in (0,1)$ such that $\rho_{E_1^*}(\mu_n) = \frac{\gamma \langle J_{E_1}^p G_n x_n, x_n - v \rangle}{2^q G_q \|x_n\|^p \|J_{E_1}^p G_n x_n\|}$, and

$$\lambda_n = \begin{cases} \frac{1}{\|A\|} \frac{1}{\|J_{E_2}^p(I - \Pi_{AK}^p B_{\delta_n}^T) A u_n\|}, & u_n \neq 0 \\ \frac{1}{\|A\|^p} \frac{\|J_{E_2}^p(I - \Pi_{AK}^p B_{\delta_n}^T) A u_n\|^{p(p-1)}}{\|J_{E_2}^p(I - \Pi_{AK}^p B_{\delta_n}^T) A u_n\|^p}, & u_n = 0, \end{cases} \tag{61}$$

and $\iota \in (0,1)$ and $\tau_n = \frac{1}{\|u_n\|^{p-1}}$ are chosen such that

$$\rho_{E_1^*}(\tau_n) = \frac{\iota}{2^q G_q \|A\|} \times \frac{\|J_{E_2}^p(I - \Pi_{AK}^p B_{\delta_n}^T) A u_n\|^p}{\|u_n\|^p \|J_{E_2}^p(I - \Pi_{AK}^p B_{\delta_n}^T) A u_n\|^{p-1}}, \tag{62}$$

and $\lim_{n \to \infty} \eta_n = 0$, for $M \geq 0$, $\sum_{n=1}^{\infty} \eta_{n-1} M < \infty$, and $\eta_n \leq \frac{1}{2}$. If $\Gamma = \Omega \cap \omega \cap \Im \neq \emptyset$, then $\{x_n\}$ converges strongly to $x^* \in \Gamma$, where $\Pi_{AK}^p B_{\delta_n}^T(x^*) = B_{\delta_n}^T(x^*)$.

3. Application to Generalized Mixed Equilibrium Problem, Split Hammerstein Integral Equations and Fixed Point Problem

Definition 4. *Let $C \subset \mathbb{R}^n$ be bounded. Let $k : C \times C \to \mathbb{R}$ and $f : C \times \mathbb{R} \to \mathbb{R}$ be measurable real-valued functions. An integral equation of Hammerstien-type has the form*

$$u(x) + \int_C k(x,y)f(y,u(y))dy = w(x),$$

where the unknown function u and non-homogeneous function w lies in a Banach space E of measurable real-valued functions. By transforming the above equation, we have that

$$u + KFu = w,$$

and therefore, without loss of generality, we have

$$u + KFu = 0. \tag{63}$$

The split Hammerstein integral equations problem is formulated as finding $x^* \in E_1$ and $y^* \in E_1^*$ such that

$$x^* + KFx^* = 0 \text{ with } Fx^* = y^* \text{ and } Ky^* + x^* = 0$$

and $Ax^* \in E_2$ and $Ay^* \in E_2^*$ such that

$$Ax^* + K'F'Ax^* = 0 \text{ with } F'Ax^* = Ay^* \text{ and } K'Ay^* + Ax^* = 0$$

where $F : E_1 \to E_1^*$, $K : E_1^* \to E_1$ and $F' : E_2 \to E_2^*$, $K' : E_2^* \to E_2$ are maximal monotone mappings.

Lemma 16 ([21]). *Let E be a Banach space. Let $F : E \to E^*$, $K : E^* \to E$ be bounded and maximal monotone operators. Let $D : E \times E^* \to E^* \times E$ be defined by $D(x,y) = (Fx - y, Ky + x)$ for all $(x,y) \in E \times E^*$. Then, the mapping D is maximal monotone.*

By Lemma 16, if K, K', and F, F' are multi-valued maximal monotone operators then, we have two resolvent mappings,

$$B_\delta^D = (J_{E_1}^p + \delta J_{E_1}^p D)^{-1} J_{E_1}^p \text{ and } B_\delta^{D'} = (J_{E_2}^p + \delta J_{E_2}^p D')^{-1} J_{E_2}^p,$$

where $F : E_1 \to E_1^*$, $K : E_1^* \to E_1$ are multi-valued and maximal monotone operators, $D : E_1 \times E_1^* \to E_1^* \times E_1$ is defined by $D(x,y) = (Fx - y, Ky + x)$ for all $(x,y) \in E_1 \times E_1^*$, and $F' : E_2 \to E_2^*$, $K' : E_2^* \to E_2$ are multi-valued and maximal monotone operators, $D' : E_2 \times E_2^* \to E_2^* \times E_2$ is defined by $D'(Ax, Ay) = (F'Ax - Ay, K'Ay + Ax)$ for all $(Ax, Ay) \in E_2 \times E_2^*$. Then D and D' are maximal monotone by Lemma 16.

When $U = D$ and $T = D'$ in Corollary 1, the algorithm (60) becomes

$$\begin{cases} g(u_n, y) + \langle J_{E_1}^p C_n u_n + J_{E_1}^p G_n x_n, y - u_n \rangle + \theta(y) - \theta(u_n) \\ + \frac{1}{r_n}\langle y - u_n, J_{E_1}^p u_n - J_{E_1}^p x_n \rangle \geq 0 \ \forall y \in K, \\ y_n = J_{E_1^*}^q \left(B_{\delta_n}^{D_n} \left(J_{E_1}^p u_n - \lambda_n A^* J_{E_2}^p (I - \Pi_{AK}^p B_{\delta_n}^{D'_n}) A u_n \right) \right) \\ x_{n+1} = J_{E_1^*}^q \left(\eta_n J_{E_1}^p (f(x_n)) + (1 - \eta_n) J_{E_1}^p (S_n(y_n)) \right) n \geq 1; \end{cases}$$

and its strong convergence is guaranteed, which solves the problem of a common solution of a system of generalized mixed equilibrium problems, split Hammerstein integral equations, and fixed-point problems for the mappings involved in this algorithm.

4. A Numerical Example

Let $i = 0$, $E_1 = E_2 = \mathbb{R}$, and $K = AK = [0, \infty)$, for $Ax = x \ \forall x \in E_1$. The generalized mixed equilibrium problem is formulated as finding a point $x \in K$ such that,

$$g_0(x, y) + \langle G_0 x, y - x \rangle + \theta_0(y) - \theta_0(x) \geq 0, \ \forall y \in K. \tag{64}$$

Let $r_0 \in (0, 1]$ and define $\theta_0 = 0$, $g_0(x, y) = \frac{y^2}{r_0} + \frac{2x^2}{r_0}$ and $G_0(x) = S_0(x) = \frac{1}{r_0} x$.

Clearly, $g_0(x, y)$ satisfies the conditions $(A1) - (A4)$ and $G_0(x) = S_0(x)$ is a Bregman asymptotically non-expansive mapping, as well as a $1-$ inverse strongly monotone mapping. Since Y_{r_0} is single-valued, therefore for $y \in K$, we have that

$$g_0(u_0, y) + \langle G_0 x, y - u_0 \rangle + \frac{1}{r_0} \langle y - u_0, u_0 - x \rangle \geq 0$$

$$\Leftrightarrow \frac{y^2}{r_0} + \frac{2u_0^2}{r_0} + \frac{1}{r_0} \langle y - u_0, u_0 \rangle \geq 0$$

$$\Leftrightarrow \frac{y^2}{r_0} + \frac{2|y u_0|}{r_0^{\frac{3}{2}}} + \frac{x^2}{r_0} \geq 0. \tag{65}$$

As (65) is a nonnegative quadratic function with respect to y variable, so it implies that the coefficient of y^2 is positive and the discriminant $\frac{4u_0^2}{r_0^3} - \frac{4x^2}{r_0^2} \leq 0$, and therefore $u_0 = x \sqrt{r_0}$. Hence,

$$Y_{r_0}(x) = x \sqrt{r_0}. \tag{66}$$

By Lemma 13 and (66), $F(Y_{r_0}) = GEP(g_0, G_0) = \{0\}$ and $F(S_0) = \{0\}$. Define

$$U_0, T_0 : \mathbb{R} \longrightarrow \mathbb{R} \ by \ U_0(x) = T_0(Ax) \begin{cases} (0, 1), x \geq 0 \\ \{1\}, x < 0, \end{cases}$$

$$P_{[0, \infty)} : \mathbb{R} \longrightarrow [0, \infty) \ by \ P_{[0, \infty)}(Ax) = \begin{cases} 0, Ax \in (-\infty, 0) \\ Ax, Ax \in [0, \infty), \end{cases}$$

$$B_\delta^{U_0} = B_\delta^T : \mathbb{R} \longrightarrow \mathbb{R} \ by \ B_\delta^T(Ay) = B_\delta^{U_0}(y) = \begin{cases} \frac{y}{1 + (0, \delta)}, y \geq 0 \\ \frac{y}{1 + \delta}, y < 0, \end{cases}$$

$$P_{[0, \infty)} B_\delta^T : \mathbb{R} \longrightarrow [0, \infty) \ by \ P_{[0, \infty)} B_\delta^T(Ay) = \begin{cases} \frac{Ay}{1 + (0, \delta)}, Ay \geq 0 \\ 0, Ay < 0. \end{cases}$$

It is clear that U_0 and T_0 are multi-valued maximal monotone mappings, such that $0 \in SOLVIP(U_0)$ and $0 \in SOLVIP(T_0)$. We define the ζ-contraction mapping by $f(x) = \frac{x}{2}$, $\delta_n = \frac{1}{2^{n+1}}$, $\eta_{n,0} = \frac{1}{n+1}$, $r_{n,0} = \frac{1}{2^{2n}}$ and $\zeta = \frac{1}{2}$. Hence, for

$$\lambda_n = \begin{cases} \frac{1 + \left(0, \frac{1}{2^{n+1}}\right)}{\left|u_{n,0}\left(1 + \left(0, \frac{1}{2^{n+1}}\right)\right) - u_{n,0}\right|}, u_{n,0} > 0, \\ 1, u_{n,0} = 0, \\ \frac{1}{|u_{n,0}|}, u_{n,0} < 0, \end{cases}$$

we get,

$$\begin{cases} u_{n,0} = \frac{1}{2^n} x_n, \\ y_n^1 = \frac{u_{n,0}}{1+\left(0,\frac{1}{2^{n+1}}\right)}(u_{n,0} - 1), \ u_{n,0} > 0, \\ y_n^2 = \left[\frac{u_{n,0}}{1+\left(0,\frac{1}{2^{n+1}}\right)}\right]^2, \ u_{n,0} = 0, \\ y_n^3 = \frac{2^{n+1} u_{n,0}}{2^{n+1}+1}(u_{n,0} + 1), \ u_{n,0} < 0, \\ x_{n+1} = \frac{x_n}{2(n+1)} + \frac{2^{2n} n y_n}{(n+1)}, \ n \geq 1, \end{cases}$$

$$x_{n+1} = \begin{cases} \frac{x_n}{2(n+1)} + \frac{n x_n^2 - 2^n x_n}{(n+1)\left(1+\left(0,\frac{1}{2^{n+1}}\right)\right)}, \ x_n > 0, \\ \frac{x_n}{2(n+1)} + \frac{n x_n^2}{(n+1)\left(1+\left(0,\frac{1}{2^{n+1}}\right)\right)}, \ x_n = 0, \\ \frac{x_n}{2(n+1)} + \frac{n 2^{n+1}(x_n^2 + x_n)}{2^{n+1}+1}, \ x_n < 0. \end{cases}$$

In particular,

$$x_{n+1} = \begin{cases} \frac{x_n}{2(n+1)} + \frac{5(n x_n^2 - 2^n x_n)}{6(n+1)}, \ x_n > 0, \\ \frac{x_n}{2(n+1)} + \frac{5 n x_n^2}{6(n+1)}, \ x_n = 0, \\ \frac{x_n}{2(n+1)} + \frac{n 2^{n+1}(x_n^2 + x_n)}{2^{n+1}+1}, \ x_n < 0. \end{cases}$$

By Theorem 1, the sequence $\{x_n\}$ converges strongly to $0 \in \Gamma$. The Figures 1 and 2 below obtained by $(MATLAB)$ software indicate convergence of $\{x_n\}$ given by (32) with $x_1 = -10.0$ and $x_1 = 10.0$, respectively.

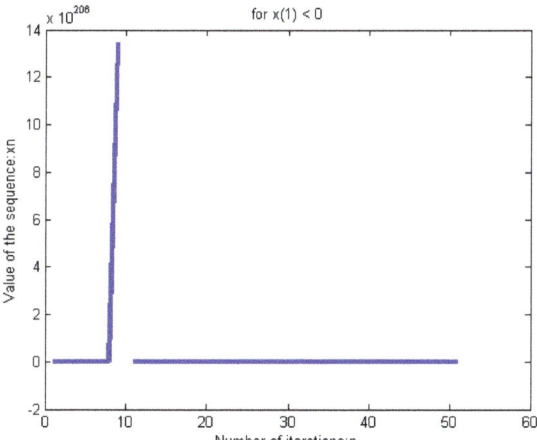

Figure 1. Sequence convergence with initial condition -10.0.

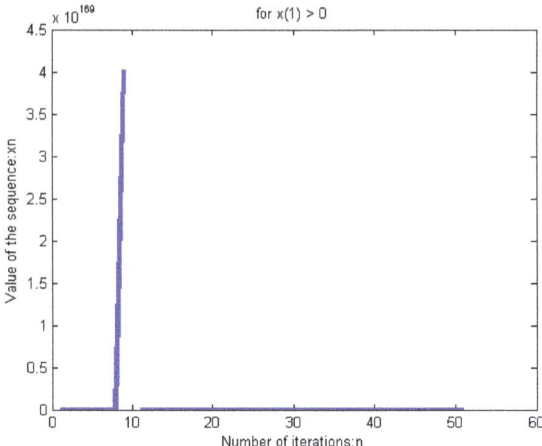

Figure 2. Sequence convergence with initial condition 10.0

Remark 1. *Our results generalize and complement the corresponding ones in [2,7,9,10,22,23].*

Author Contributions: all the authors contribute equally to all the parts of the manuscript

Funding: This work has been co-funded by the Deanship of Scientific Research (DSR) at University of Petroleum and Minerals (King Fahd University of Petroleum and Minerals KFUPM, Saudi Arabia) through Project No. IN141047 and by the Spanish Government and European Commission through Grant RTI2018-094336-B-I00 (MINECO/FEDER, UE).

Acknowledgments: The author A.R. Khan would like to acknowledge the support provided by the Deanship of Scientific Research (DSR) at University of Petroleum and Minerals (KFUPM)for funding this work through project No. IN141047.

Conflicts of Interest: The authors declare no confilct of interest.

References

1. Barbu, V. *Maximal Monotone Operators in Banach Spaces, Nonlinear Differential Equations of Monotone Types in Banach Spaces*; Springer Monographs in Mathematics; Springer: New York, NY, USA, 2011.
2. Moudafi, A. Split monotone variational inclusions. *J. Optim. Theory Appl.* **2011**, *150*, 275–283. [CrossRef]
3. Moudafi, A. Viscosity approximation methods for fixed points problems. *J. Math. Anal. Appl.* **2000**, *241*, 46–55. [CrossRef]
4. Kazmi, K.R.; Rizvi, S.H. An iterative method for split variational inclusion problem and fixed point problem for a nonexpansive mapping. *Optim. Lett.* **2014**, *8*, 1113–1124. [CrossRef]
5. Nimit, N.; Narin, P. Viscosity Approximation Methods for Split Variational Inclusion and Fixed Point Problems in Hilbert Spaces. In Proceedings of the International Multi-Conference of Engineers and Computer Scientists (IMECS 2014), Hong Kong, China, 12–14 March 2014; Volume II.
6. Schopfer, F.; Schopfer, T.; Louis, A.K. An iterative regularization method for the solution of the split feasibility problem in Banach spaces. *Inverse Probl.* **2008**, *24*, 055008. [CrossRef]
7. Chen, J.Z.; Hu, H.Y.; Ceng, L.C. Strong convergence of hybrid Bregman projection algorithm for split feasibility and fixed point problems in Banach spaces. *J. Nonlinear Sci. Appl.* **2017**, *10*, 192–204. [CrossRef]
8. Nakajo, K.; Takahashi, W. Strong convergence theorem for nonexpansive mappings and nonexpansive semigroups. *J. Math. Anal. Appl.* **2003**, *279*, 372–379. [CrossRef]
9. Takahashi, W. Split feasibility problem in Banach spaces. *J. Nonlinear Convex Anal.* **2014**, *15*, 1349–1355.
10. Wang, F.H. A new algorithm for solving multiple-sets split feasibility problem in Banach spaces. *Numer. Funct. Anal. Optim.* **2014**, *35*, 99–110. [CrossRef]

11. Payvand, M.A.; Jahedi, S. System of generalized mixed equilibrium problems, variational inequality, and fixed point problems. *Fixed Point Theory Appl.* **2016**, *2016*, 93. [CrossRef]
12. Schopfer, F. *Iterative Methods for the Solution of the Split Feasibility Problem in Banach Spaces*; der Naturwissenschaftlich-Technischen Fakultaten, Universitat des Saarlandes: Saarbrücken, Germany, 2007.
13. Martin, M.V.; Reich, S.; Sabach, S. Right Bregman nonexpansive operators in Banach spaces. *Nonlinear Anal.* **2012**, *75*, 5448–5465. [CrossRef]
14. Xu, H.K. Existence and convergence for fixed points of mappings of asymptotically nonexpansive type. *Nonlinear Anal.* **1991**, *16*, 1139–1146. [CrossRef]
15. Xu, H.K. Inequalities in Banach spaces with applications. *Nonlinear Anal.* **1991**, *16*, 1127–1138. [CrossRef]
16. Liu, L.S. Ishikawa and Mann iterative process with errors for nonlinear strongly accretive mappings in Banach spaces. *J. Math. Anal. Appl.* **1995**, *194*, 114–125. [CrossRef]
17. Xu, H.K.; Xu, Z.B. An L_p inequality and its applications to fixed point theory and approximation theory. *Proc. R. Soc. Edinb.* **1989**, *112A*, 343–351. [CrossRef]
18. Aoyamaa, K.; Yasunori, K.; Takahashi, W.; Toyoda, M. Mann Approximation of common fixed points of a countable family of nonexpansive mappings in a Banach space. *Nonlinear Anal.* **2007**, *67*, 2350–2360. [CrossRef]
19. Deng, B.C.; Chen, T.; Yin, Y.L. Srong convergence theorems for mixed equilibrium problem and asymptotically I-nonexpansive mapping in Banach spaces. *Abstr. Appl. Anal.* **2014**, *2014*, 965737.
20. Reich, S.; Sabach, S. Two strong convergence theorems for Bregman strongly nonexpansive operators in reflexive Banach spaces. *Nonlinear Anal.* **2010**, *73*, 122–135. [CrossRef]
21. Chidume, C.E.; Idu, K.O. Approximation of zeros bounded maximal monotone mappings, solutions of Hammertein integral equations and convex minimization problems. *Fixed Point Theory Appl.* **2016**, *2016*, 97. [CrossRef]
22. Khan, A.R.; Abbas, M.; Shehu, Y. A general convergence theorem for multiple-set split feasibilty problem in Hilbert spaces. *Carpathian J. Math.* **2015**, *31*, 349–357.
23. Ogbuisi, F.U.; Mewomo, O.T. On split generalized mixed equilibrium problem and fixed point problems with no prioir knowledge of operator norm. *J. Fixed Point Theory Appl.* **2017**, *19*, 2109–2128. [CrossRef]

© 2019 by the authors. Licensee MDPI, Basel, Switzerland. This article is an open access article distributed under the terms and conditions of the Creative Commons Attribution (CC BY) license (http://creativecommons.org/licenses/by/4.0/).

Article

Sixteenth-Order Optimal Iterative Scheme Based on Inverse Interpolatory Rational Function for Nonlinear Equations

Mehdi Salimi [1,2] and Ramandeep Behl [3,*]

[1] Center for Dynamics and Institute for Analysis, Department of Mathematics, Technische Universität Dresden, 01062 Dresden, Germany; mehdi.salimi@tu-dresden.de or msalimi1@yahoo.com
[2] Department of Law, Economics and Human Sciences, University Mediterranea of Reggio Calabria, 89125 Reggio Calabria, Italy; mehdi.salimi@unirc.it
[3] Department of Mathematics, King Abdulaziz University, Jeddah 21589, Saudi Arabia
* Correspondence: ramanbehl87@yahoo.in

Received: 30 April 2019; Accepted: 14 May 2019; Published: 19 May 2019

Abstract: The principal motivation of this paper is to propose a general scheme that is applicable to every existing multi-point optimal eighth-order method/family of methods to produce a further sixteenth-order scheme. By adopting our technique, we can extend all the existing optimal eighth-order schemes whose first sub-step employs Newton's method for sixteenth-order convergence. The developed technique has an optimal convergence order regarding classical Kung-Traub conjecture. In addition, we fully investigated the computational and theoretical properties along with a main theorem that demonstrates the convergence order and asymptotic error constant term. By using Mathematica-11 with its high-precision computability, we checked the efficiency of our methods and compared them with existing robust methods with same convergence order.

Keywords: simple roots; Newton's method; computational convergence order; nonlinear equations

1. Introduction

The formation of high-order multi-point iterative techniques for the approximate solution of nonlinear equations has always been a crucial problem in computational mathematics and numerical analysis. Such types of methods provide the utmost and effective imprecise solution up to the specific accuracy degree of

$$\Omega(x) = 0, \qquad (1)$$

where $\Omega : \mathbb{C} \to \mathbb{C}$ is holomorphic map/function in the neighborhood of required ζ. A certain recognition has been given to the construction of sixteenth-order iterative methods in the last two decades. There are several reasons behind this. However, some of them are advanced digital computer arithmetic, symbolic computation, desired accuracy of the required solution with in a small number of iterations, smaller residual errors, CPU time, smaller difference between two iterations, etc. (for more details please see Traub [1] and Petković et al. [2]).

We have a handful of optimal iterative methods of order sixteen [3–9]. Among these methods most probably are the improvement or extension of some classical methods e.g., Newton's method or Newton-like method, Ostrowski's method at the liability of further values of function/s and/or 1st-order derivative/s or extra numbers of sub-steps of the native schemes.

In addition, we have very few such techniques [5,10] that are applicable to every optimal 8-order method (whose first sub-step employs Newton's method) to further obtain 16-order convergence optimal scheme, according to our knowledge. Presently, optimal schemes suitable to every iterative method of particular order to obtain further high-order methods have more importance than obtaining a high-order version of a native method. Finding such general schemes are a more attractive and harder chore in the area of numerical analysis.

Therefore, in this manuscript we pursue the development of a scheme that is suitable to every optimal 8-order scheme whose first sub-step should be the classical Newton's method, in order to have further optimal 16-order convergence, rather than applying the technique only to a certain method. The construction of our technique is based on the rational approximation approach. The main advantage of the constructed technique is that it is suitable to every optimal 8-order scheme whose first sub-step employs Newton's method. Therefore, we can choose any iterative method/family of methods from [5,11–25], etc. to obtain further 16-order optimal scheme. The effectiveness of our technique is illustrated by several numerical examples and it is found that our methods execute superior results than the existing optimal methods with the same convergence order.

2. Construction of the Proposed Optimal Scheme

Here, we present an optimal 16-order general iterative scheme that is the main contribution of this study. In this regard, we consider a general 8-order scheme, which is defined as follows:

$$\begin{cases} w_r = x_r - \dfrac{\Omega(x_r)}{\Omega'(x_r)}, \\ z_r = \phi_4(x_r, w_r), \\ t_r = \psi_8(x_r, w_r, z_r), \end{cases} \quad (2)$$

where ϕ_4 and ψ_8 are optimal scheme of order four and eight, respectively.

We adopt Newton's method as a fourth sub-step to obtain a 16-order scheme, which is given by

$$x_{r+1} = t_r - \dfrac{\Omega(t_r)}{\Omega'(t_r)}, \quad (3)$$

that is non-optimal in the regard of conjecture given by Kung-Traub [5] because of six functional values at each step. We can decrease the number of functional values with the help of following $\gamma(x)$ third-order rational functional

$$\gamma(x) = \gamma(x_r) + \dfrac{(x - x_r) + b_1}{b_2(x - x_r)^3 + b_3(x - x_r)^2 + b_4(x - x_r) + b_5}, \quad (4)$$

where the values of disposable parameters $b_i (1 \leq i \leq 5)$ can be found with the help of following tangency constraints

$$\gamma(x_r) = \Omega(x_r), \ \gamma'(x_r) = \Omega'(x_r), \ \gamma(w_r) = \Omega(w_r), \ \gamma(z_r) = \Omega(z_r). \quad (5)$$

Then, the last sub-step iteration is replaced by

$$x_{r+1} = t_r - \dfrac{\Omega(t_r)}{\gamma'(t_r)}, \quad (6)$$

that does not require $\Omega'(t_r)$. Expressions (2) and (6) yield an optimal sixteenth-order scheme. It is vital to note that the $\gamma(x)$ in (4) plays a significant role in the construction of an optimal 16-order scheme.

In this paper, we adopt a different last sub-step iteration, which is defined as follows:

$$x_{r+1} = x_r - Q_\Omega(x_r, w_r, z_r), \quad (7)$$

where Q_Ω can be considered to be a correction term to be called naturally as "error corrector". The last sub-step of this type is handier for the convergence analysis and additionally in the dynamics study through basins of attraction. The easy way of obtaining such a fourth sub-step iteration with a feasible error corrector is to apply the Inverse Function Theorem [26] to (5). Since ξ is a simple root (i.e., $\gamma'(\xi) \neq 0$), then we have a unique map $\tau(x)$ satisfying $\gamma(\tau(x)) = x$ in the certain neighborhood of $\gamma(\xi)$. Hence, we adopt such an inverse map $\tau(x)$ to obtain the needed last sub-step of the form (7) instead of using $\gamma(x)$ in (5).

With the help of Inverse Function Theorem, we will yield the final sub-step iteration from the expression (5):

$$x = x_r - \frac{\varphi(x) - \varphi(x_r) + b_1}{b_2\big(\varphi(x) - \varphi(x_r)\big)^3 + b_3\big(\varphi(x) - \varphi(x_r)\big)^2 + b_4\big(\varphi(x) - \varphi(x_r)\big) + b_5}, \tag{8}$$

where $b_i, i = 1, 2, \ldots, 5$ are disposable constants. We can find them by adopting the following tangency conditions

$$\varphi(x_r) = \Omega(x_r), \quad \varphi'(x_r) = \Omega'(x_r), \quad \varphi(w_r) = \Omega(w_r), \quad \varphi(z_r) = \Omega(z_r), \quad \varphi(t_r) = \Omega(t_r). \tag{9}$$

One should note that the rational function on the right side of (8) is regarded as an error corrector. Indeed, the desired last sub-step iteration (8) is obtained using the inverse interpolatory function approach meeting the tangency constraints (9). Clearly, the last sub-step iteration (6) looks more suitable than (3) in the error analysis. It remains for us to determine parameters $b_i (1 \leq i \leq 5)$ in (8)

By using the first two tangency conditions, we obtain

$$b_1 = 0, \quad b_5 = \Omega'(x_r). \tag{10}$$

By adopting last three tangency constraints and the expression (10), we have the following three independent relations

$$b_2(\Omega(w_r) - \Omega(x_r))^2 + b_3(\Omega(w_r) - \Omega(x_r)) + b_4 = \frac{1}{w_r - x_r} - \frac{\Omega'(x_r)}{\Omega(w_r) - \Omega(x_r)},$$

$$b_2(\Omega(z_r) - \Omega(x_r))^2 + b_3(\Omega(z_r) - \Omega(x_r)) + b_4 = \frac{1}{z_r - x_r} - \frac{\Omega'(x_r)}{\Omega(z_r) - \Omega(x_r)}, \tag{11}$$

$$b_2(\Omega(t_r) - \Omega(x_r))^2 + b_3(\Omega(t_r) - \Omega(x_r)) + b_4 = \frac{1}{t_r - x_r} - \frac{\Omega'(x_r)}{\Omega(t_r) - \Omega(x_r)},$$

which further yield

$$b_2 = -\frac{\theta_1 + \theta_2\{\Omega(t_r)(t_r - x_r) + \Omega(w_r)(z_r - t_r) + \Omega(z_r)(x_r - z_r)\}}{\theta_2(\Omega(t_r) - \Omega(w_r))(\Omega(t_r) - \Omega(z_r))(\Omega(w_r) - \Omega(z_r))(t_r - x_r)(x_r - z_r)},$$

$$b_3 = \frac{b_2(\Omega(w_r) - \Omega(z_r))(\Omega(w_r) - 2\Omega(x_r) + \Omega(z_r)) + \frac{\Omega'(x_r)}{\Omega(w_r) - \Omega(x_r)} + \frac{\Omega'(x_r)(2\Omega(x_r) - \Omega(z_r))}{\Omega(x_r)(\Omega(x_r) - \Omega(z_r))} + \frac{1}{z_r - x_r}}{\Omega(z_r) - \Omega(w_r)}, \tag{12}$$

$$b_4 = \frac{b_2\Omega(x_r)(\Omega(w_r) - \Omega(x_r))^3 + b_3\Omega(x_r)(\Omega(w_r) - \Omega(x_r))^2 + \Omega'(x_r)\Omega(w_r)}{\Omega(x_r)(\Omega(x_r) - \Omega(w_r))},$$

where $\theta_1 = \Omega'(x_r)(\Omega(t_r) - \Omega(z_r))(t_r - x_r)(x_r - z_r)\big[(\Omega(x_r) - \Omega(w_r))(\Omega(t_r)(\Omega(z_r) - 2\Omega(x_r)) + \Omega(x_r)(\Omega(w_r) + 2\Omega(x_r) - 2\Omega(z_r))) + \Omega(x_r)(\Omega(t_r) - \Omega(x_r))(\Omega(x_r) - \Omega(z_r))\big]$, $\theta_2 = \Omega(x_r)(\Omega(t_r) - \Omega(x_r))(\Omega(x_r) - \Omega(w_r))(\Omega(x_r) - \Omega(z_r))$.

Let us consider that the rational Function (8) cuts the x – axis at $x = x_{r+1}$, in order to obtain the next estimation x_{r+1}. Then, we obtain

$$\varphi(x_{r+1}) = 0, \tag{13}$$

which further yield by using the above values of b_1, b_2 and b_3

$$x_{r+1} = x_r + \frac{\theta_2(\Omega(t_r) - \Omega(w_r))(\Omega(t_r) - \Omega(z_r))(\Omega(w_r) - \Omega(z_r))(t_r - x_r)(z_r - x_r)}{\theta_3 + \theta_2\Omega(w_r)\{\Omega(t_r)(t_r - x_r)(\Omega(t_r) - \Omega(w_r)) - \Omega(z_r)(\Omega(w_r) - \Omega(z_r))(x_r - z_r)\}}, \quad (14)$$

where $\theta_3 = \Omega'(x_r)\Omega(t_r)\Omega(z_r)(\Omega(t_r) - \Omega(z_r))(t_r - x_r)(x_r - z_r)[(\Omega(x_r) - \Omega(w_r))\{-\Omega(t_r)(\Omega(w_r) + 2\Omega(x_r) - 2\Omega(z_r)) + \Omega(w_r)^2 + (\Omega(w_r) + 2\Omega(x_r))(\Omega(x_r) - \Omega(z_r))\} + \Omega(x_r)(\Omega(t_r) - \Omega(x_r))(\Omega(x_r) - \Omega(z_r))]$.

Finally, by using expressions (2) and (14), we have

$$\begin{cases} w_r = x_r - \dfrac{\Omega(x_r)}{\Omega'(x_r)}, \\ z_r = \phi_4(x_r, w_r), \\ t_r = \psi_8(x_r, w_r, z_r), \\ x_{r+1} = x_r + \dfrac{\theta_2(\Omega(t_r) - \Omega(w_r))(\Omega(t_r) - \Omega(z_r))(\Omega(w_r) - \Omega(z_r))(t_r - x_r)(z_r - x_r)}{\theta_3 + \theta_2\Omega(w_r)\{\Omega(t_r)(t_r - x_r)(\Omega(t_r) - \Omega(w_r)) - \Omega(z_r)(\Omega(w_r) - \Omega(z_r))(x_r - z_r)\}}, \end{cases} \quad (15)$$

where θ_2 and θ_3 are defined earlier. We illustrate that convergence order reach at optimal 16-order without adopting any additional functional evaluations in the next Theorem 1. It is vital to note that only coefficients A_0 and B_0 from $\phi_4(x_r, w_r)$ and $\psi_8(x_r, w_r, z_r)$, respectively, contribute to its important character in the development of the needed asymptotic error constant, which can be found in Theorem 1.

Theorem 1. *Let $\Omega : \mathbb{C} \to \mathbb{C}$ be an analytic function in the region containing the simple zero ξ and initial guess $x = x_0$ is sufficiently close to ξ for guaranteed convergence. In addition, we consider that $\phi_4(x_r, w_r)$ and $\psi_8(x_r, w_r, z_r)$ are any optimal 4- and 8-order schemes, respectively. Then, the proposed scheme (15) has an optimal 16-order convergence.*

Proof. Let us consider $e_r = x_r - \xi$ be the error at rth step. With the help of the Taylor's series expansion, we expand the functions $\Omega(x_r)$ and $\Omega'(x_r)$ around $x = \xi$ with the assumption $\Omega'(\xi) \neq 0$ which leads us to:

$$\Omega(x_r) = \Omega'(\xi)\left[e_r + \sum_{k=2}^{16} c_k e_r^k + O(e_r^{17})\right] \quad (16)$$

and

$$\Omega'(x_r) = \Omega'(\xi)\left[1 + \sum_{k=2}^{16} k c_k e_r^{k-1} + O(e_r^{17})\right], \quad (17)$$

where $c_j = \dfrac{\Omega^{(j)}(\xi)}{j!\Omega'(\xi)}$ for $j = 2, 3, \ldots, 16$, respectively.

By inserting the expressions (16) and (17) in the first sub-step (15), we have

$$w_r - \xi = c_2 e_r^2 + 2(c_3 - c_2^2)e_r^3 + (4c_2^3 - 7c_3c_2 + 3c_4)e_r^4 + (20c_3c_2^2 - 8c_2^4 - 10c_4c_2 - 6c_3^2 + 4c_5)e_r^5 + \sum_{k=1}^{11} G_k e_r^{k+4} + O(e_r^{17}), \quad (18)$$

where $G_k = G_k(c_2, c_3, \ldots, c_{16})$ are given in terms of $c_2, c_3, \ldots, c_i$ with explicitly written two coefficients $G_1 = 16c_2^5 - 52c_2^3c_3 + 28c_4c_2^2 + (33c_3^2 - 13c_5)c_2 - 17c_3c_4 + 5c_6$, $G_2 = 2\{16c_2^6 - 64c_3c_2^4 + 36c_4c_2^3 + 9(7c_3^2 - 2c_5)c_2^2 + (8c_6 - 46c_3c_4)c_2 - 9c_3^3 + 6c_4^2 + 11c_3c_5 - 3c_7\}$, etc.

The following expansion of $\Omega(w_r)$ about a point $x = \xi$ with the help of Taylor series

$$\Omega(w_r) = \Omega'(\xi)[c_2 e_r^2 + 2(c_3 - c_2^2)e_r^3 + (5c_2^3 - 7c_3c_2 + 3c_4)e_r^4 + 2(6c_2^4 - 12c_3c_2^2 + 5c_4c_2 + 3c_3^2 - 2c_5)e_r^5 + \sum_{k=1}^{11} \bar{G}_k e_r^{k+4} + O(e_r^{17})]. \quad (19)$$

As in the beginning, we consider that $\phi_4(x_r, w_r)$ and $\phi_8(x_r, w_r, z_r)$ are optimal schemes of order four and eight, respectively. Then, it is obvious that they will satisfy the error equations of the following forms

$$z_r - \xi = \sum_{m=0}^{12} A_m e_r^{m+4} + O(e_r^{17}) \tag{20}$$

and

$$t_r - \xi = \sum_{m=0}^{8} B_m e_r^{m+8} + O(e_r^{17}), \tag{21}$$

respectively, where $A_0, B_0 \neq 0$. By using the Taylor series expansion, we further obtain

$$\begin{aligned}\Omega(z_r) = \Omega'(\xi)[A_0 e_r^4 + A_1 e_r^5 + A_2 e_r^6 + A_3 e_r^7 + (A_0^2 c_2 + A_4) e_r^8 + (2 A_0 A_1 c_2 + A_5) e_r^9 + \{(A_1^2 + 2 A_0 A_2) c_2 \\ + A_6\} e_r^{10} + \{2(A_1 A_2 + A_0 A_3) c_2 + A_7\} e_r^{11} + (A_0^3 c_3 + 2 A_4 A_0 c_2 + A_2^2 c_2 + 2 A_1 A_3 c_2 + A_8) e_r^{12} \\ + (3 A_1 A_0^2 c_3 + 2 A_5 A_0 c_2 + 2 A_2 A_3 c_2 + 2 A_1 A_4 c_2 + A_9) e_r^{13} + H_1 e_r^{14} + H_2 e_r^{15} + H_3 e_r^{16} + O(e_n^{17})]\end{aligned} \tag{22}$$

and

$$\begin{aligned}\Omega(t_r) = \Omega'(\xi)[B_0 e_r^8 + B_1 e_r^9 + B_2 e_r^{10} + B_3 e_r^{11} + B_4 e_r^{12} + B_5 e_r^{13} + B_6 e_r^{14} + B_7 e_r^{15} \\ + (A_2 B_0^2 + B_8) e_r^{16} + O(e_r^{17})],\end{aligned} \tag{23}$$

where $H_1 = 3 A_2 A_0^2 c_3 + 2 A_6 A_0 c_2 + 3 A_1^2 A_0 c_3 + A_3^2 c_2 + 2 A_2 A_4 c_2 + 2 A_1 A_5 c_2 + A_{10}$, $H_2 = A_1^3 c_3 + 6 A_0 A_2 A_1 c_3 + 2(A_3 A_4 + A_2 A_5 + A_1 A_6 + A_0 A_7) c_2 + 3 A_0^2 A_3 c_3 + A_{11}$ and $H_3 = A_0^4 c_4 + 3 A_4 A_0^2 c_3 + 2 A_8 A_0 c_2 + 3 A_2^2 A_0 c_3 + 6 A_1 A_3 A_0 c_3 + A_4^2 c_2 + 2 A_3 A_5 c_2 + 2 A_2 A_6 c_2 + 2 A_1 A_7 c_2 + 3 A_1^2 A_2 c_3 + A_{12}$.

With the help of expressions (16)–(23), we have

$$\begin{aligned}\frac{\theta_2(\Omega(t_r) - \Omega(w_r))(\Omega(t_r) - \Omega(z_r))(\Omega(w_r) - \Omega(z_r))(t_r - x_r)(z_r - x_r)}{\theta_3 + \theta_2 \Omega(w_r)\{\Omega(t_r)(t_r - x_r)(\Omega(t_r) - \Omega(w_r)) - \Omega(z_r)(\Omega(w_r) - \Omega(z_r))(x_r - z_r)\}} \\ = e_r - A_0 B_0 (5 c_2^4 - 10 c_2^2 c_3 + 2 c_3^2 + 4 c_2 c_4 - c_5) c_2 e_r^{16} + O(e_r^{17}).\end{aligned} \tag{24}$$

Finally, we obtain

$$e_{n+1} = A_0 B_0 (5 c_2^4 - 10 c_2^2 c_3 + 2 c_3^2 + 4 c_2 c_4 - c_5) c_2 e_r^{16} + O(e_r^{17}). \tag{25}$$

The above expression (25) claims that our scheme (15) reaches the 16-order convergence. The expression (15) is also an optimal scheme in the regard of Kung-Traub conjecture since it uses only five functional values at each step. Hence, this completes the proof. □

Remark 1. *Generally, we naturally expect that the presented general scheme (15) should contain other terms from $A_0, A_1, \ldots A_{12}$ and $B_0, B_1, \ldots, B_8$. However, there is no doubt from the expression (25) that the asymptotic error constant involves only on A_0 and B_0. This simplicity of the asymptotic error constant is because of adopting the inverse interpolatory function with the tangency constraints.*

2.1. Special Cases

This is section is devoted to the discussion of some important cases of the proposed scheme. Therefore, we consider

1. We assume an optimal eighth-order technique suggested scheme by Cordero et al. [13]. By using this scheme, we obtain the following new optimal 16-order scheme

$$\begin{cases} w_r = x_r - \dfrac{\Omega(x_r)}{\Omega'(x_r)}, \\ z_r = x_r - \dfrac{\Omega(x_r)}{\Omega'(x_r)} \left[\dfrac{\Omega(x_r) - \Omega(w_r)}{\Omega(x_r) - 2\Omega(w_r)} \right], \\ u_r = z_r - \dfrac{\Omega(z_r) \left(\dfrac{\Omega(x_r) - \Omega(w_r)}{\Omega(x_r) - 2\Omega(w_r)} + \dfrac{\Omega(z_r)}{2(\Omega(w_r) - 2\Omega(z_r))} \right)^2}{\Omega'(x_r)}, \\ t_r = u_r - \dfrac{3(b_2 + b_3)\Omega(z_r)(u_r - z_r)}{\Omega'(x_r)(b_1(u_r - z_r) + b_2(w_r - x_r) + b_3(z_r - x_r))}, \\ x_{r+1} = x_r + \dfrac{\theta_2(\Omega(t_r) - \Omega(w_r))(\Omega(t_r) - \Omega(z_r))(\Omega(w_r) - \Omega(z_r))(t_r - x_r)(z_r - x_r)}{\theta_3 + \theta_2\Omega(w_r)\{\Omega(t_r)(t_r - x_r)(\Omega(t_r) - \Omega(w_r)) - \Omega(z_r)(\Omega(w_r) - \Omega(z_r))(x_r - z_r)\}}, \end{cases} \quad (26)$$

where $b_1, b_2, b_3 \in \mathbb{R}$, provided $b_2 + b_3 \neq 0$. Let us consider $b_1 = b_2 = 1$ and $b_3 = 2$ in the above scheme, recalled by $(OM1)$.

2. Again, we consider another optimal 8-order scheme presented by Behl and Motsa in [11]. In this way, we obtain another new optimal family of 16-order methods, which is given by

$$\begin{cases} w_r = x_r - \dfrac{\Omega(x_r)}{\Omega'(x_r)}, \\ z_n = w_r - \dfrac{\Omega(x_r)\Omega(w_r)}{\Omega'(x_r)(\Omega(x_r) - 2\Omega(w_r))}, \\ t_r = z_r - \dfrac{\Omega(x_r)\Omega(w_r)\Omega(z_r)\left(1 - \dfrac{\Omega(w_r)}{2\Omega(x_r)} - \dfrac{b\Omega(x_r)(\Omega(w_r) + 4\Omega(z_r))}{2(2\Omega(w_r) - \Omega(x_r))(b\Omega(x_r) - \Omega(z_r))}\right)}{\Omega'(x_r)(-2\Omega(w_r) + \Omega(x_r))(\Omega(w_r) - \Omega(z_r))}, \\ x_{r+1} = x_r + \dfrac{\theta_2(\Omega(t_r) - \Omega(w_r))(\Omega(t_r) - \Omega(z_r))(\Omega(w_r) - \Omega(z_r))(t_r - x_r)(z_r - x_r)}{\theta_3 + \theta_2\Omega(w_r)\{\Omega(t_r)(t_r - x_r)(\Omega(t_r) - \Omega(w_r)) - \Omega(z_r)(\Omega(w_r) - \Omega(z_r))(x_r - z_r)\}}, \end{cases} \quad (27)$$

where $b \in \mathbb{R}$. We chose $b = -\frac{1}{2}$ in this expression, called by $(OM2)$.

3. Let us choose one more optimal 8-order scheme proposed by Džunić and Petković [15]. Therefore, we have

$$\begin{cases} w_r = x_r - \dfrac{\Omega(x_r)}{\Omega'(x_r)}, \\ z_n = w_r - \dfrac{\Omega(w_r)}{\Omega'(x_r)} \left[\dfrac{\Omega(x_r)}{\Omega(x_r) - 2\Omega(w_r)} \right], \\ t_r = z_r + \dfrac{\Omega(x_r)\Omega(z_r)(\Omega(x_r) + 2\Omega(z_r))(\Omega(w_r) + \Omega(z_r))}{\Omega'(x_r)\Omega(w_r)(2\Omega(x_r)\Omega(w_r) - \Omega(x_r)^2 + \Omega(w_r)^2)}, \\ x_{r+1} = x_r + \dfrac{\theta_2(\Omega(t_r) - \Omega(w_r))(\Omega(t_r) - \Omega(z_r))(\Omega(w_r) - \Omega(z_r))(t_r - x_r)(z_r - x_r)}{\theta_3 + \theta_2\Omega(w_r)\{\Omega(t_r)(t_r - x_r)(\Omega(t_r) - \Omega(w_r)) - \Omega(z_r)(\Omega(w_r) - \Omega(z_r))(x_r - z_r)\}}. \end{cases} \quad (28)$$

Let us call the above scheme by $(OM3)$.

4. Now, we pick another optimal family of eighth-order iterative methods given by Bi et al. in [12]. By adopting this scheme, we further have

$$\begin{cases} w_r = x_r - \dfrac{\Omega(x_r)}{\Omega'(x_r)}, \\ z_r = w_r - \dfrac{\Omega(w_r)}{\Omega'(x_r)} \left[\dfrac{2\Omega(x_r) - \Omega(w_r)}{2\Omega(x_r) - 5\Omega(w_r)} \right], \\ t_r = z_r - \dfrac{\Omega(x_r) + (\alpha + 2)\Omega(z_r)}{\Omega(x_r) + \alpha\Omega(z_r)} \left[\dfrac{\Omega(z_r)}{\Omega[z_r, w_r] + \dfrac{\Omega[z_r, x_r] - \Omega'(x_r)}{z_r - x_r}(z_r - w_r)} \right], \\ x_{r+1} = x_r + \dfrac{\theta_2(\Omega(t_r) - \Omega(w_r))(\Omega(t_r) - \Omega(z_r))(\Omega(w_r) - \Omega(z_r))(t_r - x_r)(z_r - x_r)}{\theta_3 + \theta_2\Omega(w_r)\{\Omega(t_r)(t_r - x_r)(\Omega(t_r) - \Omega(w_r)) - \Omega(z_r)(\Omega(w_r) - \Omega(z_r))(x_r - z_r)\}}, \end{cases} \quad (29)$$

where $\alpha \in \mathbb{R}$ and $\Omega[\cdot,\cdot]$ is finite difference of first order. Let us consider $\alpha = 1$ in the above scheme, denoted by $(OM4)$.

In similar fashion, we can develop several new and interesting optimal sixteenth-order schemes by considering any optimal eighth-order scheme from the literature whose first sub-step employs the classical Newton's method.

3. Numerical Experiments

This section is dedicated to examining the convergence behavior of particular methods which are mentioned in the Special Cases section. Therefore, we shall consider some standard test functions, which are given as follows:

$\Omega_1(x) = 10x \exp(-x^2) - 1;$ [11] $\zeta = 1.679630610428449940674920$
$\Omega_2(z) = x^5 + x^4 + 4x^2 - 15;$ [16] $\zeta = 1.347428098968304981506715$
$\Omega_3(x) = x^4 + \sin\left(\frac{\pi}{x^2}\right) - 5;$ [18] $\zeta = \sqrt{2}$
$\Omega_4(x) = \exp(-x^2 + x + 2) + x^3 - \cos(x+1) + 1;$ [20] $\zeta = -1$
$\Omega_5(x) = \cos\left(x^2 - 2x + \frac{16}{9}\right) - \log\left(x^2 - 2x + \frac{25}{9}\right) - 1;$ [4] $\zeta = 1 + \frac{\sqrt{7}}{3}i$
$\Omega_6(x) = \sin^{-1}(x^2 - 1) - \frac{x}{2} + 1;$ [12] $\zeta = 0.594810968398369177 5226562$
$\Omega_7(x) = x^3 + \log(x+1);$ [6] $\zeta = 0$
$\Omega_8(x) = \tan^{-1}(x) - x + 1;$ [2] $\zeta = 2.132267725272885131625421$

Here, we confirm the theoretical results of the earlier sections on the basis of gained results $\left|\frac{x_{r+1} - x_r}{(x_r - x_{r-1})^{16}}\right|$ and computational convergence order. We displayed the number of iteration indexes (n), approximated zeros (x_r), absolute residual error of the corresponding function $(|\Omega(x_r)|)$, error in the consecutive iterations $|x_{r+1} - x_r|$, $\left|\frac{x_{r+1} - x_r}{(x_r - x_{r-1})^{16}}\right|$, the asymptotic error constant $\eta = \lim_{n \to \infty} \left|\frac{x_{r+1} - x_r}{(x_r - x_{r-1})^{16}}\right|$ and the computational convergence order (ρ) in Table 1. To calculate (ρ), we adopt the following method

$$\rho = \left|\frac{(x_{r+1} - x_r)/\eta}{(x_r - x_{r-1})}\right|, \ n = 1, 2, 3.$$

We calculate (ρ), asymptotic error term and other remaining parameters up to a high number of significant digits (minimum 1000 significant digits) to reduce the rounding-off error. However, due to the restricted paper capacity, we depicted the values of x_r and ρ up to 25 and 5 significant figures, respectively. Additionally, we mentioned $\left|\frac{x_{r+1} - x_r}{(x_r - x_{r-1})^{16}}\right|$ and η by 10 significant figures. In addition to this, the absolute residual error in the function $|\Omega(x_r)|$ and error in the consecutive iterations $|x_{r+1} - x_r|$ are depicted up to 2 significant digits with exponent power that can be seen in Tables 1–3.

Furthermore, the estimated zeros by 25 significant figures are also mentioned in Table 1.

Now, we compare our 16-order methods with optimal 16-order families of iterative schemes that were proposed by Sharma et al. [7], Geum and Kim [3,4] and Ullah et al. [8]. Among these schemes, we pick the iterative methods namely expression (29), expression (Y1) (for more detail please see Table 1 of Geum and Kim [3]) and expression (K2) (please have look at Table 1 of Geum and Kim [4] for more details) and expression (9), respectively called by SM, $GK1$, $GK2$ and MM. The numbering and titles of the methods (used for comparisons) are taken from their original research papers.

Table 1. Convergence behavior of methods $OM1$, $OM2$, $OM3$ and $OM4$ on $\Omega_1(x)$–$\Omega_8(x)$.

Cases	$\Omega(x)$	n	x_r	$|\Omega(x_r)|$	$|x_{r+1} - x_r|$	$\left|\frac{x_{r+1} - x_r}{(x_r - x_{r-1})^p}\right|$	η	ρ
$OM1$	Ω_1	0	1.7	5.5(−2)	2.0(−2)			
		1	1.679630610428449940674920	1.0(−26)	3.8(−27)	4.299293162	3.402712013	15.940
		2	1.679630610428449940674920	1.6(−422)	5.8(−423)	3.402712013		16.000
$OM1$	Ω_2	0	1.1	7.1	2.5(−1)			
		1	1.347428099532342545074013	2.1(−8)	5.6(−10)	2.858404704	0.6398089109	14.928
		2	1.347428098968304981506715	2.5(−147)	6.7(−149)	0.6398089109		16.000
$OM2$	Ω_3	0	1.3	1.2	1.1(−1)			
		1	1.414213562373095736525797	7.8(−15)	6.9(−16)	0.8202219879	0.1956950645	15.340
		2	1.414213562373095048801689	5.5(−243)	4.9(−244)	0.1956950645		16.000
$OM2$	Ω_4	0	−0.7	1.9	3.0(−1)			
		1	−1.000000000000007093884377	4.3(−14)	7.1(−15)	1.647949998(−6)	6.821618098(−6)	17.180
		2	−1.000000000000000000000000	1.7(−231)	2.8(−232)	6.821618098(−6)		16.000
$OM3$	Ω_5	0	$0.9 + 0.8i$	2.2(−1)	1.3(−1)			
		1	$0.99999999\cdots + 0.88191710\ldots i$	5.3(−17)	3.0(−17)	0.004949317501	0.04805746878	17.111
		2	$1.0000000\cdots + 0.88191710\ldots i$	3.8(−266)	2.2(−266)	0.04805746878		16.000
$OM3$	Ω_6	0	0.5	9.8(−2)	9.5(−2)			
		1	0.594810968398369177522655	5.5(−25)	5.2(−25)	1.216280520(−8)	2.864980977(−8)	16.364
		2	0.594810968398369177522656	8.3(−397)	7.8(−397)	2.864980977(−8)		16.000
$OM4$	Ω_7	0	0.5	5.3(−1)	5.0(−1)			
		1	0.0000107256041067920231261691	1.1(−5)	1.1(−5)	0.7031544881	0.1352418133	13.6218
		2	4.148195228902998294111344(−81)	4.1(−81)	4.1(−81)	0.1352418133		16.000
$OM4$	Ω_8	0	2.2	5.6(−2)	6.8(−2)			
		1	2.132267725272885131625421	6.7(−31)	8.1(−31)	4.147660854(−12)	6.197625624(−12)	16.1492
		2	2.132267725272885131625421	1.9(−493)	2.3(−493)	6.197625624(−12)		16.000

It is straightforward to say that our proposed methods not only converge very fast towards the required zero, but they have also small asymptotic error constant.

Table 2. Comparison of residual error on the test examples $\Omega_9(x)$–$\Omega_{12}(x)$.

$\Omega(x)$	$|\Omega(x_r)|$	SM	GK1	GK2	MM	OM1	OM2	OM3	OM4
Ω_9	$|\Omega(x_0)|$	2.5(−1)	2.5(−1) *	2.5(−1) #	2.5(−1)	2.5(−1)	2.5(−1)	2.5(−1)	2.5(−1)
	$|\Omega(x_1)|$	3.7(−7)	2.8 *	7.6(−2) #	1.7(−4)	8.1(−7)	9.3(−8)	1.2(−7)	2.1(−7)
	$|\Omega(x_2)|$	8.1(−105)	2.0(−4) *	1.9(−19) #	6.1(−62)	9.3(−99)	7.7(−116)	3.9(−112)	6.5(−110)
Ω_{10}	$|\Omega(x_0)|$	1.0	1.0	1.0	1.0	1.0	1.0	1.0	1.0
	$|\Omega(x_1)|$	7.9(−14)	7.6(−13)	3.0(−12)	4.4(−12)	4.1(−14)	2.7(−14)	3.4(−14)	2.1(−13)
	$|\Omega(x_2)|$	7.3(−224)	5.3(−209)	2.1(−199)	1.1(−194)	4.0(−227)	3.7(−231)	2.1(−231)	3.1(−216)
Ω_{11}	$|\Omega(x_0)|$	3.8(−1)	3.8(−1)	3.8(−1)	3.8(−1)	3.8(−1)	3.8(1−)	3.8(−1)	3.8(−1)
	$|\Omega(x_1)|$	3.5(−12)	1.3(−12)	1.6(−10)	7.4(−11)	1.5(−12)	1.1(−12)	1.9(−12)	5.9(−13)
	$|\Omega(x_2)|$	2.1(−186)	1.9(−193)	1.3(−156)	1.4(−163)	1.5(−193)	2.4(−195)	7.6(−192)	2.7(−199)
Ω_{12}	$|\Omega(x_0)|$	1.4(−1)	1.4(−1)	1.4(−1)	1.4(−1)	1.4(−1)	1.4(−1)	1.4(−1)	1.4(−1)
	$|\Omega(x_1)|$	2.3(−11)	8.6(−5)	1.7(−4)	2.8(−7)	5.4(−11)	8.4(−12)	1.3(−11)	1.5(−11)
	$|\Omega(x_2)|$	1.4(−170)	1.2(−63)	9.3(−57)	1.0(−109)	5.5(−164)	2.6(−178)	2.2(−174)	3.1(−174)
Ω_{13}	$|\Omega(x_0)|$	1.9(+1)	1.9(+1)	1.9(+1)	1.9(+1)	1.9(+1)	1.9(+1)	1.9(+1)	1.9(+1)
	$|\Omega(x_1)|$	1.7(−26)	8.2(−26)	6.0(−23)	1.3(−28)	4.5(−26)	7.9(−28)	2.9(−26)	1.2(−29)
	$|\Omega(x_2)|$	2.8(−459)	1.3(−446)	1.3(−398)	6.7(−501)	5.4(−452)	4.8(−484)	3.8(−455)	8.2(−517)

* and # stand for converge to undesired roots −1.89549... and 0, respectively.

Table 3. Comparison of error in the consecutive iterations on the test examples $\Omega_9(x)$–$\Omega_{13}(x)$.

$\Omega(x)$	$|x_{r+1} - x_r|$	SM	GK1	GK2	MM	OM1	OM2	OM3	OM4
Ω_9	$|x_1 - x_0|$	4.0(−1)	5.5*	1.7#	4.0(−1)	4.0(−1)	4.0(−1)	4.0(−1)	4.0(−1)
	$|x_2 - x_1|$	4.5(−7)	2.1*	1.5(−1)#	2.1(−4)	9.9(−7)	1.1(−7)	1.5(−7)	2.6(−7)
	$|x_3 - x_2|$	9.9(−105)	2.4(−4)*	3.8(−19)#	7.4(−62)	1.1(−98)	9.4(−116)	4.8(−112)	8.0(−110)
Ω_{10}	$|x_1 - x_0|$	1.4(−1)	1.4(−1)	1.4(−1)	1.4(−1)	1.4(−1)	1.4(−1)	1.4(−1)	1.4(−1)
	$|x_2 - x_1|$	1.2(−14)	1.2(−13)	4.6(−13)	6.7(−13)	6.3(−15)	4.1(−15)	5.2(−15)	3.2(−14)
	$|x_3 - x_2|$	1.1(−224)	8.2(−210)	3.3(−200)	1.7(−195)	6.1(−228)	5.7(−232)	3.3(−232)	4.8(−217)
Ω_{11}	$|x_1 - x_0|$	2.8(−1)	2.8(−1)	2.8(−1)	2.8(−1)	2.8(−1)	2.8(−1)	2.8(−1)	2.8(−1)
	$|x_2 - x_1|$	3.0(−12)	1.2(−12)	1.4(−10)	6.4(−11)	1.3(−12)	9.6(−13)	1.6(−12)	5.1(−13)
	$|x_3 - x_2|$	1.8(−186)	1.7(−193)	1.2(−156)	1.3(−163)	1.3(−193)	2.1(−195)	6.6(−192)	2.4(−199)
Ω_{12}	$|x_1 - x_0|$	1.7(−1)	1.7(−1)	1.7(−1)	1.7(−1)	1.7(−1)	1.7(−1)	1.7(−1)	1.7(−1)
	$|x_2 - x_1|$	2.3(−11)	8.5(−5)	1.7(−4)	2.8(−7)	5.4(−11)	8.4(−12)	1.3(−11)	1.5(−11)
	$|x_3 - x_2|$	1.4(−170)	1.2(−63)	9.3(−57)	1.0(−109)	5.5(−164)	2.6(−178)	2.2(−174)	3.1(−174)
Ω_{13}	$|x_1 - x_0|$	5.1(−1)	5.1(−1)	5.1(−1)	5.1(−1)	5.1(−1)	5.1(−1)	5.1(−1)	5.1(−1)
	$|x_2 - x_1|$	4.5(−28)	2.2(−27)	1.6(−24)	3.5(−30)	1.2(−27)	2.1(−29)	7.9(−28)	3.1(−31)
	$|x_3 - x_2|$	7.6(−461)	3.5(−448)	3.5(−400)	1.8(−502)	1.5(−453)	1.3(−485)	1.0(−456)	2.2(−518)

* and # stand for converge to undesired roots −1.89549... and 0, respectively.

We want to demonstrate that our methods perform better than the existing ones. Therefore, instead of manipulating the results by considering self-made examples or/and cherry-picking among the starting points, we assume 4 numerical examples; the first one is taken from Sharma et al. [7]; the second one is considered from Geum and Kim [3]; the third one is picked from Geum and Kim [4] and the fourth one is considered from Ullah et al. [8] with the same starting points that are mentioned in their research articles. Additionally, we want to check what the outcomes will be if we assume different numerical examples and staring guesses that are not suggested in their articles. Therefore, we assume another numerical example from Behl et al. [27]. For the detailed information of the considered examples or test functions, please see Table 4.

We have suggested two comparison tables for every test function. The first one is associated with $(|\Omega(x_r)|)$ mentioned in Table 2. On the other hand, the second one is related to $|x_{r+1} - x_r|$ and the corresponding results are depicted in Table 3. In addition, we assume the estimated zero of considered functions in the case where exact zero is not available, i.e., corrected by 1000 significant figures to calculate $|x_r - \zeta|$. All the computations have been executed by adopting the programming package *Mathematica* 11 with multiple precision arithmetic. Finally, $b_1(\pm b_2)$ stands for $b_1 \times 10^{(\pm b_2)}$ in Tables 1–3.

Table 4. Test problems.

$\Omega(x)$	x_0	$Root(r)$
$\Omega_9(x) = \sin x - \frac{x}{2}$; [7]	1.5	1.895494267033980947144036
$\Omega_{10}(x) = \sin\left(\frac{2}{x}\right) + 3x^2 + e^{-x^2} - 3$; [3]	0.65	0.792938431630179374167859
$\Omega_{11}(x) = e^{-x}\cos(3x) + x - 2$; [4]	1.6	1.878179124117988404113719
$\Omega_{12}(x) = e^{-x} - \cos x$; [8]	$\frac{1}{6}$	0
$\Omega_{13}(z) = (x-2)^2 - \log x - 33x$; [27]	37.5	36.989473582944669865344473

4. Conclusions

We constructed a general optimal scheme of 16-order that is suitable for every optimal 8-order iterative method/family of iterative methods provided the first sub-step employs classical Newton's method, unlike the earlier studies, where researchers suggested a high-order version or extension of certain existing methods such as Ostrowski's method or King's method [28], etc. This means that we can choose any iterative method/family of methods from [5,11–21], etc. to obtain further optimal 16-order scheme. The construction of the presented technique is based on the inverse interpolatory approach. Our scheme also satisfies the conjecture of optimality of iterative methods given by Kung-Traub. In addition, we compare our methods with the existing methods with same convergence order on several of the nonlinear scalar problems. The obtained results in Tables 2 and 3 also illustrate the superiority of our methods to the existing methods, despite choosing the same test problem and same initial guess. Tables 1–3 confirm that smaller $|\Omega(x_r)|$, $|x_{r+1} - x_r|$ and simple asymptotic error terms are related to our iterative methods. The superiority of our methods over the existing robust methods may be due to the inherent structure of our technique with simple asymptotic error constants and inverse interpolatory approach.

Author Contributions: Both authors have equal contribution.

Funding: This research received no external funding

Conflicts of Interest: The authors declare no conflict of interest.

References

1. Traub, J.F. *Iterative Methods for the Solution of Equations*; Prentice-Hall: Englewood Cliffs, NJ, USA, 1964.
2. Petković, M.S.; Neta, B.; Petković, L.D.; Džunić, J. *Multipoint Methods for Solving Nonlinear Equations*; Academic Press: New York, NY, USA, 2012.
3. Geum, Y.H.; Kim, Y.I. A family of optimal sixteenth-order multipoint methods with a linear fraction plus a trivariate polynomial as the fourth-step weighting function. *Comput. Math. Appl.* **2011**, *61*, 3278–3287. [CrossRef]
4. Geum, Y.H.; Kim, Y.I. A biparametric family of optimally convergent sixteenth-order multipoint methods with their fourth-step weighting function as a sum of a rational and a generic two-variable function. *J. Comput. Appl. Math.* **2011**, *235*, 3178–3188. [CrossRef]
5. Kung, H.T.; Traub, J.F. Optimal order of one-point and multi-point iteration. *J. ACM* **1974**, *21*, 643–651. [CrossRef]
6. Neta, B. On a family of multipoint methods for non-linear equations. *Int. J. Comput. Math.* **1981**, *9*, 353–361. [CrossRef]
7. Sharma, J.R.; Guha, R.K.; Gupta, P. Improved King's methods with optimal order of convergence based on rational approximations. *Appl. Math. Lett.* **2013**, *26*, 473–480. [CrossRef]
8. Ullah, M.Z.; Al-Fhaid, A.S.; Ahmad, F. Four-Point Optimal Sixteenth-Order Iterative Method for Solving Nonlinear Equations. *J. Appl. Math.* **2013**, *2013*, 850365. [CrossRef]
9. Sharifi, S.; Salimi, M.; Siegmund, S.; Lotfi, T. A new class of optimal four-point methods with convergence order 16 for solving nonlinear equations. *Math. Comput. Simul.* **2016**, *119*, 69–90. [CrossRef]
10. Behl, R.; Amat, S.; Magreñán, Á.A.; Motsa, S.S. An efficient optimal family of sixteen order methods for nonlinear models. *J. Comput. Appl. Math.* **2019**, *354*, 271–285. [CrossRef]
11. Behl, R.; Motsa, S.S. Geometric construction of eighth-order optimal families of ostrowski's method. *Sci. World J.* **2015**, *2015*, 11. [CrossRef] [PubMed]
12. Bi, W.; Ren, H.; Wu, Q. Three-step iterative methods with eighth-order convergence for solving nonlinear equations. *J. Comput. Appl. Math.* **2009**, *255*, 105–112. [CrossRef]
13. Cordero, A.; Torregrosa, J.R.; Vassileva, M.P. Three-step iterative methods with optimal eighth-order convergence. *J. Comput. Appl. Math.* **2011**, *235*, 3189–3194. [CrossRef]
14. Cordero, A.; Hueso, J.L.; Martínez, E.; Torregrosa, J.R. New modifications of Potra-Pták's method with optimal fourth and eighth order of convergence. *J. Comput. Appl. Math.* **2010**, *234*, 2969–2976. [CrossRef]
15. Džuníc, J.; Petkovíc, M.S. A family of three point methods of Ostrowski's type for solving nonlinear equations. *J. Appl. Math.* **2012**, *2012*, 425867. [CrossRef]
16. Liu, L.; Wang, X. Eighth-order methods with high efficiency index for solving nonlinear equations. *J. Comput. Appl. Math.* **2010**, *215*, 3449–3454. [CrossRef]
17. Sharma, J.R.; Sharma, R. A new family of modified Ostrowski's methods with accelerated eighth order convergence. *Numer. Algorithms* **2010**, *54*, 445–458. [CrossRef]

18. Soleymani, F.; Vanani, S.K.; Khan, M.; Sharifi, M. Some modifications of King's family with optimal eighth-order of convergence. *Math. Comput. Model.* **2012**, *55*, 1373–1380. [CrossRef]
19. Soleymani, F.; Sharifi, M.; Mousavi, B.S. An improvement of Ostrowski's and King's techniques with optimal convergence order eight. *J. Optim. Theory Appl.* **2012**, *153*, 225–236. [CrossRef]
20. Thukral, R.; Petkovíc, M.S. A family of three point methods of optimal order for solving nonlinear equations. *J. Comput. Appl. Math.* **2010**, *233*, 2278–2284. [CrossRef]
21. Wang, X.; Liu, L. Modified Ostrowski's method with eighth-order convergence and high efficiency index. *Appl. Math. Lett.* **2010**, *23*, 549–554. [CrossRef]
22. Salimi, M.; Lotfi, T.; Sharifi, S.; Siegmund, S. Optimal Newton-Secant like methods without memory for solving nonlinear equations with its dynamics. *Int. J. Comput. Math.* **2017**, *94*, 1759–1777. [CrossRef]
23. Salimi, M.; Long, N.M.A.N.; Sharifi, S.; Pansera, B.A. A multi-point iterative method for solving nonlinear equations with optimal order of convergence. *Jpn. J. Ind. Appl. Math.* **2018** *35*, 497–509. [CrossRef]
24. Sharifi, S.; Ferrara, M.; Salimi, M.; Siegmund, S. New modification of Maheshwari method with optimal eighth order of convergence for solving nonlinear equations. *Open Math.* **2016** *14*, 443–451.
25. Lotfi, T.; Sharifi, S.; Salimi, M.; Siegmund, S. A new class of three point methods with optimal convergence order eight and its dynamics. *Numer. Algorithms* **2016**, *68*, 261–288. [CrossRef]
26. Apostol, T.M. *Mathematical Analysis*; Addison-Wesley Publishing Company, Inc.; Boston, USA. 1974.
27. Behl, R.; Cordero, A.; Motsa, S.S.; Torregrosa, J.R. Construction of fourth-order optimal families of iterative methods and their dynamics. *Appl. Math. Comput.* **2015**, *271*, 89–101. [CrossRef]
28. King, R.F. A family of fourth order methods for nonlinear equations. *SIAM J. Numer. Anal.* **1973**, *10*, 876–879. [CrossRef]

© 2019 by the authors. Licensee MDPI, Basel, Switzerland. This article is an open access article distributed under the terms and conditions of the Creative Commons Attribution (CC BY) license (http://creativecommons.org/licenses/by/4.0/).

Article

Modified Optimal Class of Newton-Like Fourth-Order Methods for Multiple Roots

Munish Kansal [1], Ramandeep Behl [2,*], Mohammed Ali A. Mahnashi [2] and Fouad Othman Mallawi [2]

[1] School of Mathematics, Thapar Institute of Engineering and Technology, Patiala 147004, India; munish.kansal@thapar.edu
[2] Department of Mathematics, King Abdulaziz University, Jeddah 21589, Saudi Arabia; love_maam@hotmail.com (M.A.A.M.); fmallawi@hotmail.com (F.O.M.)
* Correspondence: ramanbehl87@yahoo.in

Received: 24 February 2019; Accepted: 1 April 2019; Published: 11 April 2019

Abstract: Here, we propose optimal fourth-order iterative methods for approximating multiple zeros of univariate functions. The proposed family is composed of two stages and requires 3 functional values at each iteration. We also suggest an extensive convergence analysis that demonstrated the establishment of fourth-order convergence of the developed methods. It is interesting to note that some existing schemes are found to be the special cases of our proposed scheme. Numerical experiments have been performed on a good number of problems arising from different disciplines such as the fractional conversion problem of a chemical reactor, continuous stirred tank reactor problem, and Planck's radiation law problem. Computational results demonstrates that suggested methods are better and efficient than their existing counterparts.

Keywords: Multiple roots; Optimal iterative methods; Scalar equations; Order of convergence

1. Introduction

Importance of solving nonlinear problems is justified by numerous physical and technical applications over the past decades. These problems arise in many areas of science and engineering. The analytical solutions for such problems are not easily available. Therefore, several numerical techniques are used to obtain approximate solutions. When we discuss about iterative solvers for obtaining multiple roots with known multiplicity $m \geq 1$ of scalar equations of the type $g(x) = 0$, where $g : D \subseteq \mathbb{R} \to \mathbb{R}$, modified Newton's technique [1,2] (also known as Rall's method) is the most popular and classical iterative scheme, which is defined by

$$x_{s+1} = x_s - m \frac{g(x_s)}{g'(x_s)}, \quad s = 0, 1, 2, \ldots. \tag{1}$$

Given the multiplicity $m \geq 1$ in advance, it converges quadratically for multiple roots. However, modified Newton's method would fail miserably if the initial estimate x_0 is either far away from the required root or the value of the first-order derivative is very small in the neighborhood of the needed root. In order to overcome this problem, Kanwar et al. [3] considered the following one-point iterative technique

$$x_{s+1} = x_s - m \frac{g(x_s)}{g'(x_s) - \lambda g(x_s)}. \tag{2}$$

One can find the classical Newton's formula for $\lambda = 0$ and $m = 1$ in (2). The method (2) satisfies the following error equation:

$$e_{s+1} = \left(\frac{c_1 - \lambda}{m}\right) e_s^2 + O(e_s^3), \tag{3}$$

where $e_s = x_s - \alpha$, $c_j = \dfrac{m!}{(m+j)!} \dfrac{g^{(m+j)}(\alpha)}{g^{(m)}(\alpha)}$, $j = 1, 2, 3, \ldots$. Here, α is a multiple root of $g(x) = 0$ having multiplicity m.

One-point methods are not of practical interest because of their theoretical limitations regarding convergence order and efficiency index. Therefore, multipoint iterative functions are better applicants to certify as efficient solvers. The good thing with multipoint iterative methods without memory for scalar equations is that they have a conjecture related to order of convergence (for more information please have a look at the conjecture [2]). A large community of researchers from the world wide turn towards the most prime class of multipoint iterative methods and proposed various optimal fourth-order methods (they are requiring 3 functional values at each iteration) [4–10] and non-optimal methods [11,12] for approximating multiple zeros of nonlinear functions.

In 2013, Zhou et al. [13], presented a family of 4-order optimal iterative methods, defined as follows:

$$\begin{cases} w_s = x_s - m \dfrac{g(x_s)}{g'(x_s)}, \\ x_{s+1} = w_s - m \dfrac{g(x_s)}{g'(x_s)} Q(u_s), \end{cases} \quad (4)$$

where $u_s = \left(\dfrac{g(w_s)}{g(x_s)} \right)^{\frac{1}{m}}$ and $Q : \mathbb{C} \to \mathbb{C}$ is a weight function. The above family (4) requires two functions and one derivative evaluation per full iteration.

Lee et al. in [14], suggested an optimal 4-order scheme, which is given by

$$\begin{cases} w_s = x_s - m \dfrac{g(x_s)}{g'(x_s) + \lambda g(x_s)}, \\ x_{s+1} = x_s - m H_g(u_s) \dfrac{g(x_s)}{g'(x_s) + 2\lambda g(x_s)}, \end{cases} \quad (5)$$

where $u_s = \left(\dfrac{g(w_s)}{g(x_s)} \right)^{\frac{1}{m}}$, $H_g(u_s) = \dfrac{u_s(1 + (c+2)u_s + ru_s^2)}{1 + cu_s}$, λ, c, and r are free disposable parameters.

Very recently, Zafar et al. [15] proposed another class of optimal methods for multiple zeros defined by

$$\begin{cases} w_s = x_s - m \dfrac{g(x_s)}{g'(x_s) + a_1 g(x_s)}, \\ x_{s+1} = w_s - mu_s H(u_s) \dfrac{g(x_s)}{g'(x_s) + 2a_1 g(x_s)}, \end{cases} \quad (6)$$

where $u_s = \left(\dfrac{g(w_s)}{g(x_s)} \right)^{\frac{1}{m}}$ and $a_1 \in \mathbb{R}$. It can be seen that the family (5) is a particular case of (6).

We are interested in presenting a new optimal class of parametric-based iterative methods having fourth-order convergence which exploit weight function technique for computing multiple zeros. Our proposed scheme requires only three function evaluations ($g(x_s)$, $g'(x_s)$, and $g(w_s)$) at each iteration which is in accordance with the classical Kung-Traub conjecture. It is also interesting to note that the optimal fourth-order families (5) and (6) can be considered as special cases of our scheme for some particular values of free parameters. Therefore, the new scheme can be treated as more general family for approximating multiple zeros of nonlinear functions. Furthermore, we manifest that the proposed scheme shows a good agreement with the numerical results and offers smaller residual errors in the estimation of multiple zeros.

Our presentation is unfolded in what follows. The new fourth-order scheme and its convergence analysis is presented in Section 2. In Section 3, several particular cases are included based on the different choices of weight functions employed at second step of the designed family. In addition, Section 3, is also dedicated to the numerical experiments which illustrate the efficiency and accuracy of the scheme in multi-precision arithmetic on some complicated real-life problems. Section 4, presents the conclusions.

2. Construction of the Family

Here, we suggest a new fourth-order optimal scheme for finding multiple roots having known multiplicity $m \geq 1$. So, we present the two-stage scheme as follows:

$$w_s = x_s - m \frac{g(x_s)}{g'(x_s) + \lambda_1 g(x_s)},$$
$$z_n = w_s - m u_s \frac{g(x_s)}{g'(x_s) + \lambda_2 g(x_s)} Q(t_s),$$
(7)

where $Q : \mathbb{C} \to \mathbb{C}$ is the weight function and holomorphic function in the neighborhood of origin with $u_s = \left(\frac{g(w_s)}{g(x_s)}\right)^{\frac{1}{m}}$ and $t_s = \frac{u_s}{a_1 + a_2 u_s}$ and being $\lambda_1, \lambda_2, a_1$ and a_2 are free parameters.

In the following Theorem 1, we illustrate that how to construct weight function Q so that it arrives at fourth-order without consuming any extra functional values.

Theorem 1. *Let us assume that $g : \mathbb{C} \to \mathbb{C}$ is holomorphic function in the region containing the multiple zero $x = \alpha$ with multiplicity $m \geq 1$. Then, for a given initial guess x_0, the iterative expression (7) reaches 4-order convergence when it satisfies*

$$Q(0) = 1, \quad Q'(0) = 2a_1, \quad \lambda_2 = 2\lambda_1, \quad \text{and} \quad |Q''(0)| < \infty.$$
(8)

Proof. Let us assume that $x = \alpha$ is a multiple zero having known multiplicity $m \geq 1$ of $g(x)$. Adopting Taylor's series expansion of $g(x_s)$ and $g'(x_s)$ about α, we obtain

$$g(x_s) = \frac{g^{(m)}(\alpha)}{m!} e_s^m \left(1 + c_1 e_s + c_2 e_s^2 + c_3 e_s^3 + c_4 e_s^4 + O(e_s^5)\right)$$
(9)

and

$$g'(x_s) = \frac{g^m(\alpha)}{m!} e_s^{m-1} \left(m + c_1(m+1)e_s + c_2(m+2)e_s^2 + c_3(m+3)e_s^3 + c_4(m+4)e_s^4 + O(e_s^5)\right),$$
(10)

respectively. Here, $e_s = x_s - \alpha$ and $c_j = \frac{m!}{(m+j)!} \frac{g^{(m+j)}(\alpha)}{g^{(m)}(\alpha)}, j = 1, 2, 3, \ldots$.

From the Equations (9) and (10), we obtain

$$\frac{g(x_s)}{g'(x_s) + \lambda_1 g(x_s)} = \frac{e_s}{m} + \frac{(-\lambda_1 - c_1)}{m^2} e_s^2 + \frac{\left(c_1^2 + mc_1^2 - 2mc_2 + 2c_1 \lambda_1 + \lambda_1^2\right) e_s^3}{m^3} + \frac{L_1}{m^4} e_s^4 + O(e_s^5),$$
(11)

where $L_1 = \left(-c_1^3 - 2mc_1^3 - m^2 c_1^3 + 4mc_1 c_2 + 3m^2 c_1 c_2 - 3m^2 c_3 - 3c_1^2 \lambda_1 - 2mc_1^2 \lambda_1 + 4mc_2 \lambda_1 - 3c_1 \lambda_1^2 - \lambda_1^3\right)$.

Now, substituting (11) in the first substep of scheme (7), we get

$$w_s - \alpha = \frac{(\lambda_1 + c_1)}{m^2} e_s^2 - \frac{\left(c_1^2 + mc_1^2 - 2mc_2 + 2c_1 \lambda_1 + \lambda_1^2\right) e_s^3}{m^2} + \frac{L_1}{m^3} e_s^4 + O(e_s^5).$$
(12)

Using again Taylor's series, we yield

$$g(w_s) = g^{(m)}(\alpha) e_s^{2m} \left[\frac{\left(\frac{\lambda_1 + c_1}{m}\right)^m}{m!} - \frac{\left(\frac{c_1 + \lambda_1}{m}\right)^m \left((1+m)c_1^2 - 2mc_2 + 2c_1 \lambda_1 + \lambda_1^2\right) e_s}{m!(c_1 + \lambda_1)} + \frac{1}{m!} \left(\frac{c_1(c_1 + \lambda_1)\left(\frac{\lambda_1 + c_1}{m}\right)^m}{m} \right. \right.$$
$$\left. \left. \left(\frac{\lambda_1 + c_1}{m}\right)^m \frac{B_1}{2m(c_1 + \lambda_1)^2} + \frac{B_2}{m(c_1 + \lambda_1)} \right) e_s^2 + O(e_s^3) \right],$$
(13)

where

$$B_1 = (-1+m)\left((1+m)c_1^2 - 2mc_2 + 2c_1\lambda_1 + \lambda_1^2\right)^2,$$
$$B_2 = (1+m)^2 c_1^3 + 3m^2 c_3 + (3+2m)c_1^2 \lambda_1 - 4mc_2\lambda_1 + \lambda_1^3 + c_1\left(-m(4+3m)c_2 + 3\lambda_1^2\right).$$
(14)

Moreover,

$$u_s = \frac{(c_1 + \lambda_1)e_s}{m} - \frac{\left((2+m)c_1^2 - 2mc_2 + 3c_1\lambda_1 + \lambda_1^2\right)e_s^2}{m^2} + \frac{\gamma_1 e_s^3}{2m^3} + O(e_s^4),$$
(15)

where

$$\gamma_1 = \left(\left(7 + 7m + 2m^2\right)c_1^3 + 5(3+m)c_1^2\lambda_1 - 2c_1\left(m(7+3m)c_2 - 5\lambda_1^2\right) + 2\left(3m^2 c_3 - 5mc_2\lambda_1 + \lambda_1^3\right)\right).$$

Now, using the above expression (15), we get

$$t_s = \frac{(c_1 + \lambda_1)}{ma_1}e_s + \sum_{i=1}^{2} \Theta_j e_s^{j+1} + O(e_s^4).$$
(16)

where $\Theta_j = \Theta_j(a_1, a_2, m, c_1, c_2, c_3, c_4)$.

Due to the fact that $t_s = \dfrac{u_s}{a_1 + a_2 u_s} = O(e_s)$, therefore, it suffices to expand weight function $Q(t_s)$ around the origin by Taylor's series expansion up to 3-order term as follows:

$$Q(t_s) \approx Q(0) + Q'(0)t_s + \frac{1}{2!}Q''(0)t_s^2 + \frac{1}{3!}Q^{(3)}(0)t_s^3,$$
(17)

where $Q^{(k)}$ represents the k-th derivative.

Adopting the expressions (9)–(17) in (7), we have

$$e_{s+1} = \frac{\Omega_1}{m}e_s^2 + \frac{\Omega_2}{ma_1^2}e_s^3 + \frac{\Omega_3}{2m^3 a_1^2}e_s^4 + O(e_s^5).$$
(18)

where

$$\begin{aligned}\Omega_1 &= (-1 + Q(0))(c_1 + \lambda_1),\\ \Omega_2 &= (-a_1(1+m) + a_1(3+m)Q(0) - Q'(0))c_1^2 - 2a_1 m(-1+Q(0))c_2 \\ &\quad + c_1\left((-2a_1 + 4a_1 Q(0) - 2Q'(0))\lambda_1 + a_1 Q(0)\lambda_2\right) + \lambda_1\left((a_1(-1+Q(0))\right.\\ &\quad \left. -Q'(0))\lambda_1 + a_1 Q(0)\lambda_2\right).\end{aligned}$$
(19)

It is clear from error Equation (18) that in order to have at least 4-order convergence. The coefficients of e_s^2 and e_s^3 must vanish simultaneously. Therefore, inserting $Q(0) = 1$ in (19), we have

$$\Omega_2 = (2a_1 - Q'(0))c_1 - Q'(0)\lambda_1 + a_1 \lambda_2.$$
(20)

Similarly, $\Omega_2 = 0$ implies that $Q'(0) = 2a_1$ and $\lambda_2 = 2\lambda_1$.

Finally, using Equations (19) and (20) in the proposed scheme (7), we have

$$e_{s+1} = \frac{(c_1 + \lambda_1)\left((4a_1 a_2 + a_1^2(9+m) - Q''(0))c_1^2 - 2a_1^2 mc_2 + 2\left(7a_1^2 + 4a_1 a_2 - Q''(0)\right)c_1\lambda_1 + (4a_1(a_1 + a_2) - Q''(0))\lambda_1^2\right)e_s^4}{2a_1^2 m^3}$$
$$+ O(e_s^5).$$
(21)

The consequence of the above error analysis is that the family (7) acquires 4-order convergence by consuming only 3 functional values (viz. $g(x_s)$, $g'(x_s)$, and $g(w_s)$) per full iteration. Hence, the proof is completed. □

Some Particular Cases of the Suggested Class

We suggest some interesting particulars cases of (7) by choosing different forms of weight function $Q(t_s)$ that satisfy the constrains of Theorem 1.

Let us assume the following optimal class of fourth-order methods by choosing weight function directly from the Theorem 1:

$$w_s = x_s - m\frac{g(x_s)}{g'(x_s) + \lambda_1 g(x_s)},$$
$$x_{s+1} = w_s - mu_s \frac{g(x_s)}{g'(x_s) + 2\lambda_1 g(x_s)} \left[1 + 2a_1 t_s + \frac{1}{2}t_s^2 Q''(0) + \frac{1}{3!}t_s^3 Q^{(3)}(0)\right], \qquad (22)$$

where $t_s = \frac{u_s}{a_1 + a_2 u_s}$, $\lambda_1, a_1, a_2, Q''(0)$ and $Q^{(3)}(0)$ are free disposable variables.

Sub cases of the given scheme (22):

1. We assume that $Q(t_s) = 1 + 2a_1 t_s + \frac{\mu}{2} t_s^2$, in expression (22), we obtain

$$w_s = x_s - m\frac{g(x_s)}{g'(x_s) + \lambda_1 g(x_s)},$$
$$x_{s+1} = w_s - mu_s \left[1 + 2a_1 t_s + \frac{\mu}{2} t_s^2\right] \frac{g(x_s)}{g'(x_s) + 2\lambda_1 g(x_s)}, \qquad (23)$$

where $\mu \in \mathbb{R}$.

2. Considering the weight function $Q(t_s) = 1 + \alpha_1 t_s + \alpha_2 t_s^2 + \frac{\alpha_3 t_s^2}{1 + \alpha_4 t_s + \alpha_5 t_s^2}$ in expression (22), one gets

$$w_s = x_s - m\frac{g(x_s)}{g'(x_s) + \lambda_1 g(x_s)},$$
$$x_{s+1} = w_s - mu_s \left[1 + \alpha_1 t_s + \alpha_2 t_s^2 + \frac{\alpha_3 t_s^2}{1 + \alpha_4 t_s + \alpha_5 t_s^2}\right] \frac{g(x_s)}{g'(x_s) + 2\lambda_1 g(x_s)}, \qquad (24)$$

where $\alpha_1 = 2a_1, \alpha_2, \alpha_3, \alpha_4$ and α_5 are free parameters.

Case 2A: Substituting $\alpha_2 = \alpha_3 = 1$, $\alpha_4 = 15$ and $\alpha_5 = 10$ in (24), we obtain

$$w_s = x_s - m\frac{g(x_s)}{g'(x_s) + \lambda_1 g(x_s)},$$
$$x_{s+1} = w_s - mu_s \left[1 + 2a_1 t_s + t_s^2 + \frac{t_s^2}{1 + 15t_s + 10t_s^2}\right] \frac{g(x_s)}{g'(x_s) + 2\lambda_1 g(x_s)}. \qquad (25)$$

Case 2B: Substituting $\alpha_2 = \alpha_3 = 1$, $\alpha_4 = 2$ and $\alpha_5 = 1$, in (24), we have

$$w_s = x_s - m\frac{g(x_s)}{g'(x_s) + \lambda_1 g(x_s)},$$
$$x_{s+1} = w_s - mu_s \left[1 + 2a_1 t_s + \alpha_2 t_s^2 + \frac{t_s^2}{1 + \alpha_4 t_s + t_s^2}\right] \frac{g(x_s)}{g'(x_s) + 2\lambda_1 g(x_s)}. \qquad (26)$$

Remark 1. *It is worth mentioning here that the family (6) can be captured as a special case for $a_1 = 1$ and $a_2 = 0$ in the proposed scheme (22).*

Remark 2. *Furthermore, it is worthy to record that $Q(t_s)$ weight function plays a great character in the development of fourth-order schemes. Therefore, it is customary to display different choices of weight functions, provided they must assure all the constrains of Theorem 1. Hence, we have mentioned above some special cases* of new fourth-order schemes (23), (24), (25) and (26) having simple body structures so that they can be easily implemented in the numerical experiments.

3. Numerical Experiments

Here, we verify the computational aspects of the following methods: expression (23) for ($a_1 = 1$, $a_2 = 1$, $\lambda_1 = 0$, $\mu = 13$), and expression (25) for ($a_1 = 1$, $a_2 = -1$, $\lambda_1 = 0$) denoted by (MM1) and (MM2), respectively, with some already existing techniques of the same convergence order.

In this regard, we consider several test functions coming from real life problems and linear algebra that are depicted in Examples 1–5. We make a contrast of them with existing optimal 4-order methods, namely method (6) given by Zafar et al. [15] for $H(u_s) = (1 + 2u_s + \frac{k}{2}u_s^2)$ with $k = 11$ and $a_1 = 0$ denoted by (ZM). Also, family (5) proposed by Lee et al. [14] is compared by taking $H_g(u_s) = u_s(1 + u_s)^2$ for ($c = 0$, $\lambda = \frac{m}{2}$, $r = 1$), and $H_g(u_s) = \frac{u_s(1-u_s^2)}{1-2u_s}$ for ($c = -2$, $\lambda = \frac{m}{2}$, $r = -1$). We denote these methods by (LM1) and (LM2), respectively.

We compare our iterative methods with the exiting optimal 4-order methods on the basis of x_n (approximated roots), $|g(x_s)|$ (residual error of the considered function), $|x_{s+1} - x_s|$ (absolute error between two consecutive iterations), and the estimations of asymptotic error constants according to the formula $\left|\frac{x_{s+1} - x_s}{(x_s - x_{s-1})^4}\right|$ are depicted in Tables 1–5. In order to minimize the round off errors, we have considered 4096 significant digits. The whole numerical work have been carried out with Mathematica 7 programming package. In Tables 1–5, the $k_1(\pm k_2)$ stands for $k_1 \times 10^{(\pm k_2)}$.

Example 1. We assume a 5×5 matrix, which is given by

$$A = \begin{bmatrix} 29 & 14 & 2 & 6 & -9 \\ -47 & -22 & -1 & -11 & 13 \\ 19 & 10 & 5 & 4 & -8 \\ -19 & -10 & -3 & -2 & 8 \\ 7 & 4 & 3 & 1 & -3 \end{bmatrix}.$$

We have the following characteristic equation of the above matrix:

$$g_1(x) = (x - 2)^4(x + 1).$$

It is straightforward to say that the function $g_1(x)$ has a multiple zero at $x = 2$ having four multiplicity.

The computational comparisons depicted in Table 1 illustrates that the new methods (MM1), (MM2) and (ZM) have better results in terms of precision in the calculation of the multiple zero of $g_1(x)$. On the other hands, the methods (LM1) and (LM2) fail to converge.

Example 2. (Chemical reactor problem):
We assume the following function (for more details please, see [16])

$$g_2(x) = -5 \log \left[\frac{0.4(1-x)}{0.4-0.5x}\right] + \frac{x}{1-x} + 4.45977. \quad (27)$$

The variable x serve as the fractional transformation of the specific species B in the chemical reactor. There will be no physical benefits of the above expression (27) for either $x < 0$ or $x > 1$. Therefore, we are looking for a bounded solution in the interval $0 \leq x \leq 1$ and approximated zero is $\alpha \approx 0.7573962462537538794596412979 29$.

Table 1. Convergence study of distinct iterative functions on $g_1(x)$.

Methods	n	x_s	$\|g(x_s)\|$	$\|x_{s+1} - x_s\|$	$\left\|\frac{x_{s+1}-x_s}{(x_s-x_{s-1})^4}\right\|$
LM1	0	0.5	*	*	
	1		*	*	*
	2		*	*	*
	3		*	*	*
LM2	0	0.5	*	*	
	1		*	*	*
	2		*	*	*
	3		*	*	*
ZM	0	0.5	7.6(+0)	4.3(+0)	
	1	4.772004872217151226361127	3.4(+2)	2.8(+0)	4.613629651(−2)
	2	2.012505802018268992557295	7.4(−8)	1.3(−2)	2.156703982(−4)
	3	2.000000000014282551529598	1.2(−43)	1.4(−11)	5.839284100(−4)
MM1	0	0.5	7.6(+0)	3.8(+0)	
	1	4.260708441594529218722064	1.4(+2)	2.3(+0)	8.619134726(−2)
	2	2.009364265674733970271046	2.3(−8)	9.4(−3)	3.645072355(−4)
	3	2.000000000008902251900730	1.9(−44)	8.9(−12)	1.157723834(−3)
MM2	0	0.5	7.6(+0)	4.4(+0)	
	1	4.907580957752597082443289	4.2(+2)	2.9(+0)	4.04329177(−2)
	2	2.017817158679257202528994	3.0(−7)	1.8(−2)	2.554989145(−4)
	3	2.000000000144841263588776	4.3(−39)	1.4(−10)	1.437270553(−3)

*: denotes the case of failure.

We can see that the new methods possess minimal residual errors and minimal errors difference between the consecutive approximations in comparison to the existing ones. Moreover, the numerical results of convergence order that coincide with the theoretical one in each case.

Example 3. *(Continuous stirred tank reactor (CSTR))*:
In our third example, we assume a problem of continuous stirred tank reactor (CSTR). We observed the following reaction scheme that develop in the chemical reactor (see [17] for more information):

$$\begin{aligned} K_1 + P &\rightarrow K_2 \\ K_2 + P &\rightarrow K_3 \\ K_3 + P &\rightarrow K_4 \\ K_4 + P &\rightarrow K_5, \end{aligned} \tag{28}$$

where the components T and K_1 are fed at the amount of $q\text{-}Q$ and Q, respectively, to the chemical reactor. The above model was studied in detail by Douglas [18] in order to find a good and simple system that can control feedback problem. Finally, he transferred the above model to the following mathematical expression:

$$K_H \frac{2.98(t+2.25)}{(t+1.45)(t+4.35)(t+2.85)^2} = -1, \tag{29}$$

where K_H denotes for the gaining proportional controller. The suggested control system is balanced with the values of K_H. If we assume $K_H = 0$, we obtain the poles of the open-loop transferred function as the solutions of following uni-variate equation:

$$g_3(x) = x^4 + 11.50x^3 + 47.49x^2 + 83.06325x + 51.23266875 = 0 \tag{30}$$

given as: $x = -2.85, -1.45, -4.35, -2.85$. It is straightforward to say that we have one multiple root $x = -2.85$, having known multiplicity 2. The computational results for Example 3 are displayed in Table 3.

Table 2. Convergence study of distinct iterative functions on $g_2(x)$.

Methods	n	x_s	$\|g(x_s)\|$	$\|x_{s+1} - x_s\|$	$\left\|\frac{x_{s+1}-x_s}{(x_s-x_{s-1})^4}\right\|$
LM1	0	0.76	$2.2(-1)$	$2.6(-3)$	
	1	0.7573968038178290616303393	$4.4(-5)$	$5.6(-7)$	$5.769200720(+18)$
	2	0.7573962462537538794608754	$9.8(-20)$	$1.2(-21)$	$1.27693413(+4)$
	3	0.7573962462537538794596413	$2.4(-78)$	$-3.0(-80)$	$1.277007736(+4)$
LM2	0	0.76	$2.2(-1)$	$2.6(-3)$	
	1	0.7573964149978655308754320	$1.3(-5)$	$1.7(-7)$	$2.018201446(+20)$
	2	0.7573962462537538794596446	$2.6(-22)$	$3.3(-24)$	$4.015765605(+3)$
	3	0.7573962462537538794596413	$3.6(-89)$	$4.5(-91)$	$4.15789304(+3)$
ZM	0	0.76	$2.2(-1)$	$2.6(-3)$	
	1	0.7573960528543158818682498	$1.5(-5)$	$1.9(-7)$	$1.382402073(+20)$
	2	0.7573962462537538794596326	$6.9(-22)$	$8.7(-24)$	$6.198672349(+3)$
	3	0.7573962462537538794596413	$2.8(-87)$	$3.5(-89)$	$6.198509111(+3)$
MM1	0	0.76	$2.2(-1)$	$2.6(-3)$	
	1	0.7573962756803076928764181	$2.3(-6)$	$2.9(-8)$	$3.924476992(+22)$
	2	0.7573962462537538794596413	$1.3(-25)$	$1.7(-27)$	$2.210559498(+3)$
	3	0.7573962462537538794596413	$1.3(-102)$	$1.7(-104)$	$2.210597968(+3)$
MM2	0	0.76	$2.2(-1)$	$2.6(-3)$	
	1	0.7573963165291620208634917	$5.6(-6)$	$7.0(-8)$	$2.881309229(+21)$
	2	0.7573962462537538794596413	$4.2(-25)$	$5.3(-113)$	$2.165675770(+2)$
	3	0.7573962462537538794596413	$1.3(-101)$	$1.7(-103)$	$2.166423965(+2)$

Table 3. Convergence study of distinct iterative functions on $g_3(x)$.

Methods	n	x_s	$\|g(x_s)\|$	$\|x_{s+1} - x_s\|$	$\left\|\frac{x_{s+1}-x_s}{(x_s-x_{s-1})^4}\right\|$
LM1	0	-3.0	$4.7(-2)$	$1.8(-1)$	
	1	$-2.817626610201641938500885$	$2.2(-3)$	$3.2(-2)$	$2.947357688(+4)$
	2	$-2.849999804254456528880326$	$8.0(-14)$	$2.0(-7)$	$1.782172267(-1)$
	3	$-2.850000000000000000000000$	$1.9(-55)$	$3.0(-28)$	$2.052962276(-1)$
LM2	0	-3.0	$4.7(-2)$	$1.8(-1)$	
	1	$-2.817286067962330455509242$	$2.2(-3)$	$3.3(-2)$	$2.856287648(+4)$
	2	$-2.850000137877462427734760$	$4.0(-16)$	$1.4(-8)$	$1.203820035(-2)$
	3	$-2.849999999999999818950521$	$6.9(-32)$	$1.8(-16)$	$5.009843150(+15)$
ZM	0	-3.0	$4.7(-2)$	$1.5(-1)$	
	1	$-2.847808068144375821316837$	$1.0(-5)$	$2.2(-3)$	$9.49657081(+7)$
	2	$-2.850000238882998104304060$	$1.2(-13)$	$2.4(-7)$	$1.034398150(+4)$
	3	$-2.850000000000000000000000$	$7.2(-58)$	$1.8(-29)$	$5.668907061(-3)$
MM1	0	-3.0	$4.7(-2)$	$1.5(-1)$	
	1	$-2.847808129810423347656086$	$1.0(-5)$	$2.2(-3)$	$9.497358601(+7)$
	2	$-2.850000238869272754660930$	$1.2(-13)$	$2.4(-7)$	$1.034455136(+4)$
	3	$-2.850000000000000000000000$	$7.2(-58)$	$1.9(-29)$	$5.682404331(-3)$
MM2	0	-3.0	$4.7(-2)$	$1.5(-1)$	
	1	$-2.847808157132816544128276$	$1.0(-5)$	$2.2(-3)$	$9.497713760(+7)$
	2	$-2.850000238863191571180946$	$1.2(-13)$	$2.4(-7)$	$1.034480386(+4)$
	3	$-2.850000000000000000000000$	$7.2(-58)$	$1.9(-29)$	$5.689152966(-3)$

Example 4. *We consider another uni-variate function from [14], defined as follows:*

$$g_4(x) = \left(\sin^{-1}\left(\frac{1}{x} - 1\right) + e^{x^2} - 3\right)^2.$$

The function g_4 has a multiple zero at $x = 1.05655361033535$, having known multiplicity $m = 2$.

Table 4 demonstrates the computational results for problem g_4. It can be concluded from the numerical tests that results are very good for all the methods, but lower residuals error belongs to newly proposed methods.

Table 4. Convergence study of distinct iterative functions on $g_4(x)$.

Methods	n	x_s	$|g(x_s)|$	$|x_{s+1} - x_s|$	$\left|\frac{x_{s+1}-x_s}{(x_s-x_{s-1})^4}\right|$
LM1	0	1.3	4.8(+0)	2.2(−1)	
	1	1.08451403224800705967719	2.7(−2)	2.8(−2)	4.571366396(+4)
	2	1.05657438534171491408426	1.3(−8)	2.1(−5)	3.409239504(+1)
	3	1.05655361033535490274866 7	2.0(−33)	8.1(−18)	4.373385687(+1)
LM2	0	1.3	4.8(+0)	2.3(−1)	
	1	1.06539295483200133241306 4	2.5(−3)	8.8(−3)	1.447890326(+6)
	2	1.05655369487354453280418 4	2.2(−13)	8.5(−8)	1.384807214(+1)
	3	1.05655361033535489460195 4	1.9(−53)	7.8(−28)	1.519374809(+1)
ZM	0	1.3	4.8(+0)	2.3(−1)	
	1	1.06713597931183077967712 5	3.6(−3)	1.1(−2)	8.438277939(+5)
	2	1.05655354878012151604779 0	1.2(−13)	6.2(−8)	4.908212761(+0)
	3	1.05655361033536948835807 3	6.6(−27)	1.5(−14)	1.016498504(+15)
MM1	0	1.3	4.8(+0)	2.3(−1)	
	1	1.07375319366800043877143 1	9.8(−3)	1.7(−2)	1.965340386(+5)
	2	1.05655394471263474954945 3	3.5(−12)	3.3(−7)	3.821191729(+00)
	3	1.05655361033535489460195 4	2.7(−54)	3.0(−28)	2.376288351(−2)
MM2	0	1.3	4.8(+0)	2.3(−1)	
	1	1.06912174202955052317548 2	5.1(−3)	1.3(−2)	5.037130656(+5)
	2	1.05655374390921349892295 9	5.5(−13)	1.3(−7)	5.353737131(+00)
	3	1.05655361033535489460195 4	3.9(−53)	1.1(−27)	3.547176655(+00)

Example 5. *(Planck's radiation law problem):*

Here, we chosen the well-known Planck's radiation law problem [19], that addresses the density of energy in an isothermal blackbody, which is defined as follows:

$$\Omega(\delta) = \frac{8\pi ch\delta^{-5}}{e^{\frac{ch}{\delta BT}} - 1}, \tag{31}$$

where the parameters δ, T, h and c denote as the wavelength of the radiation, absolute temperature of the blackbody, Planck's parameter and c is the light speed, respectively. In order to find the wavelength δ, then we have to calculate the maximum energy density of $\Omega(\delta)$.

In addition, the maximum value of a function exists on the critical points ($\Omega'(\delta) = 0$), then we have

$$\frac{\frac{ch}{\delta BT} e^{\frac{ch}{\delta BT}}}{e^{\frac{ch}{\delta BT}} - 1} = 5, \tag{32}$$

where B is the Boltzmann constant. If $x = \frac{ch}{\delta BT}$, then (32) is satisfied when

$$g_5(x) = \frac{x}{5} - 1 + e^{-x} = 0. \tag{33}$$

Therefore, the roots of $g_5(x) = 0$, provide the maximum wavelength of radiation δ by adopting the following technique:

$$\delta \approx \frac{ch}{\alpha BT}, \tag{34}$$

where α is a solution of (33). Our desired root is $x = 4.9651142317442$ with multiplicity $m = 1$.

The computational results for $g_5(x) = 0$, displayed in Table 5. We concluded that methods (MM1) and (MM2) have small values of residual errors in comparison to the other methods.

Table 5. Convergence study of distinct iterative functions on $g_5(x)$.

Methods	n	x_s	$|g(x_s)|$	$|x_{s+1} - x_s|$	$\left|\frac{x_{s+1}-x_s}{(x_s-x_{s-1})^4}\right|$
LM1	0	5.5	1.0(−1)	5.3(−1)	
	1	4.970872146931603546368908	1.1(−3)	5.8(−3)	5.238466809(+6)
	2	4.965114231914843999162688	3.3(−11)	1.7(−10)	1.55180113(−1)
	3	4.965114231744276303698759	2.6(−41)	1.3(−40)	1.567247236(−1)
LM2	0	5.5	1.0(−1)	5.4(−1)	
	1	4.956468415831016632868463	1.7(−3)	8.6(−3)	1.547326676(+6)
	2	4.965114231063736461886677	1.3(−10)	6.8(−10)	1.217950829(−1)
	3	4.965114231744276303698759	5.0(−39)	2.6(−38)	1.213771079(−1)
ZM	0	5.5	1.0(−1)	5.3(−1)	
	1	4.965118934170088855124237	9.1(−7)	4.7(−6)	9.616878784(−15)
	2	4.965114231744276303698759	1.0(−26)	5.2(−26)	1.059300624(−4)
	3	4.965114231744276303698759	1.5(−106)	7.6(−106)	1.059306409(−4)
MM1	0	5.5	1.0(−1)	5.3(−1)	
	1	4.965119103136732738326681	9.4(−7)	4.9(−6)	8.650488681(+15)
	2	4.965114231744276303698759	1.2(−26)	6.3(−26)	1.118336091(−4)
	3	4.965114231744276303698759	3.4(−106)	1.8(−105)	1.118342654(−4)
MM2	0	5.5	1.0(−1)	5.3(−1)	
	1	4.965119178775304742593802	9.5(−7)	4.9(−6)	8.259734679(+15)
	2	4.965114231744276303698759	1.3(−26)	6.9(−26)	1.147853783(−4)
	3	4.965114231744276303698759	4.9(−106)	2.6(−105)	1.147860777(−4)

4. Conclusions

In this study, we proposed a wide general optimal class of iterative methods for approximating multiple zeros of nonlinear functions numerically. Weight functions based on function-to-function ratios and free parameters are employed at second step of the family which enable us to achieve desired convergence order four. In the numerical section, we have incorporated variety of real life problems to confirm the efficiency of the proposed technique in comparison to the existing robust methods. The computational results demonstrates that the new methods show better performance in terms of precision and accuracy for the considered test functions. Finally, we point out that high convergence order of the proposed class, makes it not only interesting from theoretical point of view but also in practice.

Author Contributions: All the authors have equal contribution for this paper.

Funding: "This research received no external funding".

Acknowledgments: We would like to express our gratitude to the anonymous reviewers for their constructive suggestions which improved the readability of the paper.

Conflicts of Interest: The authors declare no conflict of interest.

References

1. Ostrowski, A.M. *Solution of Equations and Systems of Equations*; Academic Press: New York, NY, USA, 1960.
2. Traub, J.F. *Iterative Methods for the Solution of Equations*; Prentice-Hall: Englewood Cliffs, NJ, USA, 1964.
3. Kanwar, V.; Bhatia, S.; Kansal, M. New optimal class of higher-order methods for multiple roots, permitting $f'(x_n) = 0$. *Appl. Math. Comput.* **2013**, *222*, 564–574. [CrossRef]
4. Behl, R.; Cordero, A.; Motsa, S.S.; Torregrosa, J.R. On developing fourth-order optimal families of methods for multiple roots and their dynamics. *Appl. Math. Comput.* **2015**, *265*, 520–532. [CrossRef]
5. Behl, R.; Cordero, A.; Motsa, S.S.; Torregrosa, J.R.; Kanwar, V. An optimal fourth-order family of methods for multiple roots and its dynamics. *Numer. Algor.* **2016**, *71*, 775–796. [CrossRef]
6. Li, S.; Liao, X.; Cheng, L. A new fourth-order iterative method for finding multiple roots of nonlinear equations. *Appl. Math. Comput.* **2009**, *215*, 1288–1292.
7. Neta, B.; Chun, C.; Scott, M. On the development of iterative methods for multiple roots. *Appl. Math. Comput.* **2013**, *224*, 358–361. [CrossRef]

8. Sharma, J.R.; Sharma, R. Modified Jarratt method for computing multiple roots. *Appl. Math. Comput.* **2010**, *217*, 878–881. [CrossRef]
9. Zhou, X.; Chen, X.; Song, Y. Constructing higher-order methods for obtaining the multiple roots of nonlinear equations. *Comput. Appl. Math.* **2011**, *235*, 4199–4206. [CrossRef]
10. Kim, Y.I.; Geum, Y.H. A triparametric family of optimal fourth-order multiple-root finders and their Dynamics. *Discret. Dyn. Nat. Soc.* **2016**, *2016*, 8436759. [CrossRef]
11. Li, S.; Cheng, L.; Neta, B. Some fourth-order nonlinear solvers with closed formulae for multiple roots. *Comput. Math. Appl.* **2010**, *59*, 126–135. [CrossRef]
12. Neta, B. Extension of Murakami's high-order non-linear solver to multiple roots. *Int. J. Comput. Math.* **2010**, *87*, 1023–1031. [CrossRef]
13. Zhou, X.; Chen, X.; Song, Y. Families of third and fourth order methods for multiple roots of nonlinear equations. *Appl. Math. Comput.* **2013**, *219*, 6030–6038. [CrossRef]
14. Lee, M.Y.; Kim, Y.I.; Magrenan, A.A. On the dynamics of tri-parametric family of optimal fourth-order multiple-zero finders with a weight function of the principal mth root of a function-function ratio. *Appl. Math. Comput.* **2017**, *315*, 564–590.
15. Zafar, F.; Cordero, A.; Torregrosa, J.R. Stability analysis of a family of optimal fourth-order methods for multiple roots. *Numer. Algor.* **2018**. [CrossRef]
16. Shacham, M. Numerical solution of constrained nonlinear algebraic equations. *Int. J. Numer. Method Eng.* **1986**, *23*, 1455–1481. [CrossRef]
17. Constantinides, A., Mostoufi, N. *Numerical Methods for Chemical Engineers with MATLAB Applications*; Prentice Hall PTR: Upper Saddle River, NJ, USA, 1999.
18. Douglas, J.M. *Process Dynamics and Control*; Prentice Hall: Englewood Cliffs, NJ, USA, 1972; Volume 2.
19. Jain, D. Families of Newton-like method with fourth-order convergence. *Int. J. Comput. Math.* **2013**, *90*, 1072–1082. [CrossRef]

© 2019 by the authors. Licensee MDPI, Basel, Switzerland. This article is an open access article distributed under the terms and conditions of the Creative Commons Attribution (CC BY) license (http://creativecommons.org/licenses/by/4.0/).

Article

An Efficient Class of Traub-Steffensen-Like Seventh Order Multiple-Root Solvers with Applications

Janak Raj Sharma [1,*], Deepak Kumar [1] and Ioannis K. Argyros [2,*]

1. Department of Mathematics, Sant Longowal Institute of Engineering and Technology Longowal Sangrur 148106, India; deepak.babbi@gmail.com
2. Department of Mathematical Sciences, Cameron University, Lawton, OK 73505, USA
* Correspondence: jrshira@yahoo.co.in (J.R.S.); iargyros@cameron.edu (I.K.A.)

Received: 20 March 2019; Accepted: 8 April 2019; Published: 10 April 2019

Abstract: Many higher order multiple-root solvers that require derivative evaluations are available in literature. Contrary to this, higher order multiple-root solvers without derivatives are difficult to obtain, and therefore, such techniques are yet to be achieved. Motivated by this fact, we focus on developing a new family of higher order derivative-free solvers for computing multiple zeros by using a simple approach. The stability of the techniques is checked through complex geometry shown by drawing basins of attraction. Applicability is demonstrated on practical problems, which illustrates the efficient convergence behavior. Moreover, the comparison of numerical results shows that the proposed derivative-free techniques are good competitors of the existing techniques that require derivative evaluations in the iteration.

Keywords: nonlinear equations; multiple-root solvers; Traub–Steffensen method; fast algorithms

MSC: 65H05; 41A25; 49M15

1. Introduction

Solving nonlinear equations is an important task in numerical analysis and has numerous applications in engineering, mathematical biology, physics, chemistry, medicine, economics, and other disciplines of applied sciences [1–3]. Due to advances in computer hardware and software, the problem of solving the nonlinear equations by computational techniques has acquired an additional advantage of handling the lengthy and cumbersome calculations. In the present paper, we consider iterative techniques for computing multiple roots, say α, with multiplicity m of a nonlinear equation $f(x) = 0$, that is $f^{(j)}(\alpha) = 0, j = 0, 1, 2, \ldots, m-1$ and $f^{(m)}(\alpha) \neq 0$. The solution α can be calculated as a fixed point of some function $M : D \subset \mathbb{C} \to \mathbb{C}$ by means of the fixed point iteration:

$$x_{n+1} = M(x_n), \quad n \geq 0 \qquad (1)$$

where $x \in D$ is a scalar.

Many higher order techniques, based on the quadratically-convergent modified Newton's scheme (see [4]):

$$x_{n+1} = x_n - m \frac{f(x_n)}{f'(x_n)} \qquad (2)$$

have been proposed in the literature; see, for example, [5–20] and the references therein. The techniques based on Newton's or the Newton-like method require the evaluations of derivatives of first order. There is another class of multiple-root techniques involving derivatives of both the first and second order; see [5,21]. However, higher order derivative-free techniques to handle the case of multiple roots are yet to be explored. The main problem of developing such techniques is the difficulty in finding their convergence order. Derivative-free techniques are important in the situations when the derivative of the function f is difficult to compute or is expensive to obtain. One such derivative-free technique is the classical Traub–Steffensen method [22]. The Traub–Steffensen method actually replaces the derivative in the classical Newton's method with a suitable approximation based on the difference quotient,

$$f'(x_n) \simeq \frac{f(x_n + \beta f(x_n)) - f(x_n)}{\beta f(x_n)}, \quad \beta \in \mathbb{R} - \{0\}$$

or writing more concisely:

$$f'(x_n) \simeq f[x_n, t_n]$$

where $t_n = x_n + \beta f(x_n)$ and $f[x_n, t_n] = \dfrac{f(t_n) - f(x_n)}{t_n - x_n}$ is a first order divided difference. In this way, the modified Newton's scheme (2) assumes the form of the modified Traub–Steffensen scheme:

$$x_{n+1} = x_n - m \frac{f(x_n)}{f[x_n, t_n]}. \tag{3}$$

The Traub–Steffensen scheme (3) is a noticeable improvement of Newton's scheme, since it maintains the quadratic convergence without using any derivative.

The aim of the present contribution is to develop derivative-free multiple-root iterative techniques with high computational efficiency, which means the techniques that may attain a high convergence order using as small a number of function evaluations as possible. Consequently, we develop a family of derivative-free iterative methods of seventh order convergence that requires only four function evaluations per full iteration. The scheme is composed of three steps, out of which the first step is the classical Traub–Steffensen iteration (3) and the last two steps are Traub–Steffensen-like iterations. The methodology is based on the simple approach of using weight functions in the scheme. Many special cases of the family can be generated depending on the different forms of weight functions. The efficacy of the proposed methods is tested on various numerical problems of different natures. In the comparison with existing techniques requiring derivative evaluations, the new derivative-free methods are observed to be computationally more efficient.

We summarize the contents of the rest of paper. In Section 2, the scheme of the seventh order multiple-root solvers is developed and its order of convergence is determined. In Section 3, the basins of attractors are presented to check the stability of new methods. To demonstrate the performance and comparison with existing techniques, the new techniques are applied to solve some practical problems in Section 4. Concluding remarks are given in Section 5.

2. Development of the Family of Methods

Given a known multiplicity $m \geq 1$, we consider a three-step iterative scheme with the first step as the Traub–Steffensen iteration (3) as follows:

$$\begin{cases} y_n = x_n - m \dfrac{f(x_n)}{f[x_n, t_n]} \\ z_n = y_n - m u H(u) \dfrac{f(x_n)}{f[x_n, t_n]} \\ x_{n+1} = z_n - m v G(u, w) \dfrac{f(x_n)}{f[x_n, t_n]} \end{cases} \tag{4}$$

where $u = \left(\dfrac{f(y_n)}{f(x_n)}\right)^{\frac{1}{m}}$, $v = \left(\dfrac{f(z_n)}{f(x_n)}\right)^{\frac{1}{m}}$, and $w = \left(\dfrac{f(z_n)}{f(y_n)}\right)^{\frac{1}{m}}$. The function $H(u) : \mathbb{C} \to \mathbb{C}$ is analytic in a neighborhood of 0, and the function $G(u, w) : \mathbb{C} \times \mathbb{C} \to \mathbb{C}$ is holomorphic in a neighborhood of

(0, 0). Note that second and third steps are weighted by the factors $H(u)$ and $G(u,w)$, so these factors are called weight factors or weight functions.

We shall find conditions under which the scheme (4) achieves convergence order as high as possible. In order to do this, let us prove the following theorem:

Theorem 1. *Let $f : \mathbb{C} \to \mathbb{C}$ be an analytic function in a region enclosing a multiple zero α with multiplicity m. Assume that initial guess x_0 is sufficiently close to α, then the iteration scheme defined by (4) possesses the seventh order of convergence, provided that the following conditions are satisfied:*

$$\begin{cases} H(0) = 1, \ H'(0) = 2 \ H''(0) = -2, \ and \ |H'''(0)| < \infty \\ G(0,0) = 1, \ G_{10}(0,0) = 2, \ G_{01}(0,0) = 1, \ G_{20}(0,0) = 0 \ and \ |G_{11}(0,0)| < \infty \end{cases}$$

where $G_{ij}(0,0) = \dfrac{\partial^{i+j}}{\partial u^i \partial w^j} G(u,w)|_{(0,0)}$.

Proof. Let the error at the nth iteration be $e_n = x_n - \alpha$. Using the Taylor expansion of $f(x_n)$ about α, we have that:

$$f(x_n) = \frac{f^{(m)}(\alpha)}{m!} e_n^m + \frac{f^{(m+1)}(\alpha)}{(m+1)!} e_n^{m+1} + \frac{f^{(m+2)}(\alpha)}{(m+2)!} e_n^{m+2} + \frac{f^{(m+3)}(\alpha)}{(m+3)!} e_n^{m+3} + \frac{f^{(m+4)}(\alpha)}{(m+4)!} e_n^{m+4}$$
$$+ \frac{f^{(m+5)}(\alpha)}{(m+5)!} e_n^{m+5} + \frac{f^{(m+6)}(\alpha)}{(m+6)!} e_n^{m+6} + \frac{f^{(m+7)}(\alpha)}{(m+7)!} e_n^{m+7} + O(e_n^{m+8})$$

or:

$$f(x_n) = \frac{f^{(m)}(\alpha)}{m!} e_n^m \left(1 + C_1 e_n + C_2 e_n^2 + C_3 e_n^3 + C_4 e_n^4 + C_5 e_n^5 + C_6 e_n^6 + C_7 e_n^7 + O(e_n^8)\right) \tag{5}$$

where $C_k = \dfrac{m!}{(m+k)!} \dfrac{f^{(m+k)}(\alpha)}{f^{(m)}(\alpha)}$ for $k \in \mathbb{N}$.

Using (5) in $t_n = x_n + \beta f(x_n)$, we obtain that:

$$\begin{aligned} t_n - \alpha &= x_n - \alpha + \beta f(x_n) \\ &= e_n + \frac{\beta f^{(m)}(\alpha)}{m!} e_n^m \left(1 + C_1 e_n + C_2 e_n^2 + C_3 e_n^3 + C_4 e_n^4 + C_5 e_n^5 + C_6 e_n^6 + C_7 e_n^7 + O(e_n^8)\right). \end{aligned} \tag{6}$$

Taylor's expansion of $f(t_n)$ about α is given as:

$$f(t_n) = \frac{f^{(m)}(\alpha)}{m!} (t_n - \alpha)^m \left(1 + C_1(t_n - \alpha) + C_2(t_n - \alpha)^2 + C_3(t_n - \alpha)^3 + C_4(t_n - \alpha)^4 \right. \\ \left. + C_5(t_n - \alpha)^5 + C_6(t_n - \alpha)^6 + C_7(t_n - \alpha)^7 + O((t_n - \alpha)^8)\right). \tag{7}$$

By using Equations (5)–(7) in the first step of (4), after some simple calculations, it follows that:

$$y_n - \alpha = \frac{C_1}{m} e_n^2 + \frac{2mC_2 - (m+1)C_1^2}{m^2} e_n^3 + \frac{1}{m^3}\left((m+1)^2 C_1^3 + m(4+3m)C_1 C_2 - 3m^2 C_3\right) e_n^4 + \sum_{i=1}^{3} \omega_i e_n^{i+4} + O(e_n^8) \tag{8}$$

where $\omega_i = \omega_i(m, C_1, C_2, \ldots, C_7)$ are given in terms of $m, C_1, C_2, \ldots, C_7$. The expressions of ω_i are very lengthy, so they are not written explicitly.

Taylor's expansion of $f(y_n)$ about α is given by:

$$f(y_n) = \frac{f^{(m)}(\alpha)}{m!} \left(\frac{C_1}{m}\right)^m e_n^{2m} \left(1 + \frac{2C_2 m - C_1^2(m+1)}{C_1} e_n + \frac{1}{2mC_1^2}\left((3 + 3m + 3m^2 + m^3)C_1^4 \right.\right. \\ \left.\left. - 2m(2 + 3m + 2m^2)C_1^2 C_2 + 4(-1+m)m^2 C_2^2 + 6m^2 C_1 C_3\right) e_n^2 + \sum_{i=1}^{4} \tilde{\omega}_i e_n^{i+2} + O(e_n^8)\right) \tag{9}$$

where $\bar{w}_i = \bar{w}_i(m, C_1, C_2, \ldots, C_7)$.

By using (5) and (9), we get the expression of u as:

$$u = \frac{C_1}{m} e_n + \frac{2C_2 m - C_1^2(m+2)}{m^2} e_n^2 + \sum_{i=1}^{5} \eta_i e_n^{i+2} + O(e_n^8) \tag{10}$$

where $\eta_i = \eta_i(m, C_1, C_2, \ldots, C_7)$ are given in terms of $m, C_1, C_2, \ldots, C_7$ with one explicitly-written coefficient $\eta_1 = \frac{1}{2m^3}(C_1^3(2m^2 + 7m + 7) + 6C_3 m^2 - 2C_2 C_1 m(3m+7))$.

We expand the weight function $H(u)$ in the neighborhood of 0 by the Taylor series, then we have that:

$$H(u) \approx H(0) + uH'(0) + \frac{1}{2}u^2 H''(0) + \frac{1}{6}u^3 H'''(0). \tag{11}$$

Inserting Equations (5), (9), and (11) in the second step of Scheme (4) and simplifying,

$$z_n - \alpha = -\frac{A}{m} C_1 e_n^2 + \frac{1}{m^2}\left(-2mAC_2 + C_1^2(-1 + mA + 3H(0) - H'(0))\right) e_n^3$$
$$+ \frac{1}{2m^3}\left(-6Am^2 C_3 + 2mC_1 C_2(-4 + 3Am + 11H(0) - 4H'(0)) + C_1^3(2 - 2Am^2\right. \tag{12}$$
$$\left. - 13H(0) + 10H'(0) + m(4 - 11H(0) + 4H'(0)) - H''(0))\right) e_n^4 + \sum_{i=1}^{3} \gamma_i e_n^{i+4} + O(e_n^8)$$

where $A = -1 + H(0)$ and $\gamma_i = \gamma_i(m, H(0), H'(0), H''(0), H'''(0), C_1, C_2, \ldots, C_7)$.

In order to attain higher order convergence, the coefficients of e_n^2 and e_n^3 should be simultaneously equal to zero. That is possible only for the following values of $H(0)$ and $H'(0)$:

$$H(0) = 1, \quad H'(0) = 2. \tag{13}$$

By using the above values in (12), we obtain that:

$$z_n - \alpha = \frac{-2mC_1 C_2 + C_1^3(9 + m - H''(0))}{2m^3} e_n^4 + \sum_{i=1}^{3} \gamma_i e_n^{i+4} + O(e_n^8). \tag{14}$$

Expansion of $f(z_n)$ about α leads us to the expression:

$$f(z_n) = \frac{f^{(m)}(\alpha)}{m!}(z_n - \alpha)^m \left(1 + C_1(z_n - \alpha) + C_2(z_n - \alpha)^2 + O((z_n - \alpha)^3)\right). \tag{15}$$

From (5), (9), and (15), we get the expressions of v and w as:

$$v = \frac{(9+m)C_1^3 - 2mC_1 C_2}{2m^3} e_n^3 + \sum_{i=1}^{4} \tau_i e_n^{i+3} + O(e_n^8) \tag{16}$$

and:

$$w = \frac{(9 + m - H''(0))C_1^2 - 2mC_2}{2m^3} e_n^2 + \sum_{i=1}^{5} \varsigma_i e_n^{i+2} + O(e_n^8) \tag{17}$$

where τ_i and ς_i are some expressions of $m, H''(0), H'''(0), C_1, C_2, \ldots, C_7$.

Expanding the function $G(u, w)$ in the neighborhood of origin $(0, 0)$ by Taylor series:

$$G(u, w) \approx G_{00}(0, 0) + uG_{10}(0, 0) + \frac{1}{2}u^2 G_{20}(0, 0) + w(G_{01}(0, 0) + uG_{11}(0, 0)) \tag{18}$$

where $G_{ij} = \dfrac{\partial^{i+j}}{\partial u^i \partial w^j} G(u,w)|_{(0,0)}$.

Then by substituting (5), (16), (17), and (18) into the last step of Scheme (4), we obtain the error equation:

$$e_{n+1} = \dfrac{1}{2m^3}\left((-1+G_{00}(0,0))C_1(2mC_1 - (9+m-H''(0))C_1^2)\right)e_n^4 + \sum_{i=1}^{3}\xi_i e_n^{i+4} + O(e_n^8) \quad (19)$$

where $\xi_i = \xi_i(m, H''(0), H'''(0), G_{00}, G_{01}, G_{10}, G_{20}, G_{11}, C_1, C_2, \ldots, C_7)$.

It is clear from Equation (19) that we will obtain at least fifth order convergence if we have $G_{00}(0,0) = 1$. Moreover, we can use this value in $\xi_1 = 0$ to obtain:

$$G_{10}(0,0) = 2. \quad (20)$$

By using $G_{00} = 1$ and (20) in $\xi_2 = 0$, the following equation is obtained:

$$C_1(2mC_2 - C_1^2(9+m-H''(0)))\left(-2mC_2(-1+G_{01}(0,0)) + C_1^2(-11+m(-1+G_{01}(0,0)) - (-9+H''(0))G_{01}(0,0) + G_{20}(0,0))\right) = 0$$

which further yields:

$$G_{01}(0,0) = 1, \quad G_{20}(0,0) = 0 \quad \text{and} \quad H''(0) = -2.$$

Using the above values in (19), the final error equation is given by:

$$\begin{aligned}e_{n+1} = \dfrac{1}{360m^6} &\Big(360m^3\left((47+5m)C_2^3 - 6mC_3^2 - 10mC_2C_4\right) + 120m^3C_1\left((623+78m)C_2C_3\right.\\ &\left. - 12mC_5\right) - 60m^2C_1^3C_3\left(1861 + 1025m + 78m^2 + 12H'''(0)\right) + 10mC_1^4C_2(32383\\ &+ 9911m^2 + 558m^3 + 515H'''(0) + 396G_{11}(0,0) + 36m(900 + 6H'''(0) + G_{11}(0,0)))\\ &- 60m^2C_1^2\left(-6m(67+9m)C_4 + C_2^2(3539 + 1870m + 135m^2 + 24H'''(0) + 6G_{11}(0,0))\right)\\ &- C_1^6\left(95557 + 20605m + 978m^4 + 2765H'''(0) + 10890G_{11}(0,0) + m^2(90305 + 600H'''(0)\right.\\ &\left. + 90G_{11}(0,0)) + 5m(32383 + 515H'''(0) + 396G_{11}(0,0))\right)\Big)e_n^7 + O(e_n^8).\end{aligned}$$

Hence, the seventh order convergence is established. □

Forms of the Weight Function

Numerous special cases of the family (4) are generated based on the forms of weight functions $H(u)$ and $G(u,w)$ that satisfy the conditions of Theorem 1. However, we restrict ourselves to simple forms, which are given as follows:

I. *Some particular forms of $H(u)$*

Case I(a). When $H(u)$ is a polynomial weight function, e.g.,

$$H(u) = A_0 + A_1 u + A_2 u^2.$$

By using the conditions of Theorem 1, we get $A_0 = 1$, $A_1 = 2$ and $A_2 = -1$. Then, $H(u)$ becomes:

$$H(u) = 1 + 2u - u^2.$$

Case I(b). When $H(u)$ is a rational weight function, e.g.,

$$H(u) = \frac{1 + A_0 u}{A_1 + A_2 u}.$$

Using the conditions of Theorem 1, we get that $A_0 = \frac{5}{2}$, $A_1 = 1$, and $A_2 = \frac{1}{2}$. Therefore,

$$H(u) = \frac{2 + 5u}{2 + u}.$$

Case I(c). When $H(u)$ is a rational weight function, e.g.,

$$H(u) = \frac{1 + A_0 u + A_1 u^2}{1 + A_2 u}.$$

Using the conditions of Theorem 1, then we get $A_0 = 3$, $A_1 = 2$, and $A_2 = 1$. $H(u)$ becomes:

$$H(u) = \frac{1 + 3u + u^2}{1 + u}.$$

Case I(d). When $H(u)$ is a rational function of the form:

$$H(u) = \frac{1 + A_0 u}{1 + A_1 u + A_2 u^2}.$$

Using the conditions of Theorem 1, we get $A_0 = 1$, $A_1 = -1$, and $A_2 = 1$. Then,

$$H(u) = \frac{1 + u}{1 - u + 3u^2}.$$

II. *Some particular forms of $G(u, w)$*

Case II(a). When $G(u, w)$ is a polynomial weight function, e.g.,

$$G(u, w) = A_0 + A_1 u + A_2 u^2 + (A_3 + A_4 u + A_5 u^2) w.$$

Using the conditions of Theorem 1, then we get $A_0 = 1$, $A_1 = 2$, $A_2 = 0$, and $A_3 = 1$. Therefore, $G(u, w)$ becomes:

$$G(u, w) = 1 + 2u + (1 + A_4 u + A_5 u^2) w$$

where A_4 and A_5 are free parameters.

Case II(b). When $G(u, w)$ is a rational weight function, e.g.,

$$G(u, w) = \frac{B_0 + B_1 u + B_2 w + B_3 uw}{1 + A_1 u + A_2 w + A_3 uw}.$$

Using the conditions of Theorem 1, we have $B_0 = 1$, $A_1 = 2$, $B_1 = 2$, $A_2 = 0$, and $B_2 = 1$. Then,

$$G(u, w) = \frac{1 + 2u + w + B_3 uw}{1 + A_3 uw}$$

where A_3 and B_3 are free parameters.

Case II(c). When $G(u, w)$ is the sum of two weight functions $H_1(u)$ and $H_2(w)$. Let $H_1(u) = A_0 + uA_1 + u^2 A_2$ and $H_2(u) = B_0 + wB_1 + w^2 B_2$, then $G(u, w)$ becomes:

$$G(u, w) = A_0 + A_1 u + A_2 u^2 + B_0 + B_1 w + B_2 w^2.$$

By using the conditions of Theorem 1, we get:

$$G(u,w) = 1 + 2u + w + B_2 w^2$$

where B_2 is a free parameter.

Case II(d). When $G(u,w)$ is the sum of two rational weight functions, that is:

$$G(u,w) = \frac{A_0 + A_1 u}{1 + A_2 u} + \frac{B_0 + B_1 w}{1 + B_2 w}.$$

By using the conditions of Theorem 1, we obtain that:

$$G(u,w) = 2u + \frac{1}{1-w}.$$

Case II(e). When $G(u,w)$ is product of two weight functions, that is:

$$G(u,w) = \frac{A_0 + A_1 u}{1 + A_2 u} \times \frac{B_0 + B_1 w}{1 + B_2 w}.$$

Using the conditions of Theorem 1, then we get:

$$G(u,w) = (1 + 2u)(1 + w).$$

3. Complex Dynamics of Methods

Here, our aim is to analyze the complex dynamics of the proposed methods based on the visual display of the basins of attraction of the zeros of a polynomial $p(z)$ in the complex plane. Analysis of the complex dynamical behavior gives important information about the convergence and stability of an iterative scheme. Initially, Vrscay and Gilbert [23] introduced the idea of analyzing complex dynamics. Later on, many researchers used this concept in their work, for example (see [24–26] and the references therein). We choose some of the special cases (corresponding to the above forms of $H(u)$ and $G(u,w)$) of family (4) to analyze the basins. Let us choose the combinations of special Cases II(c) (for $B_2 = 1$) and II(d) with I(a), I(b), I(c), and I(d) in (4) and denote the corresponding new methods by NM-i(j), i = 1, 2 and j = a, b, c, d.

We take the initial point as $z_0 \in D$, where D is a rectangular region in $\mathbb{C}$ containing all the roots of $p(z) = 0$. The iterative methods starting at a point z_0 in a rectangle either converge to the zero of the function $p(z)$ or eventually diverge. The stopping criterion considered for convergence is 10^{-3} up to a maximum of 25 iterations. If the desired tolerance is not achieved in 25 iterations, we do not continue and declare that the iterative method starting at point z_0 does not converge to any root. The strategy adopted is the following: A color is assigned to each starting point z_0 in the basin of attraction of a zero. If the iteration starting from the initial point z_0 converges, then it represents the basins of attraction with that particular color assigned to it, and if it fails to converge in 25 iterations, then it shows the black color.

To view complex geometry, we analyze the attraction basins for the methods NM-1(a–d) and NM-2(a–d) on the following two polynomials:

Problem 1. *In the first example, we consider the polynomial $p_1(z) = (z^2 - 1)^2$, which has zeros $\{\pm 1\}$ with multiplicity two. In this case, we use a grid of 400×400 points in a rectangle $D \in \mathbb{C}$ of size $[-2,2] \times [-2,2]$ and assign the color green to each initial point in the basin of attraction of zero ' -1' and the color red to each point in the basin of attraction of zero '1'. Basins obtained for the methods NM-1(a–d) and NM-2(a–d) are shown in Figures 1–4 corresponding to $\beta = 0.01, 0.002$. Observing the behavior of the methods, we see that the method NM-2(d) possesses a lesser number of divergent points and therefore has better stability than the* remaining methods. Notice that there is a small difference in the basins for the rest of the methods with the same value of β. Notice also that the basins are becoming wider as parameter β assumes smaller values.

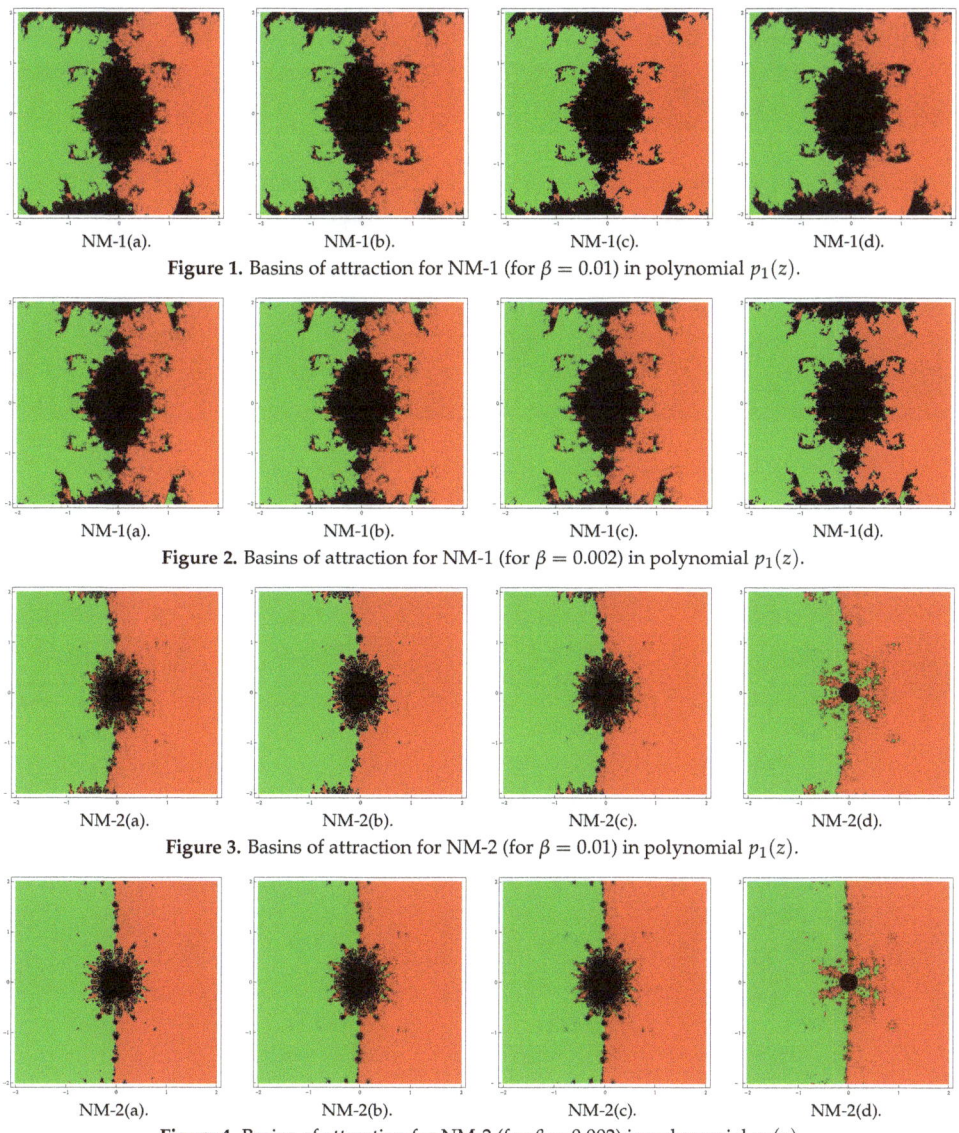

NM-1(a). NM-1(b). NM-1(c). NM-1(d).
Figure 1. Basins of attraction for NM-1 (for $\beta = 0.01$) in polynomial $p_1(z)$.

NM-1(a). NM-1(b). NM-1(c). NM-1(d).
Figure 2. Basins of attraction for NM-1 (for $\beta = 0.002$) in polynomial $p_1(z)$.

NM-2(a). NM-2(b). NM-2(c). NM-2(d).
Figure 3. Basins of attraction for NM-2 (for $\beta = 0.01$) in polynomial $p_1(z)$.

NM-2(a). NM-2(b). NM-2(c). NM-2(d).
Figure 4. Basins of attraction for NM-2 (for $\beta = 0.002$) in polynomial $p_1(z)$.

Problem 2. Let us take the polynomial $p_2(z) = (z^3 + z)^3$ having zeros $\{0, \pm i\}$ with multiplicity three. To see the dynamical view, we consider a rectangle $D = [-2, 2] \times [-2, 2] \in \mathbb{C}$ with 400×400 grid points and allocate the colors red, green, and blue to each point in the basin of attraction of $-i$, 0, and i, respectively. Basins for this problem are shown in Figures 5–8 corresponding to parameter choices $\beta = 0.01$, 0.002 in the proposed methods. Observing the behavior, we see that again, the method NM-2(d) has better convergence behavior due to a lesser number of divergent points. Furthermore, observe that in each case, the basins are becoming larger with the smaller values of β. The basins in the remaining methods other than NM-2(d) are almost the same.

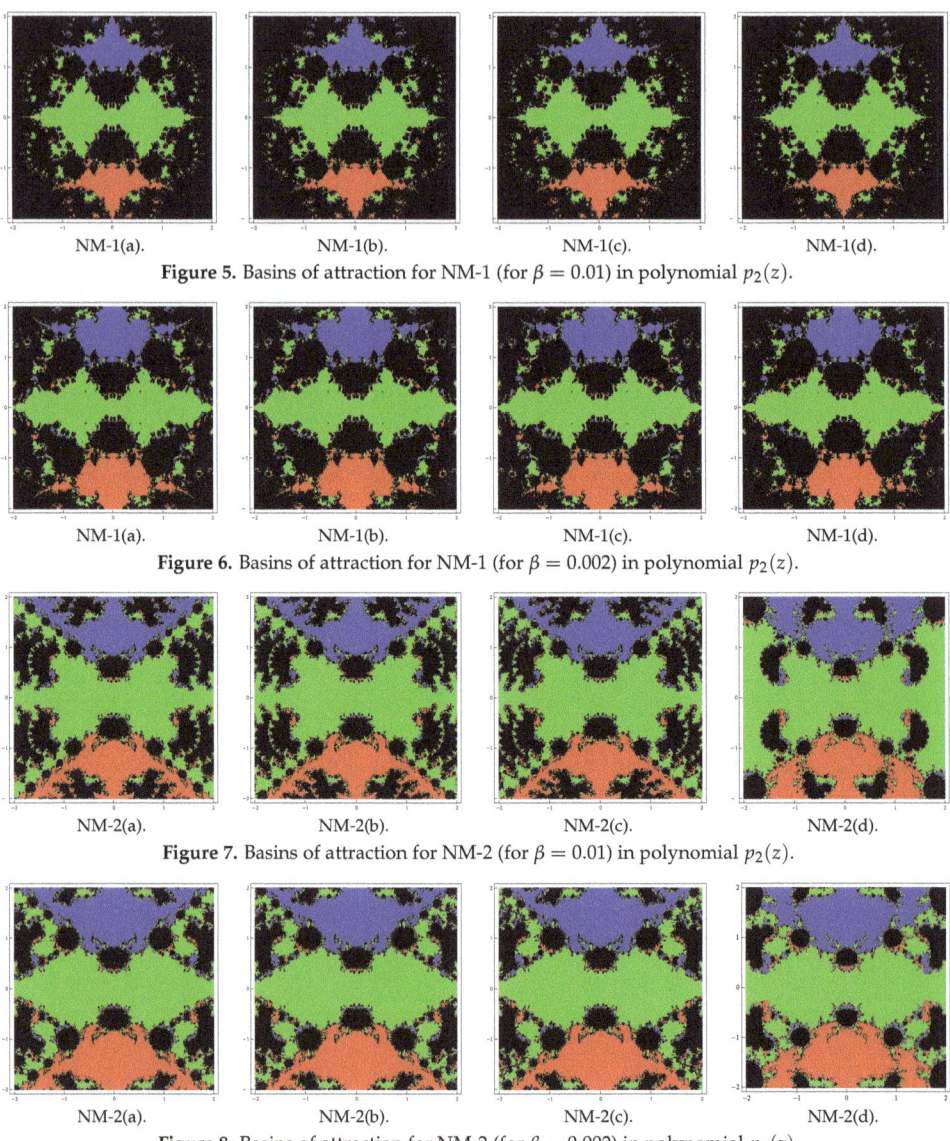

Figure 5. Basins of attraction for NM-1 (for $\beta = 0.01$) in polynomial $p_2(z)$.

Figure 6. Basins of attraction for NM-1 (for $\beta = 0.002$) in polynomial $p_2(z)$.

Figure 7. Basins of attraction for NM-2 (for $\beta = 0.01$) in polynomial $p_2(z)$.

Figure 8. Basins of attraction for NM-2 (for $\beta = 0.002$) in polynomial $p_2(z)$.

From the graphics, we can easily observe the behavior and applicability of any method. If we choose an initial guess z_0 in a region wherein different basins of attraction touch each other, it is difficult to predict which root is going to be attained by the iterative method that starts from z_0. Therefore, the choice of z_0 in such a region is not a good one. Both black regions and the regions with different colors are not suitable to assume the initial guess as z_0 when we are required to achieve a particular root. The most intricate geometry is between the basins of attraction, and this corresponds to the cases where the method is more demanding with respect to the initial point. We conclude this section with a remark that the convergence behavior of the proposed techniques depends on the value of parameter β. The smaller the value of β is, the better the convergence of the method.

4. Numerical Examples and Discussion

In this section, we implement the special cases NM-1(a–d) and NM-2(a–d) that we have considered in the previous section of the family (4), to obtain zeros of nonlinear functions. This will not only illustrate the methods' practically, but also serve to test the validity of theoretical results. The theoretical order of convergence is also confirmed by calculating the computational order of convergence (COC) using the formula (see [27]):

$$\text{COC} = \frac{\ln|(x_{n+1}-\alpha)/(x_n-\alpha)|}{\ln|(x_n-\alpha)/(x_{n-1}-\alpha)|}.$$

The performance is compared with some well-known higher order multiple-root solvers such as the sixth order methods by Geum et al. [8,9], which are expressed below:

First method by Geum et al. [8]:

$$\begin{cases} y_n = x_n - m\dfrac{f(x_n)}{f'(x_n)} \\ x_{n+1} = y_n - Q_f(u,s)\dfrac{f(y_n)}{f'(y_n)} \end{cases} \quad (21)$$

where $u = \left(\dfrac{f(y_n)}{f(x_n)}\right)^{\frac{1}{m}}$ and $s = \left(\dfrac{f'(y_n)}{f'(x_n)}\right)^{\frac{1}{m-1}}$, and $Q_f : \mathbb{C}^2 \to \mathbb{C}$ is a holomorphic function in the neighborhood of origin (0, 0). The authors have also studied various forms of the function Q_f leading to sixth order convergence of (21). We consider the following four special cases of function $Q_f(u,s)$ in the formula (21) and denote the corresponding methods by GKN-1(j), j = a, b, c, d:

(a) $Q_f(u,s) = m(1 + 2(m-1)(u-s) - 4us + s^2)$
(b) $Q_f(u,s) = m(1 + 2(m-1)(u-s) - u^2 - 2us)$
(c) $Q_f(u,s) = \dfrac{m + au}{1 + bu + cs + dus}$,
where $a = \dfrac{2m}{m-1}, b = 2 - 2m, c = \dfrac{2(2-2m+m^2)}{m-1}$, and $d = -2m(m-1)$
(d) $Q_f(u,s) = \dfrac{m + a_1 u}{1 + b_1 u + c_1 u^2}\dfrac{1}{1 + d_1 s}$,
where $a_1 = \dfrac{2m(4m^4 - 16m^3 + 31m^2 - 30m + 13)}{(m-1)(4m^2 - 8m + 7)}, b_1 = \dfrac{4(2m^2 - 4m + 3)}{(m-1)(4m^2 - 8m + 7)}$,
$c_1 = -\dfrac{4m^2 - 8m + 3}{4m^2 - 8m + 7}$, and $d_1 = 2(m-1)$.

Second method by Geum et al. [9]:

$$\begin{cases} y_n = x_n - m\dfrac{f(x_n)}{f'(x_n)} \\ z_n = x_n - mG_f(u)\dfrac{f(x_n)}{f'(x_n)} \\ x_{n+1} = x_n - mK_f(u,v)\dfrac{f(x_n)}{f'(x_n)} \end{cases} \quad (22)$$

where $u = \left(\dfrac{f(y_n)}{f(x_n)}\right)^{\frac{1}{m}}$ and $v = \left(\dfrac{f(z_n)}{f(x_n)}\right)^{\frac{1}{m}}$. The function $G_f : \mathbb{C} \to \mathbb{C}$ is analytic in a neighborhood of 0, and $K_f : \mathbb{C}^2 \to \mathbb{C}$ is holomorphic in a neighborhood of (0,0). Numerous cases of G_f and K_f have been proposed in [9]. We consider the following four special cases and denote the corresponding methods by GKN-2(j), j = a, b, c, d:

(a) $Q_f(u) = \dfrac{1+u^2}{1-u}, K_f(u,v) = \dfrac{1+u^2-v}{1-u+(u-2)v}$

(b) $Q_f(u) = 1 + u + 2u^2$, $K_f(u,v) = 1 + u + 2u^2 + (1+2u)v$

(c) $Q_f(u) = \dfrac{1+u^2}{1-u}$, $K_f(u,v) = 1 + u + 2u^2 + 2u^3 + 2u^4 + (2u+1)v$

(d) $Q_f(u) = \dfrac{(2u-1)(4u-1)}{1-7u+13u^2}$, $K_f(u,v) = \dfrac{(2u-1)(4u-1)}{1-7u+13u^2-(1-6u)v}$

Computations were carried out in the programming package *Mathematica* with multiple-precision arithmetic. Numerical results shown in Tables 1–4 include: (i) the number of iterations (n) required to converge to the solution, (ii) the values of the last three consecutive errors $e_n = |x_{n+1} - x_n|$, (iii) the computational order of convergence (COC), and (iv) the elapsed CPU time (CPU-time). The necessary iteration number (n) and elapsed CPU time are calculated by considering $|x_{n+1} - x_n| + |f(x_n)| < 10^{-350}$ as the stopping criterion.

The convergence behavior of the family of iterative methods (4) is tested on the following problems:

Example 1. (Eigenvalue problem). *Finding eigenvalues of a large square matrix is one of the difficult tasks in applied mathematics and engineering. Finding even the roots of the characteristic equation of a square matrix of order greater than four is a big challenge. Here, we consider the following* 6×6 *matrix:*

$$M = \begin{bmatrix} 5 & 8 & 0 & 2 & 6 & -6 \\ 0 & 1 & 0 & 0 & 0 & 0 \\ 6 & 18 & -1 & 1 & 13 & -9 \\ 3 & 6 & 0 & 4 & 6 & -6 \\ 4 & 14 & -2 & 0 & 11 & -6 \\ 6 & 18 & -2 & 1 & 13 & -8 \end{bmatrix}.$$

The characteristic equation of the above matrix (M) *is given as follows:*

$$f_1(x) = x^6 - 12x^5 + 56x^4 - 130x^3 + 159x^2 - 98x + 24.$$

This function has one multiple zero at $\alpha = 1$ *of multiplicity three. We choose initial approximation* $x_0 = 0.25$. *Numerical results are shown in Table 1.*

Table 1. Comparison of the performance of methods for Example 1.

Methods	n	$|e_{n-3}|$	$|e_{n-2}|$	$|e_{n-1}|$	COC	CPU-Time
GKN-1(a)	4	5.46×10^{-3}	2.40×10^{-14}	1.78×10^{-82}	6.0000	0.05475
GKN-1(b)	4	5.65×10^{-3}	3.22×10^{-14}	1.13×10^{-81}	6.0000	0.05670
GKN-1(c)	4	5.41×10^{-3}	2.80×10^{-14}	5.59×10^{-82}	6.0000	0.05856
GKN-1(d)	4	7.52×10^{-3}	4.85×10^{-13}	3.78×10^{-74}	6.0000	0.05504
GKN-2(a)	4	2.85×10^{-3}	1.57×10^{-16}	4.32×10^{-96}	6.0000	0.07025
GKN-2(b)	4	9.28×10^{-3}	1.58×10^{-12}	4.13×10^{-71}	6.0000	0.05854
GKN-2(c)	4	7.11×10^{-3}	1.87×10^{-13}	6.53×10^{-77}	6.0000	0.06257
GKN-2(d)	5	1.03×10^{-5}	3.87×10^{-30}	1.07×10^{-176}	6.0000	0.07425
NM-1(a)	4	1.62×10^{-3}	1.79×10^{-19}	3.58×10^{-131}	7.0000	0.04675
NM-1(b)	4	1.62×10^{-3}	1.85×10^{-19}	4.63×10^{-131}	6.9990	0.05073
NM-1(c)	4	1.62×10^{-3}	1.96×10^{-19}	5.92×10^{-131}	6.9998	0.05355
NM-1(d)	4	1.60×10^{-3}	1.02×10^{-19}	4.36×10^{-133}	6.9990	0.05077
NM-2(a)	4	1.37×10^{-3}	5.56×10^{-20}	1.02×10^{-134}	6.9997	0.05435
NM-2(b)	4	1.37×10^{-3}	5.77×10^{-20}	1.35×10^{-134}	6.9998	0.05454
NM-2(c)	4	1.38×10^{-3}	5.98×10^{-20}	1.77×10^{-134}	6.9996	0.05750
NM-2(d)	4	1.34×10^{-3}	2.97×10^{-20}	8.00×10^{-137}	6.9998	0.05175

Example 2. (Kepler's equation). *Let us consider Kepler's equation:*

$$f_2(x) = x - \alpha \sin(x) - K = 0, \quad 0 \leq \alpha < 1 \text{ and } 0 \leq K \leq \pi.$$

A numerical study, for different values of the parameters α and K, has been performed in [28]. As a particular example, let us take $\alpha = \frac{1}{4}$ and $K = \frac{\pi}{5}$. Consider this particular case four times with same values of the parameters, then the required nonlinear function is:

$$f_2(x) = \left(x - \frac{1}{4}\sin x - \frac{\pi}{5}\right)^4.$$

This function has one multiple zero at $\alpha = 0.80926328\ldots$ of multiplicity four. The required zero is calculated using initial approximation $x_0 = 1$. Numerical results are displayed in Table 2.

Table 2. Comparison of the performance of methods for Example 2.

Methods	n	$|e_{n-3}|$	$|e_{n-2}|$	$|e_{n-1}|$	COC	CPU-Time
GKN-1(a)	5	1.90×10^{-25}	2.55×10^{-76}	1.19×10^{-228}	3.0018	2.1405
GKN-1(b)	5	1.74×10^{-25}	1.94×10^{-76}	5.28×10^{-229}	3.0018	2.1445
GKN-1(c)	5	2.31×10^{-25}	4.56×10^{-76}	6.79×10^{-228}	3.0018	2.1835
GKN-1(d)	5	1.75×10^{-25}	1.97×10^{-76}	5.51×10^{-229}	3.0018	2.1797
GKN-2(a)	5	3.52×10^{-19}	5.28×10^{-117}	1.22×10^{-233}	1.1923	1.8047
GKN-2(b)	4	9.33×10^{-9}	6.67×10^{-53}	8.88×10^{-318}	6.0000	1.4452
GKN-2(c)	4	3.74×10^{-9}	9.33×10^{-56}	2.27×10^{-335}	6.0000	1.4415
GKN-2(d)	4	1.64×10^{-8}	2.04×10^{-51}	7.50×10^{-309}	6.0000	1.4492
NM-1(a)	3	1.91×10^{-1}	5.70×10^{-10}	6.59×10^{-70}	7.0000	0.9845
NM-1(b)	3	1.91×10^{-1}	6.02×10^{-10}	1.04×10^{-69}	7.0000	0.9650
NM-1(c)	3	1.91×10^{-1}	6.32×10^{-10}	1.58×10^{-69}	7.0000	0.9570
NM-1(d)	3	1.90×10^{-1}	9.62×10^{-11}	2.48×10^{-76}	7.0540	0.8590
NM-2(a)	3	1.91×10^{-1}	5.70×10^{-10}	6.59×10^{-70}	7.0000	0.9842
NM-2(b)	3	1.91×10^{-1}	6.02×10^{-10}	1.04×10^{-69}	7.0000	0.9607
NM-2(c)	3	1.91×10^{-1}	6.32×10^{-10}	1.58×10^{-69}	7.0000	0.9767
NM-2(d)	3	1.91×10^{-1}	9.68×10^{-11}	2.63×10^{-76}	7.0540	0.7460

Example 3. *(Manning's equation)*. Consider the isentropic supersonic flow around a sharp expansion corner. The relationship between the Mach number before the corner (i.e., M_1) and after the corner (i.e., M_2) is given by (see [3]):

$$\delta = b^{1/2}\left(\tan^{-1}\left(\frac{M_2^2 - 1}{b}\right)^{1/2} - \tan^{-1}\left(\frac{M_1^2 - 1}{b}\right)^{1/2}\right) - \left(\tan^{-1}(M_2^2 - 1)^{1/2} - \tan^{-1}(M_1^2 - 1)^{1/2}\right)$$

where $b = \frac{\gamma + 1}{\gamma - 1}$ and γ is the specific heat ratio of the gas.

As a particular case study, the equation is solved for M_2 given that $M_1 = 1.5$, $\gamma = 1.4$, and $\delta = 10^0$. Then, we have that:

$$\tan^{-1}\left(\frac{\sqrt{5}}{2}\right) - \tan^{-1}(\sqrt{x^2 - 1}) + \sqrt{6}\left(\tan^{-1}\left(\sqrt{\frac{x^2 - 1}{6}}\right) - \tan^{-1}\left(\frac{1}{2}\sqrt{\frac{5}{6}}\right)\right) - \frac{11}{63} = 0$$

where $x = M_2$.

Consider this particular case three times with the same values of the parameters, then the required nonlinear function is:

$$f_3(x) = \left(\tan^{-1}\left(\frac{\sqrt{5}}{2}\right) - \tan^{-1}(\sqrt{x^2 - 1}) + \sqrt{6}\left(\tan^{-1}\left(\sqrt{\frac{x^2 - 1}{6}}\right) - \tan^{-1}\left(\frac{1}{2}\sqrt{\frac{5}{6}}\right)\right) - \frac{11}{63}\right)^3.$$

This function has one multiple zero at $\alpha = 1.8411027704\ldots$ of multiplicity three. The required zero is calculated using initial approximation $x_0 = 1.5$. Numerical results are shown in Table 3.

Table 3. Comparison of the performance of methods for Example 3.

Methods	n	$|e_{n-3}|$	$|e_{n-2}|$	$|e_{n-1}|$	COC	CPU-Time
GKN-1(a)	4	2.17×10^{-8}	4.61×10^{-25}	1.01×10^{-151}	6.0000	1.3047
GKN-1(b)	4	2.17×10^{-8}	4.60×10^{-25}	2.27×10^{-151}	6.0000	1.2852
GKN-1(c)	4	2.11×10^{-8}	4.21×10^{-25}	1.03×10^{-151}	6.0000	1.3203
GKN-1(d)	4	1.77×10^{-8}	2.48×10^{-25}	2.68×10^{-151}	6.0000	1.2970
GKN-2(a)	4	4.83×10^{-7}	1.36×10^{-41}	6.84×10^{-249}	6.0000	1.2382
GKN-2(b)	4	4.90×10^{-7}	2.89×10^{-41}	1.21×10^{-246}	6.0000	1.2440
GKN-2(c)	4	4.88×10^{-7}	2.22×10^{-41}	1.98×10^{-247}	6.0000	1.2422
GKN-2(d)	4	4.89×10^{-7}	3.22×10^{-41}	2.62×10^{-246}	6.0000	1.2577
NM-1(a)	4	7.85×10^{-9}	1.56×10^{-60}	0	7.0000	1.0274
NM-1(b)	4	7.85×10^{-9}	1.58×10^{-60}	0	7.0000	1.0272
NM-1(c)	4	7.89×10^{-9}	1.60×10^{-60}	0	7.0000	1.0231
NM-1(d)	4	7.84×10^{-9}	1.31×10^{-60}	0	7.0000	1.0235
NM-2(a)	4	7.69×10^{-9}	1.35×10^{-60}	0	7.0000	1.0398
NM-2(b)	4	7.69×10^{-9}	1.37×10^{-60}	0	7.0000	1.0742
NM-2(c)	4	7.69×10^{-9}	1.38×10^{-60}	0	7.0000	1.0467
NM-2(d)	4	7.68×10^{-9}	1.13×10^{-60}	0	7.0000	1.0192

Example 4. Next, consider the standard nonlinear test function:

$$f_4(x) = \left(-\sqrt{1-x^2} + x + \cos\frac{\pi x}{2} + 1\right)^4$$

which has a multiple zero at $\alpha = -0.72855964390156\ldots$ of multiplicity four. Numerical results are shown in Table 4 with initial guess $x_0 = -0.5$.

Table 4. Comparison of the performance of methods for Example 4.

Methods	n	$|e_{n-3}|$	$|e_{n-2}|$	$|e_{n-1}|$	COC	CPU-Time
GKN-1(a)	4	7.20×10^{-6}	1.80×10^{-30}	4.39×10^{-178}	6.0000	0.1017
GKN-1(b)	4	7.21×10^{-6}	1.85×10^{-30}	5.32×10^{-178}	5.9999	0.0977
GKN-1(c)	4	7.42×10^{-6}	2.52×10^{-30}	3.84×10^{-177}	5.9999	0.1055
GKN-1(d)	4	8.83×10^{-6}	1.30×10^{-29}	1.34×10^{-172}	5.9999	0.1015
GKN-2(a)	4	2.15×10^{-5}	8.22×10^{-28}	2.60×10^{-162}	5.9999	0.1132
GKN-2(b)	4	2.39×10^{-5}	4.22×10^{-27}	1.27×10^{-157}	5.9999	0.1052
GKN-2(c)	4	2.33×10^{-5}	2.57×10^{-27}	4.61×10^{-159}	5.9999	0.1055
GKN-2(d)	4	2.43×10^{-5}	5.31×10^{-27}	5.83×10^{-157}	5.9999	0.1095
NM-1(a)	4	2.87×10^{-6}	1.03×10^{-37}	8.12×10^{-258}	6.9999	0.0720
NM-1(b)	4	2.88×10^{-6}	1.06×10^{-37}	9.60×10^{-258}	6.9999	0.0724
NM-1(c)	4	2.88×10^{-6}	1.08×10^{-37}	1.13×10^{-257}	6.9999	0.0722
NM-1(d)	4	2.83×10^{-6}	7.39×10^{-38}	6.09×10^{-259}	6.9999	0.0782
NM-2(a)	4	2.80×10^{-6}	8.55×10^{-38}	2.15×10^{-258}	6.9999	0.0732
NM-2(b)	4	2.80×10^{-6}	8.74×10^{-37}	2.54×10^{-258}	6.9999	0.0723
NM-2(c)	4	2.80×10^{-6}	8.93×10^{-38}	3.00×10^{-258}	6.9999	0.0746
NM-2(d)	4	2.76×10^{-6}	6.09×10^{-38}	1.56×10^{-259}	6.9999	0.0782

Example 5. Consider the standard function, which is given as (see [8]):

$$f_5(x) = \left(x - \sqrt{3}x^3 \cos\frac{\pi x}{6} + \frac{1}{x^2+1} - \frac{11}{5} + 4\sqrt{3}\right)(x-2)^4.$$

The multiple zero of function f_5 is $\alpha = 2$ with multiplicity five. We choose the initial approximation $x_0 = 1.5$ for obtaining the zero of the function. Numerical results are exhibited in Table 5.

Table 5. Comparison of the performance of methods for Example 5.

Methods	n	$|e_{n-3}|$	$|e_{n-2}|$	$|e_{n-1}|$	COC	CPU-Time
GKN-1(a)	4	1.20×10^{-5}	6.82×10^{-31}	2.31×10^{-182}	6.0000	0.5820
GKN-1(b)	4	1.20×10^{-5}	6.86×10^{-31}	2.40×10^{-182}	6.0000	0.5860
GKN-1(c)	4	1.21×10^{-5}	7.72×10^{-31}	5.18×10^{-182}	6.0000	0.5937
GKN-1(d)	4	1.58×10^{-5}	1.00×10^{-29}	6.51×10^{-175}	6.0000	0.5832
GKN-2(a)	4	3.17×10^{-5}	1.64×10^{-28}	3.21×10^{-168}	6.0000	0.7120
GKN-2(b)	4	3.50×10^{-5}	6.90×10^{-28}	4.05×10^{-164}	6.0000	0.6992
GKN-2(c)	4	3.41×10^{-5}	4.42×10^{-28}	2.09×10^{-165}	6.0000	0.6915
GKN-2(d)	4	3.54×10^{-5}	8.45×10^{-28}	1.56×10^{-163}	6.0000	0.6934
NM-1(a)	4	2.35×10^{-6}	1.81×10^{-40}	2.92×10^{-279}	7.0000	0.3712
NM-1(b)	4	2.35×10^{-6}	1.84×10^{-40}	3.31×10^{-279}	7.0000	0.3360
NM-1(c)	4	2.35×10^{-6}	1.87×10^{-40}	3.74×10^{-279}	7.0000	0.3555
NM-1(d)	4	2.33×10^{-6}	1.41×10^{-40}	4.23×10^{-280}	7.0000	0.3633
NM-2(a)	4	2.25×10^{-6}	1.34×10^{-40}	3.65×10^{-280}	7.0000	0.3585
NM-2(b)	4	2.25×10^{-6}	1.37×10^{-40}	4.15×10^{-280}	7.0000	0.3592
NM-2(c)	4	2.25×10^{-6}	1.39×10^{-40}	4.70×10^{-280}	7.0000	0.3791
NM-2(d)	4	2.24×10^{-6}	1.05×10^{-40}	5.20×10^{-281}	7.0000	0.3467

Example 6. Consider another standard function, which is given as:

$$f_6(x) = \sin\left(\frac{x\pi}{3}\right)\left(e^{x^2-2x-3} - \cos(x-3) + x^2 - 9\right)\left(\frac{27e^{2(x-3)} - x^3}{28(x^3+1)} + x\cos\frac{x\pi}{6}\right)$$

which has a zero $\alpha = 3$ of multiplicity three. Let us choose the initial approximation $x_0 = 3.5$ for obtaining the zero of the function. Numerical results are shown in Table 6.

Table 6. Comparison of the performance of methods for Example 6.

Methods	n	$|e_{n-3}|$	$|e_{n-2}|$	$|e_{n-1}|$	COC	CPU-Time
GKN-1(a)	4	5.04×10^{-4}	6.20×10^{-22}	2.15×10^{-129}	6.0000	3.8670
GKN-1(b)	4	9.53×10^{-4}	4.36×10^{-20}	3.98×10^{-118}	6.0000	4.1287
GKN-1(c)	4	1.37×10^{-4}	2.87×10^{-25}	2.43×10^{-149}	5.9999	3.8866
GKN-1(d)	4	2.53×10^{-3}	5.53×10^{-17}	6.03×10^{-99}	6.0000	4.5195
GKN-2(a)	5	4.22×10^{-7}	8.51×10^{-41}	9.95×10^{-81}	5.4576	5.5310
GKN-2(b)	4	7.24×10^{-3}	4.58×10^{-14}	2.94×10^{-81}	6.0000	3.9647
GKN-2(c)	4	4.43×10^{-3}	1.12×10^{-15}	2.90×10^{-91}	5.9995	3.7772
GKN-2(d)	8	8.78×10^{-10}	1.75×10^{-55}	1.09×10^{-329}	6.0000	6.2194
NM-1(a)	4	8.78×10^{-3}	1.35×10^{-15}	2.76×10^{-105}	7.0000	1.9372
NM-1(b)	4	3.50×10^{-6}	4.38×10^{-41}	2.10×10^{-285}	7.0000	1.5625
NM-1(c)	4	3.57×10^{-6}	5.15×10^{-41}	6.69×10^{-285}	7.0000	1.5662
NM-1(d)	4	1.83×10^{-6}	2.66×10^{-43}	3.70×10^{-301}	7.0000	1.5788
NM-2(a)	4	3.42×10^{-6}	3.63×10^{-41}	5.51×10^{-286}	7.0000	1.5900
NM-2(b)	4	3.50×10^{-6}	4.36×10^{-41}	2.05×10^{-285}	7.0000	1.5585
NM-2(c)	4	3.57×10^{-6}	5.13×10^{-41}	6.53×10^{-285}	7.0000	1.6405
NM-2(d)	4	1.82×10^{-6}	2.62×10^{-43}	3.30×10^{-301}	7.0000	1.3444

Example 7. Finally, considering yet another standard function:

$$f_7(x) = \left(\cos(x^2+1) - x\log(x^2 - \pi + 2) + 1\right)^3 (x^2 + 1 - \pi).$$

The zero of function f_7 is $\alpha = 1.4632625480850\ldots$ with multiplicity four. We choose the initial approximation $x_0 = 1.3$ to find the zero of this function. Numerical results are displayed in Table 7.

Table 7. Comparison of the performance of methods for Example 7.

Methods	n	$|e_{n-3}|$	$|e_{n-2}|$	$|e_{n-1}|$	COC	CPU-Time
GKN-1(a)	4	6.61×10^{-5}	8.80×10^{-25}	4.90×10^{-144}	6.0000	1.7305
GKN-1(b)	4	6.87×10^{-5}	1.15×10^{-24}	2.57×10^{-143}	6.0000	1.7545
GKN-1(c)	4	6.35×10^{-5}	7.67×10^{-25}	2.38×10^{-144}	6.0000	1.7150
GKN-1(d)	4	1.15×10^{-4}	8.83×10^{-23}	1.82×10^{-131}	6.0000	1.7852
GKN-2(a)	4	5.57×10^{-6}	8.57×10^{-32}	1.14×10^{-186}	6.0000	1.6405
GKN-2(b)	4	1.27×10^{-4}	1.23×10^{-22}	1.02×10^{-130}	6.0000	1.7813
GKN-2(c)	4	7.49×10^{-5}	2.89×10^{-24}	9.62×10^{-141}	6.0000	1.7382
GKN-2(d)	4	1.18×10^{-3}	9.34×10^{-17}	2.31×10^{-95}	6.0000	1.9150
NM-1(a)	4	5.19×10^{-5}	1.05×10^{-28}	1.42×10^{-194}	7.0000	1.0077
NM-1(b)	4	5.29×10^{-5}	1.23×10^{-28}	4.63×10^{-194}	7.0000	0.9062
NM-1(c)	4	5.37×10^{-5}	1.41×10^{-28}	1.23×10^{-193}	7.0000	1.0040
NM-1(d)	4	2.73×10^{-5}	7.07×10^{-31}	5.57×10^{-210}	7.0000	1.0054
NM-2(a)	4	5.14×10^{-5}	9.79×10^{-29}	8.91×10^{-195}	7.0000	0.8867
NM-2(b)	4	5.24×10^{-5}	1.16×10^{-28}	3.02×10^{-194}	7.0000	0.9802
NM-2(c)	4	5.33×10^{-5}	1.34×10^{-28}	8.30×10^{-194}	7.0000	0.9412
NM-2(d)	4	2.60×10^{-5}	5.06×10^{-31}	5.39×10^{-211}	7.0000	0.9142

It is clear from the numerical results shown in Tables 1–7 that the accuracy in the successive approximations increases as the iterations proceed. This shows the stable nature of the methods. Moreover, the present methods like that of existing methods show consistent convergence behavior. We display the value zero of $|e_n|$ in the iteration at which $|x_{n+1} - x_n| + |F(x_n)| < 10^{-350}$. The values of the computational order of convergence exhibited in the penultimate column in each table verify the theoretical order of convergence. However, this is not true for the existing methods GKN-1(a–d) and GKN-2(a) in Example 2. The entries in the last column in each table show that the new methods use less computing time than the time used by existing methods. This verifies the computationally-efficient nature of the new methods. Similar numerical tests, performed for many problems of different types, have confirmed the aforementioned conclusions to a large extent.

We conclude the analysis with an important problem regarding the choice of initial approximation x_0 in the practical application of iterative methods. The required convergence speed of iterative methods can be achieved in practice if the selected initial approximation is sufficiently close to the root. Therefore, when applying the methods for solving nonlinear equations, special care must be given for guessing close initial approximations. Recently, an efficient procedure for obtaining sufficiently close initial approximation has been proposed in [29]. For example, the procedure when applied to the function of Example 1 in the interval [0, 1.5] using the statements:

```
f[x_]=x^6-12x^5+56x^4-130x^3+159x^2-98x+24; a=0; b=1.5;
k=1; x0=0.5*(a+b+Sign[f[a]]*NIntegrate[Tanh[k*f[x]],{x,a,b}])
```

in programming package *Mathematica* yields a close initial approximation $x_0 = 1.04957$ to the root $\alpha = 1$.

5. Conclusions

In the present work, we have designed a class of seventh order derivative-free iterative techniques for computing multiple zeros of nonlinear functions, with known multiplicity. The analysis of convergence shows the seventh order convergence under standard assumptions for the nonlinear function, the zeros of which we have searched. Some special cases of the class were stated. They were applied to solve some nonlinear equations and also compared with existing techniques. Comparison of the numerical results showed that the presented derivative-free methods are good competitors of the existing sixth order techniques that require derivative evaluations. The paper is concluded with the remark that unlike the methods with derivatives, the methods without derivatives are rare in the literature. Moreover, such algorithms are good options to Newton-like iterations in the situation when derivatives are difficult to compute or expensive to obtain.

Author Contributions: Methodology, J.R.S.; writing, review and editing, J.R.S.; investigation, D.K.; data curation, D.K.; conceptualization, I.K.A.; formal analysis, I.K.A.

Funding: This research received no external funding.

Conflicts of Interest: The authors declare no conflict of interest.

References

1. Argyros, I.K. *Convergence and Applications of Newton-Type Iterations*; Springer: New York, NY, USA, 2008.
2. Argyros, I.K.; Magreñán, Á.A. *Iterative Methods and Their Dynamics with Applications*; CRC Press: New York, NY, USA, 2017.
3. Hoffman, J.D. *Numerical Methods for Engineers and Scientists*; McGraw-Hill Book Company: New York, NY, USA, 1992.
4. Schröder, E. Über unendlich viele Algorithmen zur Auflösung der Gleichungen. *Math. Ann.* **1870**, *2*, 317–365. [CrossRef]
5. Hansen, E.; Patrick, M. A family of root finding methods. *Numer. Math.* **1977**, *27*, 257–269. [CrossRef]
6. Behl, R.; Cordero, A.; Motsa, S.S.; Torregrosa, J.R. On developing fourth-order optimal families of methods for multiple roots and their dynamics. *Appl. Math. Comput.* **2015**, *265*, 520–532. [CrossRef]
7. Behl, R.; Cordero, A.; Motsa, S.S.; Torregrosa, J.R.; Kanwar, V. An optimal fourth-order family of methods for multiple roots and its dynamics. *Numer. Algorithms* **2016**, *71*, 775–796. [CrossRef]
8. Geum, Y.H.; Kim, Y.I.; Neta, B. A class of two-point sixth-order multiple-zero finders of modified double-Newton type and their dynamics. *Appl. Math. Comput.* **2015**, *270*, 387–400. [CrossRef]
9. Geum, Y.H.; Kim, Y.I.; Neta, B. A sixth–order family of three–point modified Newton–like multiple–root finders and the dynamics behind their extraneous fixed points. *Appl. Math. Comput.* **2016**, *283*, 120–140. [CrossRef]
10. Li, S.G.; Cheng, L.Z.; Neta, B. Some fourth-order nonlinear solvers with closed formulae for multiple roots. *Comput. Math. Appl.* **2010**, *59*, 126–135. [CrossRef]
11. Li, S.; Liao, X.; Cheng, L. A new fourth-order iterative method for finding multiple roots of nonlinear equations. *Appl. Math. Comput.* **2009**, *215*, 1288–1292.
12. Liu, B.; Zhou, X. A new family of fourth-order methods for multiple roots of nonlinear equations. *Nonlinear Anal. Model. Control* **2013**, *18*, 143–152.
13. Neta, B. Extension of Murakami's high-order nonlinear solver to multiple roots. *Int. J. Comput. Math.* **2010**, *87*, 1023–1031. [CrossRef]
14. Sharifi, M.; Babajee, D.K.R.; Soleymani, F. Finding the solution of nonlinear equations by a class of optimal methods. *Comput. Math. Appl.* **2012**, *63*, 764–774. [CrossRef]
15. Sharma, J.R.; Sharma, R. Modified Jarratt method for computing multiple roots. *Appl. Math. Comput.* **2010**, *217*, 878–881. [CrossRef]
16. Soleymani, F.; Babajee, D.K.R. Computing multiple zeros using a class of quartically convergent methods. *Alex. Eng. J.* **2013**, *52*, 531–541. [CrossRef]
17. Soleymani, F.; Babajee, D.K.R.; Lotfi, T. On a numerical technique for finding multiple zeros and its dynamics. *J. Egypt. Math. Soc.* **2013**, *21*, 346–353. [CrossRef]
18. Victory, H.D.; Neta, B. A higher order method for multiple zeros of nonlinear functions. *Int. J. Comput. Math.* **1983**, *12*, 329–335. [CrossRef]
19. Zhou, X.; Chen, X.; Song, Y. Constructing higher-order methods for obtaining the multiple roots of nonlinear equations. *J. Comput. Math. Appl.* **2011**, *235*, 4199–4206. [CrossRef]
20. Zhou, X.; Chen, X.; Song, Y. Families of third and fourth order methods for multiple roots of nonlinear equations. *Appl. Math. Comput.* **2013**, *219*, 6030–6038. [CrossRef]
21. Osada, N. An optimal multiple root-finding method of order three. *J. Comput. Appl. Math.* **1994**, *51*, 131–133. [CrossRef]
22. Traub, J.F. *Iterative Methods for the Solution of Equations*; Chelsea Publishing Company: New York, NY, USA, 1982.

23. Vrscay, E.R.; Gilbert, W.J. Extraneous fixed points, basin boundaries and chaotic dynamics for Schröder and König rational iteration functions. *Numer. Math.* **1988**, *52*, 1–16. [CrossRef]
24. Varona, J.L. Graphic and numerical comparison between iterative methods. *Math. Intell.* **2002**, *24*, 37–46. [CrossRef]
25. Scott, M.; Neta, B.; Chun, C. Basin attractors for various methods. *Appl. Math. Comput.* **2011**, *218*, 2584–2599. [CrossRef]
26. Lotfi, T.; Sharifi, S.; Salimi, M.; Siegmund, S. A new class of three-point methods with optimal convergence order eight and its dynamics. *Numer. Algorithms* **2015**, *68*, 261–288. [CrossRef]
27. Weerakoon, S.; Fernando, T.G.I. A variant of Newton's method with accelerated third-order convergence. *Appl. Math. Lett.* **2000**, *13*, 87–93. [CrossRef]
28. Danby, J.M.A.; Burkardt, T.M. The solution of Kepler's equation. I. *Celest. Mech.* **1983**, *40*, 95–107. [CrossRef]
29. Yun, B.I. A non-iterative method for solving non-linear equations. *Appl. Math. Comput.* **2008**, *198*, 691–699. [CrossRef]

© 2019 by the authors. Licensee MDPI, Basel, Switzerland. This article is an open access article distributed under the terms and conditions of the Creative Commons Attribution (CC BY) license (http://creativecommons.org/licenses/by/4.0/).

Article

Some Real-Life Applications of a Newly Constructed Derivative Free Iterative Scheme

Ramandeep Behl [1,*], M. Salimi [2], M. Ferrara [3,4], S. Sharifi [5] and Samaher Khalaf Alharbi [1]

1. Department of Mathematics, King Abdulaziz University, Jeddah 21589, Saudi Arabia; samaher271@gmail.com
2. Center for Dynamics and Institute for Analysis, Department of Mathematics, Technische Universität Dresden, 01062 Dresden, Germany; msalimi1@yahoo.com or mehdi.salimi@tu-dresden.de
3. Department of Law, Economics and Human Sciences, University Mediterranea of Reggio Calabria, 89125 Reggio Calabria, Italy; massimiliano.ferrara@unirc.it
4. ICRIOS—The Invernizzi Centre for Research on Innovation, Organization, Strategy and Entrepreneurship, Department of Management and Technology, Bocconi University, Via Sarfatti, 25, 20136 Milano, Italy
5. MEDAlics, Research Center at Università per Stranieri Dante Alighieri, 89125 Reggio Calabria, Italy; somayeh.sharifi69@yahoo.com
* Correspondence: ramanbehl87@yahoo.in

Received: 1 January 2019; Accepted: 28 January 2019; Published: 15 February 2019

Abstract: In this study, we present a new higher-order scheme without memory for simple zeros which has two major advantages. The first one is that each member of our scheme is derivative free and the second one is that the present scheme is capable of producing many new optimal family of eighth-order methods from every 4-order optimal derivative free scheme (available in the literature) whose first substep employs a Steffensen or a Steffensen-like method. In addition, the theoretical and computational properties of the present scheme are fully investigated along with the main theorem, which demonstrates the convergence order and asymptotic error constant. Moreover, the effectiveness of our scheme is tested on several real-life problems like Van der Waal's, fractional transformation in a chemical reactor, chemical engineering, adiabatic flame temperature, etc. In comparison with the existing robust techniques, the iterative methods in the new family perform better in the considered test examples. The study of dynamics on the proposed iterative methods also confirms this fact via basins of attraction applied to a number of test functions.

Keywords: scalar equations; computational convergence order; Steffensen's method; basins of attraction

1. Introduction

In the last few years, several scholars introduced the concept of how to remove derivatives from the iteration functions. The main practical difficulty associated with iterative methods involving derivatives is to calculate first and/or high-order derivatives at each step, which is quite difficult and time-consuming. Computing derivatives of standard nonlinear equations (which are generally considered for academic purposes) is an easy task. On the other hand, in regard to practical problems of calculating the derivatives of functions, it is either very expensive or requires a huge amount of time. Therefore, we need derivative free methods, software or tools which are capable of generating derivatives automatically (for a detailed explanation, please see [1]).

There is no doubt that optimal 8-order multi-point derivative free methods are one of the important classes of iterative methods. They have faster convergence towards the required root and a better efficiency index as compared to Newton/Steffensen's method. In addition, one can easily attain the desired accuracy of any specific number of digits within a small number of iterations with the help of these iterative methods.

In recent years, many scholars have proposed a big number of 8-order derivative free schemes in their research articles [2–18]. However, most of these eighth-order methods are the extensions or modifications of particularly well-known or unknown existing optimal fourth-order derivative free methods; for detailed explanations, please see [5,6,14,16,17]. However, there is no optimal derivative free scheme in a general way that is capable of producing optimal eighth-order convergence from every optimal fourth-order derivative free scheme to date, according to our knowledge.

In this paper, we present a new optimal scheme that doesn't require any derivative. In addition, the proposed scheme is capable of generating new optimal 8-order methods from the earlier optimal fourth-order schemes whose first substep employs Steffensen's or a Steffensen-type method. In this way, our scheme is giving the flexibility in the choice of a second-step to the scholars who can pick any existing optimal derivative free fourth-order method (available in the literature) unlike the earlier studies. The construction of the presented scheme is based on a technique similar to Sharma et al. [19] along with some modifications that can be seen in the next section. We tested the applicability of a newly proposed scheme on a good variety of numerical examples. The obtained results confirm that our methods are more efficient and faster as compared to existing methods in terms of minimum residual error, least asymptotic error constants, minimum error between two consecutive iterations, etc. Moreover, we investigate their dynamic behavior in the complex plane adopting basins of attraction. Dynamic behavior provides knowledge about convergence, and stability of the mentioned methods also supports the theoretical aspects.

2. Construction of the Proposed Scheme

This section is devoted to 8-order derivative free schemes for nonlinear equations. In order to obtain this scheme, we consider a general fourth-order method $\eta(v_j, x_j, y_j)$ in the following way:

$$\begin{cases} y_j = x_j - \dfrac{f(x_j)}{f[v_j, x_j]}, \\ z_j = \eta(v_j, x_j, y_j), \end{cases} \quad (1)$$

where $v_j = x_j + \lambda f(x_j)$, $\lambda \in \mathbb{R}$ and $f[v_j, x_j] = \dfrac{f(v_j) - f(x_j)}{v_j - x_j}$ are the first-order finite difference. We can simply obtain eighth-order convergence by applying the classical Newton's technique, which is given by

$$x_{j+1} = z_j - \dfrac{f(z_j)}{f'(z_j)}. \quad (2)$$

The above scheme is non optimal because it does not satisfy the Kung–Traub conjecture [7]. Thus, we have to reduce the number of evaluations of functions or their derivatives. In this regard, we some approximation of the first-order derivative For this purpose, we need a suitable approximation approach of functions that can approximate the derivatives. Therefore, we choose the following rational functional approach

$$\Omega(x) = \Omega(x_j) - \dfrac{(x - x_j) + \theta_1}{\theta_2(x - x_j)^2 + \theta_3(x - x_j) + \theta_4}, \quad (3)$$

where $\theta_i, i = 1, 2, 3, 4$ are free parameters. This approach is similar to Sharma et al. [19] along with some modifications. We can determine these disposable parameters θ_i by adopting the following tangency constraints

$$\Omega(x_j) = f(x_j), \quad \Omega(v_j) = f(v_j), \quad \Omega(y_j) = f(y_j), \quad \Omega(z_j) = f(z_j). \quad (4)$$

The number of tangency conditions depends on the number of undetermined parameters. If we increase the number of undetermined parameters in the above rational function, then we can also attain high-order convergence (for the detailed explanation, please see Jarratt and Nudds [20]).

By imposing the first tangency condition, we have

$$\theta_1 = 0. \tag{5}$$

The last three tangency conditions provide us with the following three linear equations:

$$\begin{aligned}
\theta_2(v_j - x_j)^2 + \theta_3(v_j - x_j) + \theta_4 &= \frac{1}{f[v_j, x_j]}, \\
\theta_2(y_j - x_j)^2 + \theta_3(y_j - x_j) + \theta_4 &= \frac{1}{f[y_j, x_j]}, \\
\theta_2(z_j - x_j)^2 + \theta_3(z_j - x_j) + \theta_4 &= \frac{1}{f[z_j, x_j]},
\end{aligned} \tag{6}$$

with three unknowns θ_2, θ_3 and θ_4.

After some simplification, we further yield

$$\begin{aligned}
\theta_2 &= \frac{f(v_j) + \theta_4 f[x_j, v_j] f[x_j, y_j](y_j - v_j) - f(y_j)}{(f(v_j) - f(x_j))(f(x_j) - f(y_j))(v_j - y_j)}, \\
\theta_3 &= -\frac{\theta_2 (f(v_j) - f(x_j))(v_j - x_j) + \theta_4 f[x_j, v_j] - 1}{f(v_j) - f(x_j)}, \\
\theta_4 &= \frac{(f(x_j) - f(y_j))(f(x_j) - f(z_j))(y_j - z_j) - a}{f[x_j, v_j] f[x_j, y_j] f[x_j, z_j](v_j - y_j)(v_j - z_j)(y_j - z_j)},
\end{aligned} \tag{7}$$

where $a = (f(v_j) - f(x_j))[(f(x_j) - f(y_j))(v_j - y_j) + (f(z_j) - f(x_j))(v_j - z_j)]$ and $f[\cdot, \cdot]$ are the finite difference of first order. Now, we differentiate the expression (3) with respect to x at the point $x = z_j$, which further provides

$$f'(z_j) \approx \Omega'(z_j) = \frac{\theta_4 - \theta_2(z_j - x_j)^2}{\left[\theta_2(z_j - x_j)^2 + \theta_3(z_j - x_j) + \theta_4\right]^2}. \tag{8}$$

Finally, by using the expressions (1), (2) and (8), we have

$$\begin{cases}
y_j = x_j - \dfrac{f(x_j)}{f[v_j, x_j]}, \\
z_j = \eta(v_j, x_j, y_j), \\
x_{j+1} = z_j - \dfrac{f(z_j)\left[\theta_2(z_j - x_j)^2 + \theta_3(z_j - x_j) + \theta_4\right]^2}{\theta_4 - \theta_2(z_j - x_j)^2},
\end{cases} \tag{9}$$

where v_j and θ_i, $i = 2, 3, 4$ was already explained earlier in the same section. Now, we demonstrate in the next Theorem 1 how a rational function of the form (2) plays an important role in the development of a new derivative free technique. In addition, we confirm the eighth-order of convergence of (9) without considering any extra functional evaluation/s.

3. Convergence Analysis

Theorem 1. *We assume that the function* $f : \mathbb{C} \to \mathbb{C}$ *is analytic in the neighborhood of simple zero* ξ. *In addition, we consider that* $\eta(v_j, x_j, y_j)$ *is any 4-order optimal derivative free iteration function and initial guess* $x = x_0$ *is close enough to the required zero* ξ *for the ensured convergence. The scheme* (9) *reaches an eighth-order convergence.*

Proof. We assume that $e_j = x_j - \xi$ is the error at jth point. We expand the function $f(x_j)$ around the point $x = \xi$ by adopting Taylor's series expansion. Then, we have

$$f(x_j) = c_1 e_j + c_2 e_j^2 + c_3 e_j^3 + c_4 e_j^4 + c_5 e_j^5 + c_6 e_j^6 + c_7 e_j^7 + c_8 e_j^8 + O(e_j^9), \tag{10}$$

where $c_n = \frac{f^{(n)}(\xi)}{n!}$ for $n = 1, 2, \ldots, 8$.

By using the above expression (10), we further obtain

$$v_j - \xi = (1 + \lambda c_1) e_j + \lambda (c_2 e_j^2 + c_3 e_j^3 + c_4 e_j^4 + c_5 e_j^5 + c_6 e_j^6 + c_7 e_j^7 + c_8 e_j^8) + O(e_j^9). \tag{11}$$

Again, we have the following expansion of $f(v_j)$ by adopting the Taylor's series expansion

$$f(v_j) = c_1 (1 + \lambda c_1) e_j + c_2 \left\{ (1 + \lambda c_1)^2 + \lambda c_1 \right\} e_j^2 + \sum_{m=1}^{6} G_m e_j^{m+2} + O(e_j^9), \tag{12}$$

where $G_m = G_m(\lambda, c_1, c_2, \ldots, c_8)$.

By using the expressions (10) and (12), we have

$$y_j - \xi = \left(\frac{1}{c_1} + \lambda \right) c_2 e_j^2 + \frac{c_1 c_3 (\lambda^2 c_1^2 + 3\lambda c_1 + 2) - c_2^2 (\lambda^2 c_1^2 + 2\lambda c_1 + 2)}{c_1^2} e_j^3 \\ + \sum_{m=1}^{5} \bar{G}_m e_j^{m+3} + O(e_j^9). \tag{13}$$

Once again, the Taylor's series expansion of $f(y_j)$ about $x = \xi$ provide

$$f(y_j) = c_2 (1 + \lambda c_1) e_j^2 + \frac{c_1 c_3 (\lambda^2 c_1^2 + 3\lambda c_1 + 2) - c_2^2 (\lambda^2 c_1^2 + 2\lambda c_1 + 2)}{c_1} e_j^3 \\ + \sum_{m=1}^{5} \tilde{G}_m e_j^{m+3} + O(e_j^9). \tag{14}$$

With the help of of expressions (10)–(14), we further obtain

$$\frac{f(x_j) - f(v_j)}{x_j - v_j} = c_1 + c_2 (2 + \lambda c_1) e_j + \left\{ c_3 (\lambda^2 c_1^2 + 3\lambda c_1 + 3) + \lambda c_2^2 \right\} e_j^2 \\ + \sum_{i=1}^{6} H_i e_j^{i+2} + O(e_j^9) \tag{15}$$

and

$$\frac{f(x_j) - f(y_j)}{x_j - y_j} = c_1 + c_2 e_j + \left(c_2^2 \left(\frac{1}{c_1} + \lambda \right) + c_3 \right) e_j^2 + \sum_{i=1}^{6} \bar{H}_i e_j^{i+2} + O(e_j^9), \tag{16}$$

where H_i and $\bar{H}_i$ are the constant functions of some constants λ and c_i, $1 \leq i \leq 8$.

Since we assumed earlier that $\eta(v_j, x_j, y_j)$ is any 4-order optimal derivative free scheme, it is therefore undeniable that it will satisfy the error equation of the following form

$$z_j - \xi = \tau_1 e_j^4 + \tau_2 e_j^5 + \tau_3 e_j^6 + \tau_4 e_j^7 + \tau_5 e_j^8 + O(e_j^9), \tag{17}$$

where $\tau_1 \neq 0$ and τ_i ($1 \leq i \leq 4$) are asymptotic error constants which may depend on some constants λ and c_i, $1 \leq i \leq 8$.

Now, we obtain the following expansion of $f(z_j)$ about $z = \xi$

$$f(z_j) = c_1 \tau_1 e_j^4 + c_1 \tau_2 e_j^5 + c_1 \tau_2 e_j^6 + c_1 \tau_3 e_j^7 + (c_2 \tau_1^2 + c_1 \tau_4) e_j^8 + O(e_j^9). \tag{18}$$

By using (10), (17) and (18), we obtain

$$\frac{f(x_j) - f(z_j)}{x_j - z_j} = c_1 + c_2 e_j + c_3 e_j^2 + c_4 e_j^3 + (c_2 \tau_1 + c_5) e_j^4 + (c_3 \tau_1 + c_2 \tau_2)$$
$$+ c_6) e_j^5 + (c_4 \tau_1 + (c_2 + c_3) \tau_2 + c_7) e_j^6 + (c_5 \tau_1 + c_3 \tau_2 \qquad (19)$$
$$+ c_4 \tau_2 + c_2 \tau_3 + c_8) e_j^7 + O(e_j^8).$$

By using the expressions (10)–(19), we have

$$\frac{f(z_j) \left[(z_j - x_j)^2 \theta_2 + (z_j - x_j) \theta_3 + \theta_4 \right]^2}{\theta_4 - (z_j - x_j)^2 \theta_2} = \tau_1 e_j^4 + \tau_2 e_j^5 + \tau_3 e_j^6 + \tau_4 e_j^7$$
$$- \frac{c_2 \tau_1 \left[c_1^3 \tau_1 + (1 + \lambda c_1)^2 (c_1^2 c_4 + c_2^3 - 2 c_1 c_2 c_3) \right]}{c_1^4} e_j^8 + O(e_j^9). \qquad (20)$$

Finally, by inserting the expressions (17) and (20) in the last sub step of scheme (9), we have

$$e_{j+1} = \frac{c_2 \tau_1 \left[c_1^3 \tau_1 + (1 + \lambda c_1)^2 (c_1^2 c_4 + c_2^3 - 2 c_1 c_2 c_3) \right]}{c_1^4} e_j^8 + O(e_j^9). \qquad (21)$$

It is straightforward to say from the expression (21) that the scheme (9) has 8-order convergence. Since the scheme (9) uses only four values of function (viz. $f(x_j)$, $f(v_j)$, $f(y_j)$ and $f(z_j)$) per step, this is therefore an optimal scheme according to the Kung–Traub conjecture. A single coefficient τ_1 from $\eta(x_j, v_j, y_j)$ occurs in the above error equation and also plays an important role in the development of our scheme. Hence, this completes the proof. □

Remark 1. *In general, it is quite obvious that one thinks that the asymptotic error constant in the error equation of scheme (9) may rely on some other constants λ, c_i, $1 \leq i \leq 8$ and τ_j, $1 \leq j \leq 5$. There is no doubt that the expression (21) confirms that the asymptotic error constant is dependent only on λ, c_1, c_2, c_3, c_4 and τ_1. This clearly demonstrates that our current rational function approach with the tangency constraints contributes a significant role in the construction of a new scheme with 8-order convergence.*

4. Numerical Examples

Here, we checked the effectiveness, convergence behavior and efficiency of our schemes with the other existing optimal eighth-order schemes without derivatives. Therefore, we assume that, out of five problems, four of them are from real-life problems, e.g., a fractional conversion problem of the chemical reactor, Van der Waal's problem, the chemical engineering problem and the adiabatic flame temperature problem. The fifth one is a standard nonlinear problem of a piecewise continuous function, which is displayed in the following Examples (1)–(5). The desired solutions are available up to many significant digits (minimum thousand), but, due to the page restriction, only 30 significant places are also listed in the corresponding example.

For comparison purposes, we require the second sub-step in the presented technique. We can choose any optimal derivative free method from the available literature whose first sub-step employs Steffensen's or a Steffensen-type method. Now, we assume some special cases of our scheme that are given as below:

1. We choose an optimal derivative free fourth-order method (6) suggested by Cordero and Torregrosa [3]. Then, we have

$$\begin{cases} y_j = x_j - \dfrac{f(x_j)^2}{f(x_j + f(x_j)) - f(x_j)}, \\ z_j = y_j - \dfrac{f(y_j)}{\dfrac{af(y_j) - bf(v_j)}{y_j - v_j} + \dfrac{cf(y_j) - df(x_j)}{y_j - x_j}}, \\ x_{j+1} = z_j - \dfrac{f(z_j)\left[\theta_2(z_j - x_j)^2 + \theta_3(z_j - x_j) + \theta_4\right]^2}{\theta_4 - \theta_2(z_j - x_j)^2}, \end{cases} \quad (22)$$

where $a, b, c, d \in \mathbb{R}$ such that $a = c = 1$ and $b + d = 1$. We consider $a = b = c = 1$ and $d = 0$ in expression (26) for checking the computational behavior, denoted by $(PM1_8)$.

2. We consider another 4-order optimal method (11) presented by Liu et al. in [8]. Then, we obtain the following new optimal 8-order derivative free scheme

$$\begin{cases} y_j = x_j - \dfrac{f(x_j)^2}{f(x_j + f(x_j)) - f(x_j)}, \\ z_j = y_j - \dfrac{f[y_j, x_j] - f[v_j, y_j] + f[v_j, x_j]}{(f[y_j, x_j])^2} f(y_j), \\ x_{j+1} = z_j - \dfrac{f(z_j)\left[\theta_2(z_j - x_j)^2 + \theta_3(z_j - x_j) + \theta_4\right]^2}{\theta_4 - \theta_2(z_j - x_j)^2}, \end{cases} \quad (23)$$

Let us call the above expression $(PM2_8)$ for computational experimentation.

3. Once again, we pick expression (12) from a scheme given by Ren et al. in [10]. Then, we obtain another interesting family

$$\begin{cases} y_j = x_j - \dfrac{f(x_j)^2}{f(x_j + f(x_j)) - f(x_j)}, \\ z_j = y_j - \dfrac{f(y_j)}{f[y_j, x_j] + f[v_j, y_j] - f[v_j, x_j] + a(y_j - x_j)(y_j - v_j)}, \\ x_{j+1} = z_j - \dfrac{f(z_j)\left[(z_j - x_j)^2 \theta_2 + (z_j - x_j)\theta_3 + \theta_4\right]^2}{\theta_4 - (z_j - x_j)^2 \theta_2}, \end{cases} \quad (24)$$

where $a \in \mathbb{R}$. We choose $a = 1$ in (30), known as $(PM3_8)$.

4. Now, we assume another 4-order optimal method (12), given by Zheng et al. in [18], which further produces

$$\begin{cases} y_j = x_j - \dfrac{f(x_j)^2}{f(x_j + f(x_j)) - f(x_j)}, \\ z_j = y_j - \left[\dfrac{f[y_j, x_j] + (p-1)f[v_j, y_j] - (p-1)f[v_j, x_j] - b(y_j - x_j)(y_j - v_j)}{f[y_j, x_j] + pf[v_j, y_j] - pf[v_j, x_j] + a(y_j - x_j)(y_j - v_j)}\right] \times \dfrac{f(y_j)}{f[y_j, x_j]}, \\ x_{j+1} = z_j - \dfrac{f(z_j)\left[\theta_2(z_j - x_j)^2 + \theta_3(z_j - x_j) + \theta_4\right]^2}{\theta_4 - \theta_2(z_j - x_j)^2}, \end{cases} \quad (25)$$

where $a, b, p \in \mathbb{R}$. We choose $p = 2$ and $a = b = 0$ in (31), called $(PM4_8)$.

Now, we compare them with iterative methods presented by Kung–Traub [7]. Out of these, we considered an optimal eighth-order method, called KT_8. We also compare them with a derivative free optimal family of 8-order iterative functions given by Kansal et al. [5]. We have picked expression

(23) out of them, known as KM_8. Finally, we contrast them with the optimal derivative free family of 8-order methods suggested by Soleymani and Vanani [14], out of which we have chosen the expression (21), denoted by SV_8.

We compare our methods with existing methods on the basis of approximated zeros (x_j), absolute residual error $(|f(x_j)|)$, error difference between two consecutive iterations $|x_{j+1} - x_j|$, $\left|\frac{e_{j+1}}{e_j^8}\right|$, asymptotic error constant $\eta = \lim_{n\to\infty}\left|\frac{e_{j+1}}{e_j^8}\right|$ and computational convergence order $\rho \approx \frac{\ln|\check{e}_{j+1}/\check{e}_j|}{\ln|\check{e}_j/\check{e}_{n-1}|}$, where $\check{e}_j = x_j - x_{n-1}$ (for the details, please see Cordero and Torregrosa [21]) and the results are mentioned in Tables 1–5.

The values of all above-mentioned parameters are available for many significant digits (with a minimum of a thousand digits), but, due to the page restrictions, results are displayed for some significant digits (for the details, please see Tables 1–5). The values of all these parameters have been calculated by adopting programming package *Mathematica* 9 for multiple precision arithmetic. Finally, the meaning of $a_1(\pm a_2)$ is $a_1 \times 10^{(\pm a_2)}$ in the following Tables 1–5.

Table 1. Convergence performance of distinct 8-order optimal derivative free methods for $f_1(x)$.

Cases	j	x_j	$|f(x_j)|$	$|x_{j+1} - x_j|$	ρ	$\left|\frac{x_{j+1}-x_j}{(x_j-x_{n-1})^8}\right|$	η
KT_8	1	0.75742117642117592668	2.0(−3)	2.5(−5)			
	2	0.75739624625375387946	1.0(−19)	1.3(−21)		8.387722076(+15)	8.409575862(+15)
	3	0.75739624625375387946	4.0(−150)	5.1(−152)	7.9999	8.409575862(+15)	
KM_8	1	0.75739472392262620965	1.2(−4)	1.5(−6)			
	2	0.75739624625375387946	1.2(−34)	1.5(−36)		5.252005934(+10)	2.765111335(+10)
	3	0.75739624625375387946	6.1(−275)	7.7(−277)	8.0093	2.765111335(+10)	
SV_8	1	0.75726839017571335554	1.0(−2)	1.3(−4)			
	2	0.75739624625375406009	1.4(−14)	1.8(−16)		2.529459671(+15)	1.540728199(+14)
	3	0.75739624625375387946	1.4(−110)	1.7(−112)	8.1026	1.540728199(+14)	
$PM1_8$	1	0.75739624679631343572	4.3(−8)	5.4(−10)			
	2	0.75739624625375387946	7.9(−60)	9.9(−62)		1.318011692(+13)	1.318013290(+13)
	3	0.75739624625375387946	9.7(−474)	1.2(−475)	8.0000	1.318013290(+13)	
$PM2_8$	1	0.75739624527627277118	7.8(−8)	9.8(−10)			
	2	0.75739624625375387946	5.3(−58)	6.7(−60)		8.002563231(+12)	8.002546457(+12)
	3	0.75739624625375387946	2.5(−459)	3.1(−461)	8.0000	8.002546457(+12)	
$PM3_8$	1	0.75739624669712714014	3.5(−8)	4.4(−10)			
	2	0.75739624625375387946	1.6(−60)	2.0(−62)		1.316590806(+13)	1.316592111(+13)
	3	0.75739624625375387946	2.3(−479)	2.9(−481)	8.0000	1.316592111(+13)	
$PM4_8$	1	0.75739625664695918279	8.3(−7)	1.0(−8)			
	2	0.75739624625375387946	1.7(−49)	2.1(−51)		1.522844707(+13)	1.522886893(+13)
	3	0.75739624625375387946	4.1(−391)	5.2(−393)	8.0000	1.522886893(+13)	

Table 2. Convergence performance of distinct 8-order optimal derivative free methods for $f_2(x)$.

Cases	j	x_j	$\|f(x_j)\|$	$\|e_j\|$	ρ	$\left\|\dfrac{e_{j+1}}{e_j^8}\right\|$	η
KT_8	1	1.9299358075659180242	7.7(−6)	9.0(−5)			
	2	1.9298462428478622185	9.2(−28)	1.1(−26)		2.570367432(+6)	2.580781373(+6)
	3	1.9298462428478622185	3.7(−203)	4.3(−202)	7.9999	2.580781373(+6)	
KM_8	1	1.9300063313329939091	1.4(−5)	1.6(−4)			
	2	1.9298462428478622185	7.0(−26)	8.1(−25)		1.872886840(+6)	1.859196359(+6)
	3	1.9298462428478622185	2.9(−188)	3.4(−187)	8.0002	1.859196359(+6)	
SV_8	1	1.9299298655571245217	7.2(−6)	8.4(−5)			
	2	1.9298462428478622185	2.6(−30)	3.0(−29)		1.272677056(+4)	5.345691399(+3)
	3	1.9298462428478622185	3.4(−226)	3.9(−225)	8.0154	5.345691399(+3)	
$PM1_8$	1	1.9298703396056890283	2.1(−6)	2.4(−5)			
	2	1.9298462428478622185	3.2(−33)	3.7(−32)		3.292189981(+5)	3.294743419(+5)
	3	1.9298462428478622185	1.1(−247)	1.3(−246)	8.0000	3.294743419(+5)	
$PM2_8$	1	1.9299039277100182896	5.0(−6)	5.8(−5)			
	2	1.9298462428478622185	1.5(−29)	1.7(−28)		1.415845181(+6)	1.419322205(+6)
	3	1.9298462428478622185	1.0(−217)	1.2(−216)	8.0000	1.419322205(+6)	
$PM3_8$	1	1.9298835516272248348	3.2(−6)	3.7(−5)			
	2	1.9298462428478622185	2.0(−31)	2.3(−30)		6.132728979(+5)	6.140666943(+5)
	3	1.9298462428478622185	4.2(−233)	4.8(−232)	8.0000	6.140666943(+5)	
$PM4_8$	1	1.9298454768935056951	6.6(−8)	7.7(−7)			
	2	1.9298462428478622185	1.6(−46)	1.9(−45)		1.600600022(+4)	1.600542542(+4)
	3	1.9298462428478622185	2.3(−355)	2.7(−354)	8.0000	1.600542542(+4)	

Table 3. Convergence performance of distinct 8-order optimal derivative free methods for $f_3(x)$.

Cases	j	x_j	$\|f(x_j)\|$	$\|e_j\|$	ρ	$\left\|\dfrac{e_{j+1}}{e_j^8}\right\|$	η
KT_8	1	$3.94856259325568 + 0.31584953607444i$	2.8(−3)	2.7(−4)			
	2	$3.94854244556204 + 0.31612357089701i$	1.1(−21)	1.1(−22)		3.278944412(+6)	3.291035449(+6)
	3	$3.94854244556204 + 0.31612357089701i$	5.5(−169)	5.5(−170)	7.999	3.291035449(+6)	
KM_8	1	$3.94541341953964 + 0.28830540896626i$	2.7(−1)	2.8(−2)			
	2	$3.94854253806613 + 0.31612376121596i$	2.1(−6)	2.1(−7)		5.611004628(+5)	1.267588109(+4)
	3	$3.94854244556204 + 0.31612357089701i$	5.2(−49)	5.1(−50)	8.3214	1.267588109(+4)	
SV_8	1	$3.94857741336794 + 0.31574108761478i$	3.9(−3)	3.8(−4)			
	2	$3.94854244556204 + 0.31612357089701i$	9.1(−21)	9.0(−22)		1.895162520(+6)	1.896706799(+6)
	3	$3.94854244556204 + 0.31612357089701i$	8.1(−162)	8.0(−163)	8.0000	1.896706799(+6)	
$PM1_8$	1	$3.94848048827814 + 0.31602117152370i$	1.2(−3)	1.2(−4)			
	2	$3.94854244556204 + 0.31612357089701i$	2.5(−25)	2.5(−26)		5.923125406(+5)	5.903970786(+5)
	3	$3.94854244556204 + 0.31612357089701i$	8.9(−199)	8.8(−200)	8.0001	5.903970786(+5)	
$PM2_8$	1	$3.94846874984553 + 0.31601667713734i$	1.3(−3)	1.3(−4)			
	2	$3.94854244556204 + 0.31612357089701i$	5.1(−25)	5.0(−26)		6.241093912(+5)	6.214835024(+5)
	3	$3.94854244556204 + 0.31612357089701i$	2.6(−196)	2.6(−197)	8.0001	6.214835024(+5)	
$PM3_8$	1	$3.94848290176499 + 0.31601668833975i$	1.2(−3)	1.2(−4)			
	2	$3.94854244556204 + 0.31612357089701i$	3.1(−25)	3.1(−26)		6.078017700(+5)	6.059534898(+5)
	3	$3.94854244556204 + 0.31612357089701i$	4.6(−198)	4.6(−199)	8.0001	6.059534898(+5)	
$PM4_8$	1	$3.94849208916059 + 0.31602400692668i$	1.1(−3)	1.1(−4)			
	2	$3.94854244556204 + 0.31612357089701i$	1.4(−25)	1.4(−26)		5.704624073(+5)	5.691514905(+5)
	3	$3.94854244556204 + 0.31612357089701i$	7.1(−201)	7.1(−202)	8.0000	5.691514905(+5)	

Table 4. Convergence performance of distinct 8-order optimal derivative free methods for $f_4(x)$.

Cases	j	x_j	$|f(x_j)|$	$|e_j|$	ρ	$\left|\dfrac{e_{j+1}}{e_j^8}\right|$	η
KT_8	1	4305.3099136661255630	3.3(−19)	1.5(−20)			
	2	4305.3099136661255630	1.1(−179)	4.8(−181)		2.234387851(−22)	2.234387851(−22)
	3	4305.3099136661255630	1.4(−1463)	6.3(−1465)	8.0000	2.234387851(−22)	
KM_8	1	4305.4966166546986926	4.2	1.9(−1)			
	2	4305.3099136647999238	3.0(−8)	1.3(−9)		8.978735581(−4)	1.132645694(−16)
	3	4305.3099136661255630	2.4(−86)	1.1(−87)	9.5830	1.132645694(−16)	
SV_8	1	4305.3099136661255630	1.5(−19)	6.9(−21)			
	2	4305.3099136661255630	1.2(−182)	5.4(−184)		1.038308478(−22)	1.038308478(−22)
	3	4305.3099136661255630	1.6(−1487)	7.1(−1489)	8.0000	1.038308478(−22)	
$PM1_8$	1	4305.3099136661255630	3.5(−20)	1.6(−21)			
	2	4305.3099136661255630	2.1(−188)	9.3(−190)		2.393094045(−23)	2.393094045(−23)
	3	4305.3099136661255630	3.1(−1534)	1.4(−1535)	8.0000	2.393094045(−23)	
$PM2_8$	1	4305.3099136661255630	4.0(−20)	1.8(−21)			
	2	4305.3099136661255630	5.8(−188)	2.6(−189)		2.683028981(−23)	2.683028981(−23)
	3	4305.3099136661255630	1.3(−1530)	5.6(−1532)	8.0000	2.683028981(−23)	
$PM3_8$	1	4305.3099136690636946	6.6(−8)	2.8(−9)			
	2	4305.3099136661255630	8.8(−77)	3.9(−78)		7.055841652(−10)	7.055841652(−10)
	3	4305.3099136661255630	8.8(−628)	3.9(−629)	8.0000	7.055841652(−10)	
$PM4_8$	1	4305.3099136661255630	4.0(−20)	1.8(−21)			
	2	4305.3099136661255630	5.8(−188)	2.6(−189)		2.119306545(−23)	2.119306545(−23)
	3	4305.3099136661255630	1.3(−1530)	5.6(−1532)	8.0000	2.119306545(−23)	

Table 5. Convergence performance of distinct 8-order optimal derivative free methods for $f_5(x)$.

Cases	j	x_j	$|f(x_j)|$	$|e_j|$	ρ	$\left|\dfrac{e_{j+1}}{e_j^8}\right|$	η
KT_8	1	1.4142135646255204265	6.4(−9)	2.3(−9)			
	2	1.4142135623730950488	2.8(−69)	9.8(−70)		1.483428355	1.483428382
	3	1.4142135623730950488	3.7(−552)	1.3(−552)	8.0000	1.483428382	
KM_8	1	1.4141886104951680577	7.1(−5)	2.5(−5)			
	2	1.4142135641342028617	5.0(−9)	1.8(−9)		1.171425936(+28)	0.1339769256
	3	1.4142135623730950488	3.5(−71)	1.2(−71)	14.972	0.1339769256	
SV_8	1	1.4142135639458229191	4.4(−9)	1.6(−9)			
	2	1.4142135623730950488	8.4(−71)	3.0(−71)		0.7923194647	0.7923194693
	3	1.4142135623730950488	1.3(−564)	4.7(−564)	8.0000	0.7923194693	
$PM1_8$	1	1.4142135629037874832	1.5(−9)	5.3(−10)			
	2	1.4142135623730950488	5.3(−75)	1.9(−75)		0.2966856754	0.2966856763
	3	1.4142135623730950488	1.2(−598)	4.4(−599)	8.0000	0.2966856763	
$PM2_8$	1	1.4142135630941303743	2.0(−9)	7.2(−10)			
	2	1.4142135623730950488	8.7(−74)	3.1(−74)		0.4230499025	0.4230499045
	3	1.4142135623730950488	1.0(−588)	3.5(−589)	8.0000	0.4230499045	
$PM3_8$	1	1.4142135672540404368	1.4(−8)	4.9(−9)			
	2	1.4142135623730950488	2.5(−66)	8.8(−67)		2.742159025	2.742159103
	3	1.4142135623730950488	2.9(−528)	1.0(−528)	8.0000	2.742159103	
$PM4_8$	1	1.4142135627314914846	1.0(−9)	3.6(−10)			
	2	1.4142135623730950488	1.5(−76)	5.2(−77)		0.1905635592	0.1905635596
	3	1.4142135623730950488	2.8(−611)	1.0(−611)	8.0000	0.1905635596	

Example 1. *Chemical reactor problem:*

In regard to fraction transformation in a chemical reactor, we consider

$$f_1(x) = \frac{x}{1-x} - 5\log\left[\frac{0.4(1-x)}{0.4-0.5x}\right] + 4.45977, \tag{26}$$

where the variable x denotes a fractional transformation of a particular species A in the chemical reactor problem (for a detailed explanation, please have a look at [22]). It is important to note that, if $x \leq 0$, then the expression (26) has no physical meaning. Hence, this expression has only a bounded region $0 \leq x \leq 1$, but its derivative is approaching zero in the vicinity of this region. Therefore, we have to take care of these facts while choosing required zero and initial approximations, which we consider as $\xi = 0.75739624625375387945964129792$ and $x_0 = 0.76$, respectively:

Example 2. *Van der Waal's equation:*

$$\left(P + \frac{a_1 n^2}{V^2}\right)(V - na_2) = nRT. \tag{27}$$

The above expression interprets real and ideal gas behavior with variables a_1 and a_2, respectively. For calculating the gas volume V, we can rewrite the above expression (27) in the following way:

$$PV^3 - (na_2 P + nRT)V^2 + a_1 n^2 V - a_1 a_2 n^2 = 0. \tag{28}$$

By considering the particular values of parameters, namely a_1 and a_2, n, P and T, we can easily get the following nonlinear function:

$$f_2(x) = 0.986x^3 - 5.181x^2 + 9.067x - 5.289. \tag{29}$$

The function f_2 has three zeros and our required zero is $\xi = 1.92984624284786221848752742787$. In addition, we consider the initial guess as $x_0 = 2$.

Example 3. *If we convert the fraction of nitrogen–hydrogen to ammonia, then we obtain the following mathematical expression (for more details, please see [23,24])*

$$f_3(z) = z^4 - 7.79075z^3 + 14.7445z^2 + 2.511z - 1.674. \tag{30}$$

The f_3 has four zeros and our required zero is $\xi = 3.9485424455620457727 + 0.3161235708970163733i$. In addition, we consider the initial guess as $x_0 = 4 + 0.25i$.

Example 4. *Let us assume an adiabatic flame temperature equation, which is given by*

$$f_4(x) = \Delta H + \alpha(x - 298) + \frac{\beta}{2}(x^2 - 298^2) + \frac{\gamma}{3}(x^3 - 298^3), \tag{31}$$

where $\Delta H = -57798$, $\alpha = 7.256$, $\beta = 2.298 \times 10^{-3}$ and $\gamma = 0.283 \times 10^{-6}$. For the details of this function, please see the research articles [24,25]. This function has a simple zero $\xi = 4305.30991366612556304019892945$ and assumes the initial approximation is $x_0 = 4307$ for this problem.

Example 5. *Finally, we assume a piece-wise continuous function [5], which is defined as follows:*

$$f_5(x) = \begin{cases} -(x^2 - 2), & \text{if } x < \sqrt{2}, \\ x^2 - 2, & \text{if } x \geq \sqrt{2}. \end{cases} \tag{32}$$

The above function has a simple zero $\xi = 1.41421356237309504880168872421$ with an initial guess being $x_0 = 1.5$.

5. Graphical Comparison by Means of Attraction Basins

It is known that a good selection of initial guesses plays a definitive role in iterative methods—in other words, that all methods converge if the initial estimation is chosen suitably. We numerically approximate the domain of attraction of the zeros as a qualitative measure of how demanding the method on the initial approximation of the root is. In order to graphically compare by means of attraction basins, we investigate the dynamics of the new methods $PM1_8$, $PM2_8$, $PM3_8$ and $PM4_8$ and compare them with available methods from the literature, namely SM_8, KT_8 and KM_8. For more details and many other examples of the study of the dynamic behavior for iterative methods, one can consult [26–29].

Let $Q : \mathbb{C} \to \mathbb{C}$ be a rational map on the complex plane. For $z \in \mathbb{C}$, we define its orbit as the set $\text{orb}(z) = \{z, Q(z), Q^2(z), \dots\}$. A point $z_0 \in \mathbb{C}$ is called a periodic point with minimal period m if $Q^m(z_0) = z_0$, where m is the smallest positive integer with this property (and thus $\{z_0, Q(z_0), \dots, Q^{m-1}(z_0)\}$ is a cycle). The point having minimal period 1 is known as a fixed point. In addition, the point z_0 is called repelling if $|Q'(z_0)| > 1$, attracting if $|Q'(z_0)| < 1$, and neutral otherwise. The Julia set of a nonlinear map $Q(z)$, denoted by $J(Q)$, is the closure of the set of its repelling periodic points. The complement of $J(Q)$ is the Fatou set $F(Q)$.

In our case, the methods $PM1_8$, $PM2_8$, $PM3_8$ and $PM4_8$ and SM_8, KT_8 and KM_8 provide the iterative rational maps $Q(z)$ when they are applied to find the roots of complex polynomials $p(z)$. In particular, we are interested in the basins of attraction of the roots of the polynomials where the basin of attraction of a root z^* is the complex set $\{z_0 \in \mathbb{C} : \text{orb}(z_0) \to z^*\}$. It is well known that the basins of attraction of the different roots lie in the Fatou set $F(Q)$. The Julia set $J(Q)$ is, in general, a fractal and, in it, the rational map Q is unstable.

For a graphical point of view, we take a 512×512 grid of the square $[-3, 3] \times [-3, 3] \subset \mathbb{C}$ and assign a color to each point $z_0 \in D$ according to the simple root to which the corresponding orbit of the iterative method starting from z_0 converges, and we mark the point as black if the orbit does not converge to a root in the sense that, after at most 15 iterations, it has a distance to any of the roots that is larger than 10^{-3}. We have used only 15 iterations because we are using eighth-order methods. Therefore, if the method converges, it is usually very fast. In this way, we distinguish the attraction basins by their color.

Different colors are used for different roots. In the basins of attraction, the number of iterations needed to achieve the root is shown by the brightness. Brighter color means less iteration steps. Note that black color denotes lack of convergence to any of the roots. This happens, in particular, when the method converges to a fixed point that is not a root or if it ends in a periodic cycle or at infinity. Actually and although we have not done it in this paper, infinity can be considered an ordinary point if we consider the Riemann sphere instead of the complex plane. In this case, we can assign a new "ordinary color" for the basin of attraction of infinity. Details for this idea can be found in [30].

We have tested several different examples, and the results on the performance of the tested methods were similar. Therefore, we merely report the general observation here for two test problems in the following Table 6.

Table 6. Test problems p_1 and p_2 and their roots.

Test Problem	Roots
$p_1(z) = z^2 - 1$	$1, -1$
$p_2(z) = z^2 - z - 1/z$	$1.46557, -0.232786 \pm 0.792552i$

From Figures 1 and 2, we conclude that our methods, namely, $PM1_8$, $PM3_8$ and $PM4_8$, are showing less chaotic behavior and have less non-convergent points as compared to the existing methods, namely SM_8 and KM_8. In addition, our methods, namely, $PM1_8$, $PM3_8$ and $PM4_8$, have almost similar basins of attraction to KT_8. On the other hand, Figures 3 and 4 confirm that our methods, namely, $PM1_8$, $PM2_8$ $PM3_8$ and $PM4_8$, have less divergent points as compared to the existing methods, namely KT_8 and KM_8. There is no doubt that the SM_8 behavior is better than all other mentioned methods, namely, $PM1_8$, $PM3_8$ and $PM4_8$ in problem $p_2(z)$ in terms of chaos.

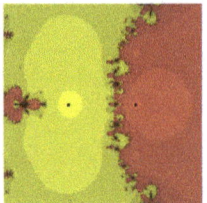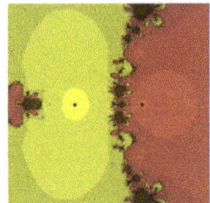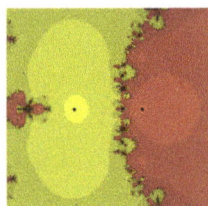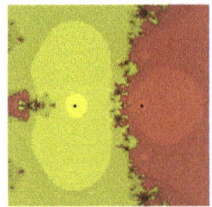

Figure 1. The dynamical behavior of our methods namely, $PM1_8$, $PM2_8$, $PM3_8$ and $PM4_8$, respectively, from left to right for test problem $p_1(z)$.

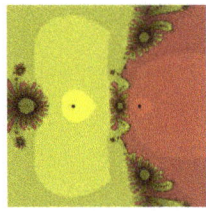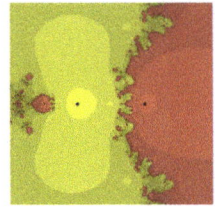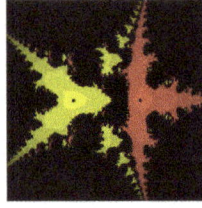

Figure 2. The dynamical behavior of methods SM_8, KT_8 and KM_8, respectively, from left to right for test problem $p_1(z)$.

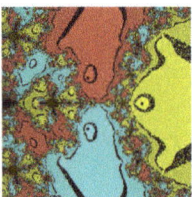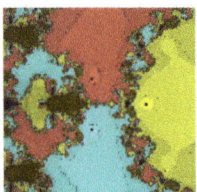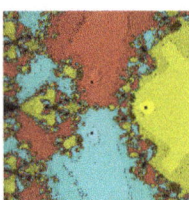

Figure 3. The dynamical behavior of our methods namely, $PM1_8$, $PM2_8$, $PM3_8$ and $PM4_8$, respectively, from left to right for test problem $p_2(z)$.

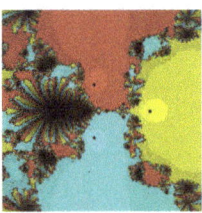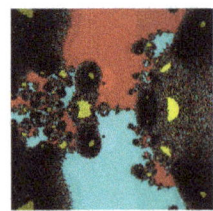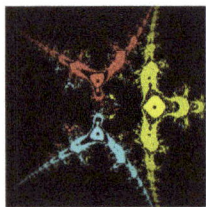

Figure 4. The dynamical behavior of methods SM_8, KT_8 and KM_8, respectively from left to right for test problem $p_2(z)$.

6. Conclusions

In this study, we present a new technique of eighth-order in a general way. The main advantages of our technique are that is a derivative free scheme, there is a choice of flexibility at the second substep, and it is capable of generating new 8-order derivative free schemes from every optimal 4-order method employing Steffensen's or Steffensen-type methods. Every member of (9) is an optimal method according to Kung–Traub conjecture. It is clear from the obtained results in Tables 1–5 that our methods have minimum residual error $|f(x_j)|$, the difference between two consecutive iterations $|x_{j+1} - x_j|$, and stable computational convergence order as compared to existing methods, namely, SM_8, KT_8 and KM_8. The dynamic study of our methods also confirms that they perform better than existing ones of similar order.

It is important to note that we are not claiming that our methods will always be superior to these methods. One may obtain different results when they rest them on distinct nonlinear functions because the computational results depend on several constraints, including initial approximation, body structure of the iterative method, the considered test problem, configuration of the used system and programming softwares, etc. In future work, we will try to obtain a new family of high-order optimal derivative free iteration functions that depend on the rational functional approach.

Author Contributions: All the authors have equal contribution to this study.

Funding: No funding for this paper.

Conflicts of Interest: The authors declare no conflict of interest.

References

1. Griewank, A.; Walther, A. *Evaluating Derivatives: Principles and Techniques of Algorithmic Differentiation*, 2nd ed.; SIAM: Philadelphia, PA, USA, 2008.
2. Behl, R.; Maroju, P.; Motsa, S.S. A family of second derivative free fourth order continuation method for solving nonlinear equations. *J. Comut. Appl. Math.* **2017**, *318*, 38–46. [CrossRef]
3. Cordero, A.; Torregrosa, J.R. A class of Steffensen type methods with optimal order of convergence. *Appl. Math. Comput.* **2011**, *217*, 7653–7659. [CrossRef]
4. Cordero, A.; Hueso, J.L.; Martínez, E.; Torregrosa, J.R. Steffensen type methods for solving nonlinear equations. *Appl. Math. Comput.* **2012**, *236*, 3058–3064. [CrossRef]
5. Kansal, M.; Kanwar, V.; Bhatia, S. An optimal eighth-order derivative-free family of Potra-Pták's method. *Algorithms* **2015**, *8*, 309–320. [CrossRef]
6. Khattri, S.K.; Steihaug, T. Algorithm for forming derivative-free optimal methods. *Numer. Algor.* **2014**, *65*, 809–824. [CrossRef]
7. Kung, H.T.; Traub, J.F. Optimal order of one-point and multi-point iteration. *J. ACM* **1974**, *21*, 643–651. [CrossRef]
8. Liu, Z.; Zheng, Q.; Zhao, P. A variant of Steffensen's method of fourth-order convergence and its applications. *Appl. Math. Comput.* **2010**, *216*, 1978–1983. [CrossRef]
9. Matthies, G.; Salimi, M.; Sharifi, S.; Varona, J.L. An optimal eighth-order iterative method with its dynamics. *Jpn. J. Ind. Appl. Math.* **2016**, *33*, 751–766. [CrossRef]
10. Ren, H.; Wu, Q.; Bi, W. A class of two-step Steffensen type methods with fourth-order convergence. *Appl. Math. Comput.* **2009**, *209*, 206–210. [CrossRef]
11. Salimi, M.; Lotfi, T.; Sharifi, S.; Siegmund, S. Optimal Newton-Secant like methods without memory for solving nonlinear equations with its dynamics. *Int. J. Comput. Math.* **2017**, *94*, 1759–1777. [CrossRef]
12. Salimi, M.; Nik Long, N.M.A.; Sharifi, S.; Pansera, B.A. A multi-point iterative method for solving nonlinear equations with optimal order of convergence. *Jpn. J. Ind. Appl. Math.* **2018**, *35*, 497–509. [CrossRef]
13. Sharifi, S.; Salimi, M.; Siegmund, S.; Lotfi, T. A new class of optimal four-point methods with convergence order 16 for solving nonlinear equations. *Math. Comput. Simul.* **2016**, *119*, 69–90. [CrossRef]
14. Soleymani, F.; Vanani, S.K. Optimal Steffensen-type methods with eighth order of convergence. *Comput. Math. Appl.* **2011**, *62*, 4619–4626. [CrossRef]
15. Traub, J.F. *Iterative Methods for the Solution of Equations*; Prentice-Hall, Englewood Cliffs: Upper Saddle River, NJ, USA, 1964.
16. Thukral, R. Eighth-order iterative methods without derivatives for solving nonlinear equations. *Int. Sch. Res. Net. Appl. Math.* **2011**, *2011*, 693787. [CrossRef]
17. Zheng, Q.; Li, J.; Huang, F. An optimal Steffensen-type family for solving nonlinear equations. *Appl. Math. Comput.* **2011**, *217*, 9592–9597. [CrossRef]
18. Zheng, Q.; Zhao, P.; Huang, F. A family of fourth-order Steffensen-type methods with the applications on solving nonlinear ODEs. *Appl. Math. Comput.* **2011**, *217*, 8196–8203. [CrossRef]
19. Sharma, J.R.; Guhaa, R.K.; Gupta, P. Improved King's methods with optimal order of convergence based on rational approximations. *Appl. Math. Lett.* **2013**, *26*, 473–480. [CrossRef]
20. Jarratt, P.; Nudds, D. The use of rational functions in the iterative solution of equations on a digital computer. *Comput. J.* **1965**, *8*, 62–65. [CrossRef]

21. Cordero, A.; Torregrosa, J.R. Variants of Newton's method using fifth-order quadrature formulas. *Appl. Math. Comput.* **2007**, *190*, 686–698. [CrossRef]
22. Shacham, M. Numerical solution of constrained nonlinear algebraic equations. *Int. J. Numer. Method Eng.* **1986**, *23*, 1455–1481. [CrossRef]
23. Balaji, G.V.; Seader, J.D. Application of interval Newton's method to chemical engineering problems. *Reliab. Comput.* **1995**, *1*, 215–223. [CrossRef]
24. Shacham, M. An improved memory method for the solution of a nonlinear equation. *Chem. Eng. Sci.* **1989**, *44*, 1495–1501. [CrossRef]
25. Shacham, M.; Kehat, E. Converging interval methods for the iterative solution of nonlinear equations. *Chem. Eng. Sci.* **1973**, *28*, 2187–2193. [CrossRef]
26. Ezquerro, J.A.; Hernández, M.A. An optimization of Chebyshev's method. *J. Complex.* **2009**, *25*, 343–361. [CrossRef]
27. Ferrara, M.; Sharifi, S.; Salimi, M. Computing multiple zeros by using a parameter in Newton-Secant method. *SeMA J.* **2017**, *74*, 361–369. [CrossRef]
28. Stewart, B.D. Attractor Basins of Various Root-Finding Methods. Master's Thesis, Naval Postgraduate School, Monterey, CA, USA, 2001.
29. Varona, J.L. Graphic and numerical comparison between iterative methods. *Math. Intell.* **2002**, *24*, 37–46. [CrossRef]
30. Hernández-Paricio, L.J.; Marañón-Grandes, M.; Rivas-Rodríguez, M.T. Plotting basins of end points of rational maps with Sage. *Tbil. Math. J.* **2012**, *5*, 71–99.

© 2019 by the authors. Licensee MDPI, Basel, Switzerland. This article is an open access article distributed under the terms and conditions of the Creative Commons Attribution (CC BY) license (http://creativecommons.org/licenses/by/4.0/).

Article
Two-Step Solver for Nonlinear Equations

Ioannis K. Argyros [1], Stepan Shakhno [2,*] and Halyna Yarmola [2]

[1] Department of Mathematics, Cameron University, Lawton, OK 73505, USA; iargyros@cameron.edu
[2] Faculty of Applied Mathematics and Informatics, Ivan Franko National University of Lviv, Universitetska Str. 1, Lviv 79000, Ukraine; halyna.yarmola@lnu.edu.ua
* Correspondence: stepan.shakhno@lnu.edu.ua

Received: 23 December 2018; Accepted: 18 January 2019; Published: 23 January 2019

Abstract: In this paper we present a two-step solver for nonlinear equations with a nondifferentiable operator. This method is based on two methods of order of convergence $1 + \sqrt{2}$. We study the local and a semilocal convergence using weaker conditions in order to extend the applicability of the solver. Finally, we present the numerical example that confirms the theoretical results.

Keywords: Nondifferentiable operator; nonlinear equation; divided difference; Lipschitz condition; convergence order; local and semilocal convergence

---

1. Introduction

A plethora of real-life applications from various areas, including Computational Science and Engineering, are converted via mathematical modeling to equations valued on abstract spaces such as n-dimensional Euclidean, Hilbert, Banach, and other spaces [1,2]. Then, researchers face the great challenge of finding a solution x_* in the closed form of the equation. However, this task is generally very difficult to achieve. This is why iterative methods are developed to provide a sequence approximating x_* under some initial conditions.

Newton's method, and its variations are widely used to approximate x_* [1–14]. There are problems with the implementation of these methods, since the invertibility of the linear operator involved is, in general, costly or impossible. That is why secant-type methods were also developed which are derivative-free. In these cases however, the order of convergence drops from 2 to $\dfrac{1+\sqrt{5}}{2}$.

Then, one considers methods that mix Newton and secant steps to increase the order of convergence. This is our first objective in this paper. Moreover, the study of iterative methods involves local convergence where knowledge about the solution x_* is used to determine upper bounds on the distances and radii of convergence. The difficulty of choosing initial points is given by local results, so they are important. In the semilocal convergence we use knowledge surrounding the initial point to find sufficient conditions for convergence. It turns out that in both cases the convergence region is small, limiting the applicability of iterative methods. That is why we use our ideas of the center-Lipschitz condition, in combination with the notion of the restricted convergence region, to present local as well as semilocal improvements leading to the extension of the applicability of iterative methods.

The novelty of the paper is that since the new Lipschitz constants are special cases of older ones, no additional cost is required for these improvements (see also the remarks and numerical examples). Our ideas can be used to improve the applicability of other iterative methods [1–14].

By E_1, E_2 we consider Banach spaces and by $\Omega \subseteq E_1$ a convex set. $F : \Omega \to E_2$ is differentiable in the Fréchet sense, $G : \Omega \to E_2$ is a continuous but its differentiability is not assumed. Then, we study equation

$$H(x) = 0, \text{ for } H(x) = F(x) + G(x). \qquad (1)$$

This problem was considered by several authors. Most of them used one-step methods for finding an approximate solution of (1), for example, Newton's type method [14], difference methods [4,5] and combined methods [1–3,11].

We proposed a two-step method [6,10,12] to numerically solve (1)

$$x_{n+1} = x_n - \left[F'\left(\frac{x_n + y_n}{2}\right) + Q(x_n, y_n)\right]^{-1}(F(x_n) + G(x_n)),$$
$$y_{n+1} = x_{n+1} - \left[F'\left(\frac{x_n + y_n}{2}\right) + Q(x_n, y_n)\right]^{-1}(F(x_{n+1}) + G(x_{n+1})), \quad n = 0, 1, \ldots \quad (2)$$

with $Q(x,y)$ a first order divided difference of the operator G at the points x and y. This method relates to methods with the order of convergence $1 + \sqrt{2}$ [7,13].

If $Q : \Omega \times \Omega \to L(E_1, E_2)$, gives $Q(x,y)(x-y) = G(x) - G(y)$ for all x, y with $x \neq y$, then, we call it a divided difference.

Two-step methods have some advantages over one-step methods. First, they usually require fewer number of iterations for finding an approximate solution. Secondly, at each iteration, they solve two similar linear problems, therefore, there is a small increase in computational complexity. That is why they are often used for solving nonlinear problems [2,6,8–10,12,13].

In [6,10,12] the convergence analysis of the proposed method was provided under classical and generalized Lipschitz conditions and superquadratic convergence order was shown. Numerical results for method (2) were presented in [10,12].

2. Local Convergence

Let $S(x_*, \rho) = \{x : \|x - x_*\| < \rho\}$.

From now on by differentiable, we mean differentiable in the Fréchet sense. Moreover, F, G are assumed as previously.

Theorem 1 ([10,12]). *Assume (1) has a solution $x_* \in \Omega$, G has a first order divided difference Q in Ω, and there exist $[T(x;y)]^{-1} = \left[F'\left(\frac{x+y}{2}\right) + Q(x,y)\right]^{-1}$ for each $x \neq y$ and $\|[T(x;y)]^{-1}\| \leq B$. Moreover, assume for each $x, y, u, v \in \Omega, x \neq y$*

$$\|F'(x) - F'(y)\| \leq 2p_1\|x - y\|, \quad (3)$$
$$\|F''(x) - F''(y)\| \leq p_2\|x - y\|^\alpha, \ \alpha \in (0, 1], \quad (4)$$
$$\|Q(x,y) - Q(u,v)\| \leq q_1(\|x - u\| + \|y - v\|). \quad (5)$$

Assume $S(x_*, r_*) \subset \Omega$, where r_* is the minimal positive zero of

$$q(r) = 1,$$
$$3B(p_1 + q_1)rq(r) = 1,$$
$$q(r) = B\left[(p_1 + q_1)r + \frac{p_2}{4(\alpha+1)(\alpha+2)}r^{1+\alpha}\right].$$

Then, the sequences $\{x_n\}_{n \geq 0}, \{y_n\}_{n \geq 0}$ for $x_0, y_0 \in S(x_*, r_*)$ remain in $S(x_*, r_*)$ with $\lim_{n \to \infty} x_n = x_*$, and

$$\|x_{n+1} - x_*\| \leq B\left[(p_1 + q_1)\|y_n - x_*\| + \frac{p_2}{4(\alpha+1)(\alpha+2)}\|x_n - x_*\|^{1+\alpha}\right]\|x_n - x_*\|, \quad (6)$$

$$\|y_{n+1} - x_*\| \leq B(p_1 + q_1)\left[\|y_n - x_*\| + \|x_n - x_*\| + \|x_{n+1} - x_*\|\right]\|x_{n+1} - x_*\|. \quad (7)$$

The condition $\|[T(x;y)]^{-1}\| \leq B$ used in [10,12] is very strong in general. That is why in what follows, we provide a weaker alternative. Indeed, assume that there exists $a > 0$ and $b > 0$ such that

$$\|F'(x_*) - F'(x)\| \leq a\|x_* - x\|, \tag{8}$$
$$\|Q(x,y) - G'(x_*)\| \leq b(\|x - x_*\| + \|y - x_*\|) \text{ for each } x, y \in \Omega. \tag{9}$$

Set $c = (a + 2b)\|T_*^{-1}\|$, $\Omega_0 = \Omega \cap S(x_*, \frac{1}{c})$ and $T_* = F'(x_*) + G'(x_*)$. It follows, for each $x, y \in S(x_*, r), r \in [0, \frac{1}{c}]$ we get in turn by (8) and (9) provided that T_*^{-1} exists

$$\begin{aligned}
\|T_*^{-1}\|\|T(x;y) - T_*\| &\leq \|T_*^{-1}\|\left[\|F'(\frac{x+y}{2}) - F'(x_*)\| + \|Q(x,y) - G'(x_*)\|\right] \\
&\leq \|T_*^{-1}\|\left[\frac{a}{2}(\|x - x_*\| + \|y - x_*\|) + b(\|x - x_*\| + \|y - x_*\|)\right] \\
&\leq \|T_*^{-1}\|(\frac{a}{2} + b)[\|x - x_*\| + \|y - x_*\|] \\
&< \|T_*^{-1}\|\left[(\frac{a}{2} + b) + (\frac{a}{2} + b)\right]\frac{1}{c} = 1.
\end{aligned} \tag{10}$$

Then, (10) and the Banach lemma on invertible operators [2] assure $T(x;y)^{-1}$ exists with

$$\|T(x;y)^{-1}\| \leq \bar{B} = \bar{B}(r) = \frac{\|T_*^{-1}\|}{1 - cr}. \tag{11}$$

Then, Theorem 1 holds but with $\bar{B}, \bar{p}_1, \bar{q}_1, \bar{p}_2, \bar{r}_1, \bar{r}_2, \bar{r}_*$ replacing $B, p_1, q_1, p_2, r_1, r_2, r_*$, respectively. Next, we provide a weaker alternative to the Theorem 1.

Theorem 2. *Assume $x_* \in \Omega$, exists with $F(x_*) + G(x_*) = 0$, $T_*^{-1} \in L(E_2, E_1)$ and together with conditions (8) and (9) following items hold for each $x, y, u, v \in \Omega_0$*

$$\begin{aligned}
\|F'(y) - F'(x)\| &\leq 2\bar{p}_1\|y - x\|, \\
\|F''(y) - F''(x)\| &\leq \bar{p}_2\|y - x\|^\alpha, \, \alpha \in (0, 1], \\
\|Q(x,y) - Q(u,v)\| &\leq \bar{q}_1(\|x - u\| + \|y - v\|).
\end{aligned}$$

Let $\bar{r}_1, \bar{r}_2$ be the minimal positive zeros of equations

$$\bar{q}(r) = 1,$$
$$3\bar{B}(\bar{p}_1 + \bar{q}_1)r\bar{q}(r) = 1,$$

respectively, where

$$\bar{q}(r) = \bar{B}\left[(\bar{p}_1 + \bar{q}_1)r + \frac{\bar{p}_2}{4(\alpha + 1)(\alpha + 2)}r^{1+\alpha}\right]$$

and set $\bar{r}_* = \min\{\bar{r}_1, \bar{r}_2\}$. Moreover, assume that $S(x_*, \bar{r}_*) \subset \Omega$.
Then, the sequences $\{x_n\}_{n\geq 0}, \{y_n\}_{n\geq 0}$ for $x_0, y_0 \in S(x_*, \bar{r}_*)$ remain in $S(x_*, \bar{r}_*)$, $\lim_{n\to\infty} x_n = x_*$, and

$$\|x_{n+1} - x_*\| \leq \bar{B}\left[(\bar{p}_1 + \bar{q}_1)\|y_n - x_*\| + \frac{\bar{p}_2}{4(\alpha + 1)(\alpha + 2)}\|x_n - x_*\|^{1+\alpha}\right]\|x_n - x_*\|, \tag{12}$$

$$\|y_{n+1} - x_*\| \leq \bar{B}(\bar{p}_1 + \bar{q}_1)\left[\|y_n - x_*\| + \|x_n - x_*\| + \|x_{n+1} - x_*\|\right]\|x_{n+1} - x_*\|. \tag{13}$$

Proof. It follows from the proof of Theorem 1, (10), (11) and the preceding replacements. □

Corollary 1. *Assume hypotheses of Theorem 2 hold. Then, the order of convergence of method (2) is $1 + \sqrt{1 + \alpha}$.*

Proof. Let

$$a_n = \|x_n - x_*\|, \quad b_n = \|y_n - x_*\|, \quad \tilde{C}_1 = \bar{B}(\bar{p}_1 + \bar{q}_1), \quad \tilde{C}_2 = \frac{\bar{B}\bar{p}_2}{4(\alpha+1)(\alpha+2)}.$$

By (12) and (13), we get

$$\begin{aligned} a_{n+1} &\leq \tilde{C}_1 a_n b_n + \tilde{C}_2 a_n^{2+\alpha}, \\ b_{n+1} &\leq \tilde{C}_1(a_{n+1} + a_n + b_n)a_{n+1} \leq \tilde{C}_1(2a_n + b_n)a_{n+1} \\ &\leq \tilde{C}_1(2a_n + \tilde{C}_1(2a_0 + b_0)a_n)a_{n+1} = \tilde{C}_1(2 + \tilde{C}_1(2a_0 + b_0))a_n a_{n+1}, \end{aligned}$$

Then, for large n and $a_{n-1} < 1$, from previous inequalities, we obtain

$$\begin{aligned} a_{n+1} &\leq \tilde{C}_1 a_n b_n + \tilde{C}_2 a_n^2 a_{n-1}^\alpha \\ &\leq \tilde{C}_1^2 (2 + \tilde{C}_1(2a_0 + b_0))a_n^2 a_{n-1} + \tilde{C}_2 a_n^2 a_{n-1}^\alpha \\ &\leq [\tilde{C}_1^2(2 + \tilde{C}_1(2a_0 + b_0)) + \tilde{C}_2] a_n^2 a_{n-1}^\alpha. \end{aligned} \quad (14)$$

From (14) we relate (2) to $t^2 - 2t - \alpha = 0$, leading to the solution $t^* = 1 + \sqrt{1+\alpha}$. □

Remark 1. *To relate Theorem 1 and Corollary 2 in [12] to our Theorem 2 and Corollary 1 respectively, let us notice that under (3)–(5) B_1 can replace B in these results, where $B_1 = B_1(r) = \frac{\|T_*^{-1}\|}{1 - c_1 r}$, $c_1 = 2(p_1 + q_1)\|T_*^{-1}\|$.*

Then, we have

$$\begin{aligned} \bar{p}_1 &\leq p_1, \\ \bar{p}_2 &\leq p_2, \\ \bar{q}_1 &\leq q_1, \\ c &\leq c_1, \\ \bar{B}(t) &\leq B_1(t) \text{ for each } t \in [0, \tfrac{1}{c_1}), \\ \tilde{C}_1 &\leq C_1, \\ \tilde{C}_2 &\leq C_2 \end{aligned}$$

and

$$\Omega_0 \subseteq \Omega$$

since $r_* \leq \bar{r}_*$, which justify the advantages claimed in the Introduction of this study.

3. Semilocal Convergence

Theorem 3 ([12]). *We assume that $S(x_0, r_0) \subset \Omega$, the linear operator $T_0 = F'\left(\frac{x_0 + y_0}{2}\right) + Q(x_0, y_0)$, where $x_0, y_0 \in \Omega$, is invertible and the Lipschitz conditions are fulfilled*

$$\|T_0^{-1}(F'(y) - F'(x))\| \leq 2p_0 \|y - x\|, \quad (15)$$
$$\|T_0^{-1}(Q(x,y) - Q(u,v))\| \leq q_0(\|x - u\| + \|y - v\|). \quad (16)$$

Let's λ, μ ($\mu > \lambda$), r_0 be non-negative numbers such that

$$\|x_0 - x_{-1}\| \leq \lambda, \quad \|T_0^{-1}(F(x_0) + G(x_0))\| \leq \mu, \quad (17)$$

$$r_0 \geq \mu/(1-\gamma), \quad (p_0 + q_0)(2r_0 - \lambda) < 1,$$

$$\gamma = \frac{(p_0+q_0)(r_0-\lambda)+0.5p_0r_0}{1-(p_0+q_0)(2r_0-\lambda)}, \quad 0 \leq \gamma < 1.$$

Then, for each $n = 0, 1, 2, \ldots$

$$\|x_n - x_{n+1}\| \leq t_n - t_{n+1}, \quad \|y_n - x_{n+1}\| \leq s_n - t_{n+1},$$
$$\|x_n - x_*\| \leq t_n - t^*, \quad \|y_n - x_*\| \leq s_n - t^*,$$

where

$$t_0 = r_0, \quad s_0 = r_0 - \lambda, \quad t_1 = r_0 - \mu,$$

$$t_{n+1} - t_{n+2} = \frac{(p_0+q_0)(s_n-t_{n+1})+0.5p_0(t_n-t_{n+1})}{1-(p_0+q_0)[(t_0-t_{n+1})+(s_0-s_{n+1})]}(t_n - t_{n+1}), \quad (18)$$

$$t_{n+1} - s_{n+1} = \frac{(p_0+q_0)(s_n-t_{n+1})+0.5p_0(t_n-t_{n+1})}{1-(p_0+q_0)[(t_0-t_n)+(s_0-s_n)]}(t_n - t_{n+1}), \quad (19)$$

$\{t_n\}_{n \geq 0}$, $\{s_n\}_{n \geq 0}$ are non-negative, decreasing sequences that converge to some t^* such that $r_0 - \mu/(1-\gamma) \leq t^* < t_0$; sequences $\{x_n\}_{n \geq 0}$, $\{y_n\}_{n \geq 0} \subseteq S(x_0, t^*)$ and converge to a solution x_* of equation (1).

Next, we present the analogous improvements in the semilocal convergence case. Assume that for all $x, y, u, v \in \Omega$

$$\|T_0^{-1}(F'(z) - F'(x))\| \leq 2\bar{p}_0 \|z - x\|, \quad z = \frac{x_0 + y_0}{2} \quad (20)$$

and

$$\|T_0^{-1}(Q(x,y) - Q(x_0,y_0))\| \leq \bar{q}_0 (\|x - x_0\| + \|y - y_0\|). \quad (21)$$

Set $\Omega_0 = \Omega \cap S(x_0, \bar{r}_0)$, where $\bar{r}_0 = \frac{1 + \lambda(\bar{p}_0 + \bar{q}_0)}{2(\bar{p}_0 + \bar{q}_0)}$. Define parameter $\bar{\gamma}$ and sequences $\{\bar{t}_n\}$, $\{\bar{s}_n\}$ for each $n = 0, 1, 2, \ldots$ by $\bar{\gamma} = \frac{(\bar{p}_0^0 + \bar{q}_0^0)(\bar{r}_0 - \lambda) + 0.5\bar{p}_0^0 \bar{r}_0}{1 - (\bar{p}_0 + \bar{q}_0)(2\bar{r}_0 - \lambda)}$,

$$\bar{t}_0 = \bar{r}_0, \quad \bar{s}_0 = \bar{r}_0 - \lambda, \quad \bar{t}_1 = \bar{r}_0 - \mu,$$

$$\bar{t}_{n+1} - \bar{t}_{n+2} = \frac{(\bar{p}_0^0 + \bar{q}_0^0)(\bar{s}_n - \bar{t}_{n+1}) + 0.5\bar{p}_0^0(\bar{t}_n - \bar{t}_{n+1})}{1 - (\bar{p}_0 + \bar{q}_0)[(\bar{t}_0 - \bar{t}_{n+1}) + (\bar{s}_0 - \bar{s}_{n+1})]}(\bar{t}_n - \bar{t}_{n+1}), \quad (22)$$

$$\bar{t}_{n+1} - \bar{s}_{n+1} = \frac{(\bar{p}_0^0 + \bar{q}_0^0)(\bar{s}_n - \bar{t}_{n+1}) + 0.5\bar{p}_0^0(\bar{t}_n - \bar{t}_{n+1})}{1 - (\bar{p}_0 + \bar{q}_0)[(\bar{t}_0 - \bar{t}_n) + (\bar{s}_0 - \bar{s}_n)]}(\bar{t}_n - \bar{t}_{n+1}). \quad (23)$$

As in the local convergence case, we assume instead of (15) and (16) the restricted Lipschitz-type conditions for each $x, y, u, v \in \Omega_0$

$$\|T_0^{-1}(F'(x) - F'(y))\| \leq 2p_0^0 \|x - y\|, \quad (24)$$
$$\|T_0^{-1}(Q(x,y) - Q(u,v))\| \leq q_0^0 (\|x - u\| + \|y - v\|). \quad (25)$$

Then, instead of the estimate in [12] using (15) and (16):

$$\begin{aligned}
\|T_0^{-1}[T_0 - T_{n+1}]\| &\leq \left\|T_0^{-1}\left[F'\left(\frac{x_0 + y_0}{2}\right) - F'\left(\frac{x_{n+1} + y_{n+1}}{2}\right)\right]\right\| + \|T_0^{-1}[Q(x_0, y_0) - Q(x_{n+1}, y_{n+1})]\| \\
&\leq 2p_0\left(\frac{\|x_0 - x_{n+1}\| + \|y_0 - y_{n+1}\|}{2}\right) + q_0(\|x_0 - x_{n+1}\| + \|y_0 - y_{n+1}\|) \\
&= (p_0 + q_0)(\|x_0 - x_{n+1}\| + \|y_0 - y_{n+1}\|) \leq (p_0 + q_0)(t_0 - t_{n+1} + s_0 - s_{n+1}) \\
&\leq (p_0 + q_0)(t_0 + s_0) = (p_0 + q_0)(2r_0 - \lambda) < 1,
\end{aligned} \qquad (26)$$

we obtain more precise results using (20) and (21)

$$\begin{aligned}
\|T_0^{-1}[T_0 - T_{n+1}]\| &\leq \leq 2\bar{p}_0\left(\frac{\|x_0 - x_{n+1}\| + \|y_0 - y_{n+1}\|}{2}\right) + \bar{q}_0(\|x_0 - x_{n+1}\| + \|y_0 - y_{n+1}\|) \\
&\leq (\bar{p}_0 + \bar{q}_0)(\|x_0 - x_{n+1}\| + \|y_0 - y_{n+1}\|) \\
&\leq (\bar{p}_0 + \bar{q}_0)(\bar{t}_0 - \bar{t}_{n+1} + \bar{s}_0 - \bar{s}_{n+1}) \\
&\leq (\bar{p}_0 + \bar{q}_0)(\bar{t}_0 + \bar{s}_0) = (\bar{p}_0 + \bar{q}_0)(2\bar{t}_0 - \lambda) < 1,
\end{aligned}$$

since

$$\begin{aligned}
\Omega_0 &\subseteq \Omega, \\
\bar{p}_0 &\leq p_0, \\
\bar{q}_0 &\leq q_0, \\
p_0^0 &\leq p_0, \\
q_0^0 &\leq q_0, \\
\bar{\gamma} &\leq \gamma, \\
\text{and } \bar{r}_0 &\geq r_0.
\end{aligned} \qquad (27)$$

Then, by replacing $p_0, q_0, r_0, \gamma, t_n, s_n$, (26) with p_0^0, q_0^0 (at the numerator in (18) and (19)), or $\bar{p}_0, \bar{q}_0$ (at the denominator in (18) and (19)), and with $\bar{r}_0, \bar{\gamma}, \bar{t}_n, \bar{s}_n$, (27) respectively, we arrive at the following improvement of Theorem 3.

Theorem 4. *Assume together with (17), (20), (21), (24), (25) that $\bar{r}_0 \geq \mu(1 - \bar{\gamma})$, $(\bar{p}_0 + \bar{q}_0)(2\bar{r}_0 - \lambda) < 1$ and $\bar{\gamma} \in [0, 1]$. Then, for each $n = 0, 1, 2, \ldots$*

$$\|x_n - x_{n+1}\| \leq \bar{t}_n - \bar{t}_{n+1}, \quad \|y_n - x_{n+1}\| \leq \bar{s}_n - \bar{t}_{n+1}, \qquad (28)$$
$$\|x_n - x_*\| \leq \bar{t}_n - t^*, \quad \|y_n - x_*\| \leq \bar{s}_n - t^*, \qquad (29)$$

with sequences $\{\bar{t}_n\}_{n \geq 0}$, $\{\bar{s}_n\}_{n \geq 0}$ given in (22) and (23) decreasing, non-negative sequences that converge to some t^* such that $r_0 - \mu/(1 - \bar{\gamma}) \leq t^* < \bar{t}_0$. Moreover, sequences $\{x_n\}_{n \geq 0}$, $\{y_n\}_{n \geq 0} \subseteq S(x_0, \bar{t}^*)$ for each $n = 0, 1, 2, \ldots,$ and $\lim_{n \to \infty} x_n = x_*$.

Remark 2. It follows (27) that by hypotheses of Theorem 3, Theorem 4, and by a simple inductive argument that the following items hold

$$t_n \leq \bar{t}_n,$$
$$s_n \leq \bar{s}_n,$$
$$0 \leq \bar{t}_n - \bar{t}_{n+1} \leq t_n - t_{n+1},$$
$$0 \leq \bar{s}_n - \bar{t}_{n+1} \leq s_n - t_{n+1},$$
$$\text{and } t^* \leq \bar{t}^*.$$

Hence, the new results extend the applicability of the method (2).

Remark 3. If we choose $F(x) = 0$, $p_1 = 0$, $p_2 = 0$. Then, the estimates (6) and (7) reduce to similar ones in [7] for the case $\alpha = 1$.

Remark 4. Section 3 contains existence results. The uniqueness results are omitted, since they can be found in [2,6] but with center-Lipschitz constants replacing the larger Lipschitz constants.

4. Numerical Experiments

Let $E_1 = E_2 = R^3$ and $\Omega = S(x_*, 1)$. Define functions F and G for $v = (v_1, v_2, v_3)^T$ on Ω by

$$F(v) = \left(e^{v_1} - 1, \frac{e-1}{2}v_2^2 + v_2, v_3\right)^T,$$
$$G(v) = \left(|v_1|, |v_2|, |v_3|\right)^T,$$

and set $H(v) = F(v) + G(v)$. Moreover, define a divided difference $Q(\cdot, \cdot)$ by

$$Q(v, \bar{v}) = \text{diag}\left(\frac{|\bar{v}_1| - |v_1|}{\bar{v}_1 - v_1}, \frac{|\bar{v}_2| - |v_2|}{\bar{v}_2 - v_2}, \frac{|\bar{v}_3| - |v_3|}{\bar{v}_3 - v_3}\right)$$

if $v_i \neq \bar{v}_i$, $i = 1, 2, 3$. Otherwise, set $Q(v, \bar{v}) = \text{diag}(1,1,1)$. Then, $T_* = 2\text{diag}(1,1,1)$, so $\|T_*^{-1}\| = 0.5$. Notice that $x_* = (0,0,0)^T$ solves equation $H(v) = 0$. Furthermore, we have $\Omega_0 = S(x_*, \frac{2}{e+1})$, so

$$p_1 = \frac{e}{2}, p_2 = e, q_1 = 1, B = B(t) = \frac{1}{2(1 - c_1 t)},$$

$b = 1$, $\alpha = 1$, $a = e - 1$, $\bar{p}_1 = \bar{p}_2 = \frac{1}{2}e^{\frac{2}{e+1}}$, $\bar{q}_1 = 1$

and Ω_0 is a strict subset of Ω. As well, the new parameters and functions are also more strict than the old ones in [12]. Hence, the aforementioned advantages hold. In particular, $r_* \approx 0.2265878$ and $\bar{r}_* \approx 0.2880938$.

Let's give results obtained by the method (2) for approximate solving the considered system of nonlinear equations. We chose initial approximations as $x_0 = (0.1; 0.1; 0.1)d$ (d is a real number) and $y_0 = x_0 + 0.0001$. The iterative process was stopped under the condition $\|x_{n+1} - x_n\| \leq 10^{-10}$ and $\|H(x_{n+1})\| \leq 10^{-10}$. We used the Euclidean norm. The obtained results are shown in Table 1.

Table 1. Value of $\|x_n - x_{n-1}\|$ for each iteration.

n	$d = 1$	$d = 10$	$d = 50$
1	0.1694750	1.4579613	5.9053855
2	0.0047049	0.3433874	1.9962504
3	0.0000005	0.0112749	1.3190118
4	4.284×10^{-16}	0.0000037	1.0454772
5		2.031×10^{-14}	0.4157737
6			0.0260385
7			0.0000271
8			1.389×10^{-12}

5. Conclusions

The convergence region of iterative methods is, in general, small under Lipschitz-type conditions, leading to a limited choice of initial points. Therefore, extending the choice of initial points without imposing additional, more restrictive, conditions than before is extremely important in computational sciences. This difficult task has been achieved by defining a convergence region where the iterates lie, that is more restricted than before, ensuring the Lipschitz constants are at least as small as in previous works. Hence, we achieve: more initial points, fewer iterations to achieve a predetermined error accuracy, and a better knowledge of where the solution lies. These are obtained without additional cost because the new Lipschitz constants are special cases of the old ones. This technique can be applied to other iterative methods.

Author Contributions: Conceptualization, Editing, I.K.A.; Investigation, H.Y. and S.S.

Funding: This research received no external funding.

Conflicts of Interest: The authors declare no conflict of interest.

References

1. Argyros, I.K. A unifying local-semilocal convergence analysis and applications for two-point Newton-like methods in Banach space. *J. Math. Anal. Appl.* **2004**, *298*, 374–397. [CrossRef]
2. Argyros, I. K.; Magrenán, Á.A. *Iterative Methods and Their Dynamics with Applications: A Contemporary Study*; CRC Press: Boca Raton, FL, USA, 2017.
3. Argyros, I.K.; Ren, H. On the convergence of modified Newton methods for solving equations containing a non-differentiable term. *J. Comp. App. Math.* **2009**, *231*, 897–906. [CrossRef]
4. Hernandez, M.A.; Rubio, M.J. A uniparametric family of iterative processes for solving nondiffrentiable operators. *J. Math. Anal. Appl.* **2002**, *275*, 821–834. [CrossRef]
5. Ren, H.; Argyros, I.K. A new semilocal convergence theorem for a fast iterative method with nondifferentiable operators. *Appl. Math. Comp.* **2010**, *34*, 39–46. [CrossRef]
6. Shakhno, S.M. Convergence of the two-step combined method and uniqueness of the solution of nonlinear operator equations. *J. Comp. App. Math.* **2014**, *261*, 378–386. [CrossRef]
7. Shakhno, S.M. On an iterative algorithm with superquadratic convergence for solving nonlinear operator equations. *J. Comp. App. Math.* **2009**, *231*, 222–235. [CrossRef]
8. Shakhno, S.M. On the difference method with quadratic convergence for solving nonlinear operator equations. *Mat. Stud.* **2006**, *26*, 105–110. (In Ukrainian)
9. Shakhno, S.M. On two-step iterative process under generalized Lipschitz conditions for the first order divided differences. *Math. Methods Phys. Fields* **2009**, *52*, 59–66. (In Ukrainian) [CrossRef]
10. Shakhno, S.; Yarmola, H. On the two-step method for solving nonlinear equations with nondifferentiable operator. *Proc. Appl. Math. Mech.* **2012**, *12*, 617–618. [CrossRef]
11. Shakhno, S.M.; Yarmola, H.P. Two-point method for solving nonlinear equation with nondifferentiable operator. *Mat. Stud.* **2009**, *36*, 213–220.
12. Shakhno, S.; Yarmola, H. Two-step method for solving nonlinear equations with nondifferentiable operator. *J. Numer. Appl. Math.* **2012**, *109*, 105–115. [CrossRef]

13. Werner, W. Über ein Verfahren der Ordnung $1 + \sqrt{2}$ zur Nullstellenbestimmung. *Numer. Math.* **1979**, *32*, 333–342. [CrossRef]
14. Zabrejko, P.P.; Nguen, D.F. The majorant method in the theory of Newton-Kantorovich approximations and the Pta'k error estimates. *Numer. Funct. Anal. Optim.* **1987**, *9*, 671–684. [CrossRef]

© 2019 by the authors. Licensee MDPI, Basel, Switzerland. This article is an open access article distributed under the terms and conditions of the Creative Commons Attribution (CC BY) license (http://creativecommons.org/licenses/by/4.0/).

Article

Local Convergence of a Family of Weighted-Newton Methods

Ramandeep Behl [1,*], Ioannis K. Argyros [2], J.A. Tenreiro Machado [3] and Ali Saleh Alshomrani [1]

1. Department of Mathematics, King Abdulaziz University, Jeddah 21589, Saudi Arabia; aszalshomrani@kau.edu.sa
2. Department of Mathematics Sciences, Cameron University, Lawton, OK 73505, USA; iargyros@cameron.edu
3. Institute of Engineering, Polytechnic of Porto Department of Electrical Engineering, 4200-072 Porto, Portugal; jtm@isep.ipp.pt
* Correspondence: ramanbehl87@yahoo.in

Received: 9 December 2018; Accepted: 12 January 2019; Published: 17 January 2019

Abstract: This article considers the fourth-order family of weighted-Newton methods. It provides the range of initial guesses that ensure the convergence. The analysis is given for Banach space-valued mappings, and the hypotheses involve the derivative of order one. The convergence radius, error estimations, and results on uniqueness also depend on this derivative. The scope of application of the method is extended, since no derivatives of higher order are required as in previous works. Finally, we demonstrate the applicability of the proposed method in real-life problems and discuss a case where previous studies cannot be adopted.

Keywords: Banach space; weighted-Newton method; local convergence; Fréchet-derivative; ball radius of convergence

PACS: 65D10; 65D99; 65G99; 47J25; 47J05

---

1. Introduction

In this work, $\mathbb{B}_1$ and $\mathbb{B}_2$ denote Banach spaces, $\mathbb{A} \subseteq \mathbb{B}_1$ stands for a convex and open set, and $\varphi : \mathbb{A} \to \mathbb{B}_2$ is a differentiable mapping in the Fréchet sense. Several scientific problems can be converted to the expression. This paper addresses the issue of obtaining an approximate solution s_* of:

$$\varphi(x) = 0, \qquad (1)$$

by using mathematical modeling [1–4]. Finding a zero s_* is a laborious task in general, since analytical or closed-form solutions are not available in most cases.

We analyze the local convergence of the two-step method, given as follows:

$$\begin{aligned} y_j &= x_j - \delta \varphi'(x_j)^{-1} \varphi(x_j), \\ x_{n+1} &= x_j - A_j^{-1}\big(c_1 \varphi(x_j) + c_2 \varphi(y_j)\big), \end{aligned} \qquad (2)$$

where $x_0 \in \mathbb{A}$ is a starting point, $A_j = \alpha \varphi'(x_j) + \beta \varphi'\left(\frac{x_j + y_j}{2}\right) + \gamma \varphi'(y_j)$, and $\alpha, \beta, \gamma, \delta, c_1, c_2 \in \mathbb{S}$, where $\mathbb{S} = \mathbb{R}$ or $\mathbb{S} = \mathbb{C}$. The values of the parameters α, γ, β, and c_1 are given as follows:

$$\alpha = -\frac{1}{3}c_2(3\delta^2 - 7\delta + 2), \quad \beta = -\frac{4}{3}c_2(2\delta - 1),$$

$$\gamma = \frac{1}{3}c_2(\delta - 2) \text{ and } c_1 = -c_2(\delta^2 - \delta + 1), \text{ for } \delta \neq 0, \; c_2 \neq 0.$$

Comparisons with other methods, proposed by Cordero et al. [5], Darvishi et al. [6], and Sharma [7], defined respectively as:

$$w_j = x_j - \varphi'(x_j)^{-1}\varphi(x_j),$$
$$x_{n+1} = w_j - B_j^{-1}\varphi(w_j), \tag{3}$$

$$w_j = x_j - \varphi'(x_j)^{-1}\varphi(x_j),$$
$$z_j = x_j - \varphi'(x_j)^{-1}(\varphi(x_j) + \varphi(w_j)), \tag{4}$$
$$x_{n+1} = x_j - C_j^{-1}\varphi(x_j),$$

$$y_j = x_j - \frac{2}{3}\varphi'(x_j)^{-1}\varphi(x_j),$$
$$x_{n+1} = x_j - \frac{1}{2}D_j^{-1}\varphi'(x_j)^{-1}\varphi(x_j), \tag{5}$$

where:

$$B_j = 2\varphi'(x_j)^{-1} - \varphi'(x_j)^{-1}\varphi'(w_j)\varphi'(x_j)^{-1},$$
$$C_j = \frac{1}{6}\varphi'(x_j) + \frac{2}{3}\varphi'\left(\frac{x_j+w_j}{2}\right) + \frac{1}{6}\varphi'(z_j),$$
$$D_j = -I + \frac{9}{4}\varphi'(y_j)^{-1}\varphi'(x_j) + \frac{3}{4}\varphi'(x_j)^{-1}\varphi'(y_j),$$

were also reported in [8]. The local convergence of Method (2) was shown in [8] for $\mathbb{B}_1 = \mathbb{B}_2 = \mathbb{R}^m$ and $\mathbb{S} = \mathbb{R}$, by using Taylor series and hypotheses reaching up to the fourth Fréchet-derivative. However, the hypothesis on the fourth derivative limits the applicability of Methods (2)–(5), particularly because only the derivative of order one is required. Let us start with a simple problem. Set $\mathbb{B}_1 = \mathbb{B}_2 = \mathbb{R}$ and $\mathbb{A} = [-\frac{5}{2}, \frac{3}{2}]$. We suggest a function $\varphi : \mathbb{A} \to \mathbb{R}$ as:

$$\varphi(x) = \begin{cases} 0, & x = 0 \\ x^3 \ln x^2 + x^5 - x^4, & x \neq 0 \end{cases},$$

which further yield:

$$\varphi'(x) = 3x^2 \ln x^2 + 5x^4 - 4x^3 + 2x^2,$$
$$\varphi''(x) = 12x \ln x^2 + 20x^3 - 12x^2 + 10x,$$
$$\varphi'''(x) = 12 \ln x^2 + 60x^2 - 12x + 22,$$

where the solution is $s_* = 1$. Obviously, the function $\varphi'''(x)$ is unbounded in the domain $\mathbb{A}$. Therefore, the results in [5–9] and Method (2) cannot be applicable to such problems or its special cases that require the hypotheses on the third- or higher order derivatives of φ. Without a doubt, some of the iterative method in Brent [10] and Petkovíc et al. [4] are derivative free and are used to locate zeros of functions. However, there have been many developments since then. Faster iterative methods have been developed whose convergence order is determined using Taylor series or with the technique introduce in our paper. The location of the initial points is a "shot in the dark" in these references; no uniqueness results or estimates on $\|x_n - x_*\|$ are available. Methods on abstract spaces derived from the ones on the real line are also not addressed.

These works do not give a radius of convergence, estimations on $\|x_j - s_*\|$, or knowledge about the location of s_*. The novelty of this study is that it provides this information, but requiring only the derivative of order one for method (2). This expands the scope of utilization of (2) and similar methods. It is vital to note that the local convergence results are very fruitful, since they give insight into the difficult operational task of choosing the starting points/guesses.

Otherwise, with the earlier approaches: (i) use the Taylor series and high-order derivative; (ii) have no clue about the choice of the starting point x_0; (iii) have no estimate in advance about the number of iterations needed to obtain a predetermined accuracy; and (iv) have no knowledge of the uniqueness of the solution.

The work is laid out as follows: we give the convergence of the iterative scheme (2) with the main Theorem 1 is given in Section 2. Six numerical problems are discussed in Section 3. The final conclusions are summarized in Section 4.

2. Convergence Study

This section starts by analyzing the convergence of Scheme (2). We assume that $L > 0$, $L_0 > 0$, $M \geq 1$ and γ, α, β, δ, c_1, $c_2 \in \mathbb{S}$. We consider some maps/functions and constant numbers. Therefore, we assume the following functions g_1, p, and h_p on the open interval $[0, \frac{1}{L_0})$ by:

$$g_1(t) = \frac{1}{2(1-L_0 t)}(Lt + 2M|1-\delta|),$$

$$p(t) = \frac{L_0}{|\alpha+\beta+\gamma|}\left(|\alpha| + \frac{|\beta|}{2}\left(\frac{|\beta|}{2}+|\gamma|\right)g_1(t)\right)t, \text{ for } \alpha+\beta+\gamma \neq 0,$$

$$h_p(t) = p(t) - 1,$$

and the values of r_1 and r_A are given as follows:

$$r_1 = \frac{2(M|1-\delta|-1)}{}, \quad r_A = \frac{2}{L+2L_0}.$$

Consider that:

$$M|1-\delta| < 1. \tag{6}$$

It is clear from the function g_1, parameters r_1 and r_A, and Equation (6), that $0 < r_1 \leq r_A < \frac{1}{L_0}$, $g_1(r_1) = 1$, and $0 \leq g_1(t) < 1$, for each $t \in [0, r_1)$ and $h_p(0) = -1$ and $h_p(t) \to +\infty$ as $t \to \frac{1}{L_0}^-$. On the basis of the classical intermediate value theorem, the function h_p has at least one zero in the open interval $\left(0, \frac{1}{L_0}\right)$. Let us call r_p as the smallest zero. We suggest some other functions g_2 and h_2 on the interval $[0, r_p)$ by means of the expressions:

$$g_2(t) = \frac{1}{2(1-L_0 t)}\left[Lt + \frac{2M^2(|\alpha-1|+|\beta|+|\gamma|)(|1-c_1|+|c_2|g_1(t))}{|\alpha+\beta+\gamma|(1-L_0 t)(1-p(t))} + \frac{2M(|1-c_1|+|c_2|g_1(t))}{1-L_0 t}\right]$$

and:

$$h_2(t) = g_2(t) - 1.$$

Suppose that:

$$M(|1-c_1|+c_2 M|1-\delta|)\left(1 + \frac{M(|\alpha-1|+|\beta|+|\gamma|)}{|\alpha+\beta+\gamma|}\right) < 1. \tag{7}$$

Then, we have by Equation (7) that $h_2(0) < 0$ and $h_2(t) \to +\infty$ as $t \to r_p^-$ by the definition of r_p. We recall r_2 as the least zero of h_2 on $(0, r_p)$.
Define:

$$r = \min\{r_1, r_2\}. \tag{8}$$

Then, notice that for all $t \in [0, r)$:

$$0 < r < r_A, \tag{9}$$

$$0 \leq g_1(t) < 1, \tag{10}$$

$$0 \leq p(t) < 1, \tag{11}$$

$$0 \leq g_2(t) < 1. \tag{12}$$

Assume that $Q(x, \delta) = \left\{ y \in \mathbb{B}_1 : \|x - y\| < \delta \right\}$. We can now proceed with the local convergence study of (2) adopting the preceding notations.

Theorem 1. *Let us assume that $\varphi : \mathbb{A} \subset \mathbb{B}_1 \to \mathbb{B}_2$ is a differentiable operator. In addition, we consider that there exist $s_* \in \mathbb{A}$, $L > 0$, $L_0 > 0$, $M \geq 1$ and the parameters α, β, γ, c_1, $c_2 \in \mathbb{S}$, with $\alpha + \beta + \gamma \neq 0$, are such that:*

$$\varphi(s_*) = 0, \quad \varphi'(s_*)^{-1} \in L(\mathbb{B}_2, \mathbb{B}_1), \tag{13}$$

$$\|\varphi'(s_*)^{-1}(\varphi'(s_*) - \varphi'(x))\| \leq L_0 \|s_* - x\|, \quad \forall \, x \in \mathbb{A}. \tag{14}$$

Set $x, y \in \mathbb{A}_0 = \mathbb{A} \cap Q\left(s_*, \frac{1}{L_0}\right)$ so that:

$$\|\varphi'(s_*)^{-1}(\varphi'(y) - \varphi'(x))\| \leq L\|y - x\|, \quad \forall \, y, \, x \in \mathbb{A}_0 \tag{15}$$

$$\|\varphi'(s_*)^{-1}\varphi'(x)\| \leq M, \quad \forall \, x \in \mathbb{A}_0, \tag{16}$$

satisfies Equations (6) and (7), the condition:

$$\bar{Q}(s_*, r) \subset \mathbb{A}, \tag{17}$$

holds, and the convergence radius r is provided by (8). The obtained sequence of iterations $\{x_j\}$ generated for $x_0 \in Q(s_*, r) - \{x^*\}$ by (2) is well defined. In addition, the sequence also converges to the required root s_*, remains in $Q(s_*, r)$ for every $n = 0, 1, 2, \ldots$, and:

$$\|y_j - s_*\| \leq g_1(\|x_j - s_*\|) \|x_j - s_*\| \leq \|x_j - s_*\| < r, \tag{18}$$

$$\|x_{n+1} - s_*\| \leq g_2(\|x_j - s_*\|) \|x_j - s_*\| < \|x_j - s_*\|, \tag{19}$$

where the g functions were described previously. Moreover, the limit point s_* of the obtained sequence $\{x_j\}$ is the only root of $\varphi(x) = 0$ in $\mathbb{A}_1 := \bar{Q}(s_*, T) \cap \mathbb{A}$, and T is defined as $T \in [r, \frac{2}{L_0})$.

Proof. We prove the estimates (18)–(19), by mathematical induction. Adopting the hypothesis $x_0 \in Q(s_*, r) - \{x^*\}$ and Equations (6) and (14), it results:

$$\|\varphi'(s_*)^{-1}(\varphi'(x_0) - \varphi'(s_*))\| \leq L_0 \|x_0 - s_*\| < L_0 r < 1. \tag{20}$$

Using Equation (20) and the results on operators by [1–3] that $\varphi'(x_0) \neq 0$, we get:

$$\|\varphi'(x_0)^{-1} \varphi'(s_*)\| \leq \frac{1}{1 - L_0 \|x_0 - s_*\|}. \tag{21}$$

Therefore, it is clear that y_0 exists. Then, by using Equations (8), (10), (15), (16), and (21), we obtain:

$$\begin{aligned}
\|y_0 - s_*\| &= \|(x_0 - s_* - \varphi'(x_0)^{-1}\varphi(x_0)) + (1 - \delta)\varphi'(x_0)^{-1}\varphi(x_0)\| \\
&\leq \|\varphi'(x_0)^{-1}\varphi'(s_*)\| \| \int_0^1 \varphi'(x^*)^{-1}[\varphi'(s_* + \theta(x_0 - s_*)) - \varphi'(x_0)](x_0 - s_*) d\theta\| \\
&\quad + \|\varphi'(x_0)^{-1}\varphi'(s_*)\| \| \int_0^1 \varphi'(x^*)^{-1}\varphi'(s_* + \theta(x_0 - s_*))(x_0 - s_*) d\theta\| \\
&\leq \frac{L\|x_0 - x^*\|^2}{2(1 - L_0\|x_0 - s_*\|)} + \frac{M|1 - \delta|\|x_0 - s_*\|}{1 - L_0\|x_0 - s_*\|} \\
&= g_1(\|x_0 - s_*\|) \|x_0 - s_*\| < \|x_0 - s_*\| < r,
\end{aligned} \tag{22}$$

illustrating that $y_0 \in Q(s_*, r)$ and Equation (18) is true for $j = 0$.

Now, we demonstrate that the linear operator A_0 is invertible. By Equations (8), (10), (14), and (22), we obtain:

$$\begin{aligned}
\|((\alpha + \beta + \gamma)\varphi'(s_*))^{-1}(A_0 - (\alpha + \beta + \gamma)\varphi'(s_*))\| \\
\leq \frac{L_0}{|\alpha + \beta + \gamma|}\left[|\alpha|\|x_0 - s_*\| + \frac{|\beta|}{2}(\|x_0 - s_*\| + \|y_0 - s_*\|) + |\gamma|\|y_0 - s_*\|\right] \\
\leq \frac{L_0}{|\alpha + \beta + \gamma|}\left[|\alpha| + \frac{|\beta|}{2}\left(\frac{|\beta|}{2} + |\gamma|\right)g_1(\|x_0 - s_*\|)\|x_0 - s_*\|\right] \\
= p(\|x_0 - s_*\|) < p(r) < 1.
\end{aligned} \tag{23}$$

Hence, $A_0^{-1} \in L(\mathbb{B}_2, \mathbb{B}_1)$,

$$\|A_0^{-1}\varphi'(s_*)\| \leq \frac{1}{|\alpha + \beta + \gamma|(1 - p(\|x_0 - s_*\|))}, \tag{24}$$

and x_1 exists. Therefore, we need the identity:

$$\begin{aligned}
x_1 - s_* = & x_0 - s_* - \varphi'(x_0)^{-1}\varphi(x_0) - \varphi'(x_0)^{-1}((1 - c_1)\varphi(x_0) + c_2\varphi(y_0)) \\
& + \varphi'(x_0)^{-1}(A_0 - \varphi'(x_0))A_0^{-1}(c_1\varphi(x_0) + c_2\varphi(y_0)).
\end{aligned} \tag{25}$$

Further, we have:

$$\begin{aligned}
\|x_1 - s_*\| \leq & \|x_0 - s_* - \varphi'(x_0)^{-1}\varphi(x_0)\| + \|\varphi'(x_0)^{-1}((1 - c_1)\varphi(x_0) + c_2\varphi(y_0))\| \\
& + \|\varphi'(x_0)^{-1}\varphi'(s_*)\|\|\varphi'(s_*)^{-1}(A_0 - \varphi'(x_0))\|\|A_0^{-1}\varphi'(s_*)\|\|\varphi'(s_*)^{-1}(c_1\varphi(x_0) + c_2\varphi(y_0))\| \\
\leq & \frac{L\|x_0 - s_*\|^2}{2(1 - L_0\|x_0 - s_*\|)} + \frac{M(|1 - c_1|\|x_0 - s_*\| + |c_2|\|y_0 - s_*\|)}{1 - L_0\|x_0 - s_*\|} \\
& + \frac{M^2(|\alpha - 1| + |\beta| + |\gamma|)(|1 - c_1| + |c_2|g_1(\|x_0 - s_*\|))\|x_0 - s_*\|}{|\alpha + \beta + \gamma|(1 - L_0\|x_0 - s_*\|)(1 - p(\|x_0 - s_*\|))} \\
\leq & g_2(\|x_0 - s_*\|)\|x_0 - s_*\| < \|x_0 - s_*\| < r,
\end{aligned} \tag{26}$$

which demonstrates that $x_1 \in Q(s_*, r)$ and (19) is true for $j = 0$, where we used (15) and (21) for the derivation of the first fraction in the second inequality. By means of Equations (21) and (16), we have:

$$\begin{aligned}
\|\varphi(s_*)^{-1}\varphi(x_0)\| &= \|\varphi'(s_*)^{-1}(\varphi(x_0) - \varphi(s_*))\| \\
&= \left\|\int_0^1 \varphi'(s_*)^{-1}\varphi'(s_* + \theta(x_0 - s_*))d\theta\right\| \leq M\|x_0 - s_*\|.
\end{aligned}$$

In the similar fashion, we obtain $\|\varphi'(s_*)^{-1}\varphi(y_0)\| \leq M\|y_0 - s_*\| \leq Mg_1(\|x_0 - s_*\|)\|x_0 - s_*\|$ (by (22)) and the definition of $\mathbb{A}$ to arrive at the second section. We reach (18) and (19), just by changing x_0, z_0, y_0, and x_1 by x_j, z_j, y_j, and x_{j+1}, respectively. Adopting the estimates $\|x_{j+1} - s_*\| \leq q\|x_j - s_*\| < r$, where $q = g_2(\|x_0 - s_*\|) \in [0, 1)$, we conclude that $x_{j+1} \in Q(s_*, r)$ and $\lim_{j \to \infty} x_j = s_*$. To illustrate the unique solution, we assume that $y_* \in \mathbb{A}_1$, satisfying $\varphi(y_*) = 0$ and $U = \int_0^1 \varphi'(y_* + \theta(s_* - y_*))d\theta$. From Equation (14), we have:

$$\begin{aligned}
\|\varphi'(s_*)^{-1}(U - \varphi'(s_*))\| &\leq \left\|\int_0^1 L_0|y_* + \theta(s_* - y_*) - s_*\|d\theta\right. \\
&\leq \int_0^1 (1 - t)\|y_* - s_*\|d\theta \leq \frac{L_0}{2}T < 1.
\end{aligned} \tag{27}$$

It follows from Equation (27) that U is invertible. Therefore, the identity $0 = \varphi(y_*) - \varphi(s_*) = U(y_* - s_*)$ leads to $y_* = s_*$. □

3. Numerical Experiments

Herein, we illustrate the previous theoretical results by means of six examples. The first two are standard test problems. The third is a counter problem where we show that the previous results are not applicable. The remaining three examples are real-life problems considered in several disciplines of science.

Example 1. We assume that $\mathbb{B}_1 = \mathbb{B}_2 = \mathbb{R}^3$, $\mathbb{A} = \bar{Q}(0, 1)$. Then, the function φ is defined on $\mathbb{A}$ for $u = (x_1, x_2, x_3)^T$ as follows:

$$\varphi(u) = \left(e_1^x - 1, \; x_2 - \frac{1}{2}(1-e)x_2^2, \; x_3\right)^T. \tag{28}$$

We yield the following Fréchet-derivative:

$$\varphi'(u) = \begin{bmatrix} e^{x_1} & 0 & 0 \\ 0 & (e-1)x_2 + 1 & 0 \\ 0 & 0 & 1 \end{bmatrix}.$$

It is important to note that we have $s_* = (0, 0, 0)^T$, $L_0 = e - 1 < L = e^{\frac{1}{L_0}}$, $\delta = 1$, $M = 2$, $c_1 = 1$, and $\varphi'(s_*) = \varphi'(s_*)^{-1} = \begin{bmatrix} 1 & 0 & 0 \\ 0 & 1 & 0 \\ 0 & 0 & 1 \end{bmatrix}$. By considering the parameter values that were defined in Theorem 1, we get the different radii of convergence that are depicted in Tables 1 and 2.

Table 1. Radii of convergence for Example 1, where $L_0 < L$.

Cases	Different Values of Parameters That Are Defined in Theorem 1						
	α	β	γ	c_2	r_1	r_2	$r = \min\{r_1, r_2\}$
1	$-\frac{2}{3}$	$\frac{4}{3}$	$\frac{1}{3}$	-1	0.382692	0.0501111	0.0501111
2	$-\frac{2}{3}$	$\frac{4}{3}$	-100	$\frac{1}{100}$	0.382692	0.334008	0.334008
3	1	1	1	0	0.382692	0.382692	0.382692
4	1	1	1	$\frac{1}{100}$	0.382692	0.342325	0.342325
5	10	$\frac{1}{10}$	$\frac{1}{10}$	$\frac{1}{100}$	0.382692	0.325413	0.325413

Table 2. Radii of convergence for Example 1, where $L_0 = L = e$ by [3,11].

Cases	Different Values of Parameters That Are Defined in Theorem 1						
	α	β	γ	c_2	r_1	r_2	$r = \min\{r_1, r_2\}$
1	$-\frac{2}{3}$	$\frac{4}{3}$	$\frac{1}{3}$	-1	0.245253	0.0326582	0.0326582
2	$-\frac{2}{3}$	$\frac{4}{3}$	-100	$\frac{1}{100}$	0.245253	0.213826	0.213826
3	1	1	1	0	0.245253	0.245253	0.245253
4	1	1	1	$\frac{1}{100}$	0.245253	0.219107	0.219107
5	10	$\frac{1}{10}$	$\frac{1}{10}$	$\frac{1}{100}$	0.245253	0.208097	0.208097

Example 2. Let us consider that $\mathbb{B}_1 = \mathbb{B}_2 = C[0, 1]$, $\mathbb{A} = \bar{Q}(0, 1)$ and introduce the space of continuous maps in $[0, 1]$ having the max norm. We consider the following function φ on $\mathbb{A}$:

$$\varphi(\phi)(x) = \varphi(x) - 5 \int_0^1 x\tau\phi(\tau)^3 d\tau, \tag{29}$$

which further yields:

$$\varphi'(\phi(\mu))(x) = \mu(x) - 15 \int_0^1 x\tau\phi(\tau)^2\mu(\tau)d\tau, \text{ for each } \mu \in \mathbb{A}.$$

We have $s_* = 0$, $L = 15$, $L_0 = 7.5$, $M = 2$, $\delta = 1$, and $c_1 = 1$. We will get different radii of convergence on the basis of distinct parametric values as mentioned in Tables 3 and 4.

Table 3. Radii of convergence for Example 2, where $L_0 < L$.

Cases	Different Values of Parameters That Are Defined in Theorem 1						
	α	β	γ	c_2	r_1	r_2	$r = \min\{r_1, r_2\}$
1	$-\frac{2}{3}$	$\frac{4}{3}$	$\frac{1}{3}$	-1	0.0666667	0.00680987	0.00680987
2	$-\frac{2}{3}$	$\frac{4}{3}$	-100	$\frac{1}{100}$	0.0666667	0.0594212	0.0594212
3	1	1	1	0	0.0666667	0.0666667	0.0666667
4	1	1	1	$\frac{1}{100}$	0.0666667	0.0609335	0.0609335
5	10	$\frac{1}{10}$	$\frac{1}{10}$	$\frac{1}{100}$	0.0666667	0.0588017	0.0588017

Table 4. Radii of convergence for Example 2, where $L_0 = L = 15$ by [3,11].

Cases	Different Values of Parameters That Are Defined in Theorem 1						
	α	β	γ	c_2	r_1	r_2	$r = \min\{r_1, r_2\}$
1	$-\frac{2}{3}$	$\frac{4}{3}$	$\frac{1}{3}$	-1	0.0444444	0.00591828	0.00591828
2	$-\frac{2}{3}$	$\frac{4}{3}$	-100	$\frac{1}{100}$	0.0444444	0.0387492	0.0387492
3	1	1	1	0	0.0444444	0.0444444	0.0444444
4	1	1	1	$\frac{1}{100}$	0.0444444	0.0397064	0.0397064
5	10	$\frac{1}{10}$	$\frac{1}{10}$	$\frac{1}{100}$	0.0444444	0.0377112	0.0377112

Example 3. Let us return to the problem from the Introduction. We have $s_* = 1$, $L = L_0 = 96.662907$, $M = 2$, $\delta = 1$, and $c_1 = 1$. By substituting different values of the parameters, we have distinct radii of convergence listed in Table 5.

Table 5. Radii of convergence for Example 3.

Cases	Different Values of Parameters That Are Defined in Theorem 1						
	α	β	γ	c_2	r_1	r_2	$r = \min\{r_1, r_2\}$
1	$-\frac{2}{3}$	$\frac{4}{3}$	$\frac{1}{3}$	-1	0.00689682	0.000918389	0.000918389
2	$-\frac{2}{3}$	$\frac{4}{3}$	-100	$\frac{1}{100}$	0.00689682	0.00601304	0.00601304
3	1	1	1	0	0.00689682	0.00689682	0.00689682
4	1	1	1	$\frac{1}{100}$	0.00689682	0.00616157	0.00616157
5	10	$\frac{1}{10}$	$\frac{1}{10}$	$\frac{1}{100}$	0.00689682	0.0133132	0.0133132

Example 4. The chemical reaction [12] illustrated in this case shows how W_1 and W_2 are utilized at rates $q_* - Q_*$ and Q_*, respectively, for a tank reactor (known as CSTR), given by:

$$W_2 + W_1 \rightarrow W_3$$
$$W_3 + W_1 \rightarrow W_4$$
$$W_4 + W_1 \rightarrow W_5$$
$$W_5 + W_1 \rightarrow W_6$$

Douglas [13] analyzed the CSTR problem for designing simple feedback control systems. The following mathematical formulation was adopted:

$$K_C \frac{2.98(x+2.25)}{(x+1.45)(x+2.85)^2(x+4.35)} = -1,$$

where the parameter K_C has a physical meaning and is described in [12,13]. For the particular value of choice $K_C = 0$, we obtain the corresponding equation:

$$\varphi(x) = x^4 + 11.50x^3 + 47.49x^2 + 83.06325x + 51.23266875. \tag{30}$$

The function φ has four zeros $s_* = (-1.45, -2.85, -2.85, -4.35)$. Nonetheless, the desired zero is $s_* = -4.35$ for Equation (30). Let us also consider $\mathbb{A} = [-4.5, -4]$.

Then, we obtain:

$$L_0 = 1.2547945, \ L = 29.610958, \ M = 2, \ \delta = 1, \ c_1 = 1.$$

Now, with the help of different values of the parameters, we get different radii of convergence displayed in Table 6.

Table 6. Radii of convergence for Example 4.

Cases	Different Values of Parameters That Are Defined in Theorem 1						
	α	β	γ	c_2	r_1	r_2	$r = \min\{r_1, r_2\}$
1	$-\frac{2}{3}$	$\frac{4}{3}$	$\frac{1}{3}$	-1	0.0622654	0.00406287	0.00406287
2	$-\frac{2}{3}$	$\frac{4}{3}$	-100	$\frac{1}{100}$	0.0622654	0.0582932	0.0582932
3	1	1	1	0	0.0622654	0.0622654	0.0622654
4	1	1	1	$\frac{1}{100}$	0.0622654	0.0592173	0.0592173
5	10	$\frac{1}{10}$	$\frac{1}{10}$	$\frac{1}{100}$	0.0622654	0.0585624	0.0585624

Example 5. Here, we assume one of the well-known Hammerstein integral equations (see pp. 19–20, [14]) defined by:

$$x(s) = 1 + \frac{1}{5} \int_0^1 F(s,t) x(t)^3 dt, \ x \in C[0,1], \ s,t \in [0,1], \tag{31}$$

where the kernel F is:

$$F(s,t) = \begin{cases} s(1-t), s \leq t, \\ (1-s)t, t \leq s. \end{cases}$$

We obtain (31) by using the Gauss–Legendre quadrature formula with $\int_0^1 \phi(t) dt \simeq \sum_{k=1}^{8} w_k \phi(t_k)$, where t_k and w_k are the abscissas and weights, respectively. Denoting the approximations of $x(t_i)$ with x_i ($i = 1,2,3,...,8$), then it yields the following 8×8 system of nonlinear equations:

$$5x_i - 5 - \sum_{k=1}^{8} a_{ik} x_k^3 = 0, \ i = 1, 2, 3 \ldots, 8,$$

$$a_{ik} = \begin{cases} w_k t_k (1 - t_i), & k \leq i, \\ w_k t_i (1 - t_k), & i < k. \end{cases}$$

The values of t_k and w_k can be easily obtained from the Gauss–Legendre quadrature formula when $k = 8$. The required approximate root is:

$$s_* = (1.002096\ldots, 1.009900\ldots, 1.019727\ldots, 1.026436\ldots, 1.026436\ldots,$$
$$1.019727\ldots, 1.009900\ldots, 1.002096\ldots)^T.$$

Then, we have:

$$L_0 = L = \frac{3}{40}, \ M = 2, \ \delta = 1, \ c_1 = 1$$

and $\mathbb{A} = Q(s_*, 0.11)$. By using the different values of the considered disposable parameters, we have different radii of convergence displayed in Table 7.

Table 7. Radii of convergence for Example 5.

Cases	Different Values of Parameters That Are Defined in Theorem 1						
	α	β	γ	c_2	r_1	r_2	$r = \min\{r_1, r_2\}$
1	$-\frac{2}{3}$	$\frac{4}{3}$	$\frac{1}{3}$	-1	8.88889	1.18366	1.18366
2	$-\frac{2}{3}$	$\frac{4}{3}$	-100	$\frac{1}{100}$	8.88889	7.74984	7.74984
3	1	1	1	0	8.88889	8.88889	8.88889
4	1	1	1	$\frac{1}{100}$	8.88889	7.94127	7.94127
5	10	$\frac{1}{10}$	$\frac{1}{10}$	$\frac{1}{100}$	8.88889	7.54223	7.54223

Example 6. *One can find the boundary value problem in [14], given as:*

$$y'' = \frac{1}{2}y^3 + 3y' - \frac{3}{2-x} + \frac{1}{2}, \ y(0) = 0, \ y(1) = 1. \tag{32}$$

We suppose the following partition of $[0, 1]$:

$$x_0 = 0 < x_1 < x_2 < x_3 < \cdots < x_j, \text{ where } x_{i+1} = x_i + h, \ h = \frac{1}{j}.$$

In addition, we assume that $y_0 = y(x_0) = 0$, $y_1 = y(x_1)$, ..., $y_{j-1} = y(x_{j-1})$ and $y_j = y(x_j) = 1$. Now, we can discretize this problem (32) relying on the first- and second-order derivatives, which is given by:

$$y'_k = \frac{y_{k+1} - y_{k-1}}{2h}, \ y''_k = \frac{y_{k-1} - 2y_k + y_{k+1}}{h^2}, \ k = 1, 2, \ldots, j-1.$$

Hence, we find the following general $(j-1) \times (j-1)$ nonlinear system:

$$y_{k+1} - 2y_k + y_{k-1} - \frac{h^2}{2}y_k^3 - \frac{3}{2-x_k}h^2 - \frac{1}{h^2} = 0, \ k = 1, 2, \ldots, j-1.$$

We choose the particular value of $j = 7$ that provides us a 6×6 nonlinear systems. The roots of this nonlinear system are $s_* = (0.07654393\ldots, 0.1658739\ldots, 0.2715210\ldots, 0.3984540\ldots, 0.5538864\ldots, 0.7486878\ldots)^T$, and the results are mentioned in Table 8.

Then, we get that:
$$L_0 = 73, \ L = 75, \ M = 2, \ \delta = 1, \ c_1 = 1,$$

and $\mathbb{A} = Q(s_*, 0.15)$.

With the help of different values of the parameters, we have the different radii of convergence listed in Table 8.

Table 8. Radii of convergence for Example 6.

Cases	Different Values of Parameters That Are Defined in Theorem 1						
	α	β	γ	c_2	r_1	r_2	$r = \min\{r_1, r_2\}$
1	$-\frac{2}{3}$	$\frac{4}{3}$	$\frac{1}{3}$	-1	0.00904977	0.00119169	0.00119169
2	$-\frac{2}{3}$	$\frac{4}{3}$	-100	$\frac{1}{100}$	0.00904977	0.00789567	0.00789567
3	1	1	1	0	0.00904977	0.00904977	0.00904977
4	1	1	1	$\frac{1}{100}$	0.00904977	0.00809175	0.00809175
5	10	$\frac{1}{10}$	$\frac{1}{10}$	$\frac{1}{100}$	0.00904977	0.00809175	0.00809175

Remark 1. *It is important to note that in some cases, the radii r_i are larger than the radius of $Q(s_*, r)$. A similar behavior for Method (2) was noticed in Table 7. Therefore, we have to choose all $r_i = 0.11$ because Expression (17) must be also satisfied.*

4. Concluding Remarks

The local convergence of the fourth-order scheme (2) was shown in earlier works [5,6,8,15] using Taylor series expansion. In this way, the hypotheses reach to four-derivative of the function φ in the particular case when $\mathbb{B}_1 = \mathbb{B}_2 = \mathbb{R}^m$ and $S = \mathbb{R}$. These hypotheses limit the applicability of methods such (2). We analyze the local convergence using only the first derivative for Banach space mapping. The convergence order can be found using the computational order of convergence (COC) or the approximate computational order of convergence ($ACOC$) (Appendix A), avoiding the computation of higher order derivatives. We found also computable radii and error bounds not given before using Lipschitz constants, expanding, therefore, the applicability of the technique. Six numerical problems were proposed for illustrating the feasibility of the new approach. Our technique can be used to study other iterative methods containing inverses of mapping such as (3)–(5) (see also [1–9,11–45]) and to expand their applicability along the same lines.

Author Contributions: All the authors have equal contribution for this paper.

Funding: This research received no external funding.

Conflicts of Interest: The authors declare no conflict of interest.

Abbreviations

The following abbreviations are used in this manuscript:

MDPI	Multidisciplinary Digital Publishing Institute
DOAJ	Directory of open access journals
TLA	Three letter acronym
LD	linear dichroism
COC	Computational order of convergence
(COC)	Approximate computational order of convergence

Appendix A

Remark

(a) The procedure of studying local convergence was already given in [1,2] for similar methods. Function $M(t) = M = 2$ or $M(t) = 1 + L_0 t$, since $0 \leq t < \frac{1}{L_0}$ can be replaced by (16). The convergence radius r cannot be bigger than the radius r_A for the Newton method given in this paper. These results are used to solve autonomous differential equations. The differential equation plays an important role in the study of network science, computer systems, social networking systems, and biochemical systems [46].

In fact, we refer the reader to [46], where a different technique is used involving discrete samples from the existence of solution spaces. The existence of intervals with common solutions, as well as disjoint intervals and the multiplicity of intervals with common solutions is also shown. However, this work does not deal with spaces that are continuous and multidimensional.

(b) It is important to note that the scheme (2) does not change if we adopt the hypotheses of Theorem 1 rather than the stronger ones required in [5–9]. In practice, for the error bounds, we adopt the following formulas [22] for the computational order of convergence (COC), when the required root is available, or the approximate computational order of convergence (ACOC), when the required root is not available in advance, which can be written as:

$$\xi = \frac{\ln \frac{\|x_{k+2} - s_*\|}{\|x_{k+1} - s_*\|}}{\ln \frac{\|x_{k+1} - s_*\|}{\|x_k - s_*\|}}, \quad k = 0, 1, 2, 3 \ldots,$$

$$\xi^* = \frac{\ln \frac{\|x_{k+2} - x_{k+1}\|}{\|x_{k+1} - x_k\|}}{\ln \frac{\|x_{k+1} - x_k\|}{\|x_k - x_{k-1}\|}}, \quad k = 1, 2, 3, \ldots,$$

respectively. By means of the above formulas, we can obtain the convergence order without using estimates on the high-order Fréchet derivative.

References

1. Argyros, I.K. *Convergence and Application of Newton-type Iterations*; Springer: Berlin, Germany, 2008.
2. Argyros, I.K.; Hilout, S. *Numerical Methods in Nonlinear Analysis*; World Scientific Publ. Comp.: Hackensack, NJ, USA, 2013.
3. Traub, J.F. *Iterative Methods for the Solution of Equations*; Prentice- Hall Series in Automatic Computation: Englewood Cliffs, NJ, USA, 1964.
4. Petkovic, M.S.; Neta, B.; Petkovic, L.; Džunič, J. *Multipoint Methods for Solving Nonlinear Equations*; Elsevier: Amsterdam, The Netherlands, 2013.
5. Cordero, A.; Martínez, E.; Torregrosa, J.R. Iterative methods of order four and five for systems of nonlinear equations. *J. Comput. Appl. Math.* **2009**, *231*, 541–551. [CrossRef]
6. Darvishi, M.T.; Barati, A. A fourth-order method from quadrature formulae to solve systems of nonlinear equations. *Appl. Math. Comput.* **2007**, *188*, 257–261. [CrossRef]
7. Sharma, J.R.; Guha, R.K.; Sharma, R. An efficient fourth order weighted-Newton method for systems of nonlinear equations. *Numer. Algorithms* **2013**, *62*, 307–323. [CrossRef]
8. Su, Q. A new family weighted-Newton methods for solving systems of nonlinear equations, to appear in. *Appl. Math. Comput.*
9. Noor, M.A.; Waseem, M. Some iterative methods for solving a system of nonlinear equations. *Comput. Math. Appl.* **2009**, *57*, 101–106. [CrossRef]
10. Brent, R.P. *Algorithms for Finding Zeros and Extrema of Functions Without Calculating Derivatives*; Report TR CS 198; DCS: Stanford, CA, USA, 1971.

11. Rheinboldt, W.C. An adaptive continuation process for solving systems of nonlinear equations. *Polish Acad. Sci. Banach Cent. Publ.* **1978**, *3*, 129–142. [CrossRef]
12. Constantinides, A.; Mostoufi, N. *Numerical Methods for Chemical Engineers with MATLAB Applications*; Prentice Hall PTR: Upper Saddle River, NJ, USA, 1999.
13. Douglas, J.M. *Process Dynamics and Control*; Prentice Hall: Englewood Cliffs, NJ, USA, 1972; Volume 2.
14. Ortega, J.M.; Rheinboldt, W.C. *Iterative Solution of Nonlinear Equations in Several Variables*; Academic Press: New York, NY, USA, 1970.
15. Chun, C. Some improvements of Jarratt's method with sixth-order convergence. *Appl. Math. Comput.* **2007**, *190*, 1432–1437. [CrossRef]
16. Candela, V.; Marquina, A. Recurrence relations for rational cubic methods I: The Halley method. *Computing* **1990**, *44*, 169–184. [CrossRef]
17. Chicharro, F.; Cordero, A.; Torregrosa, J.R. Drawing dynamical and parameters planes of iterative families and methods. *Sci. World J.* **2013**, 780153. [CrossRef]
18. Cordero, A.; García-Maimó, J.; Torregrosa, J.R.; Vassileva, M.P.; Vindel, P. Chaos in King's iterative family. *Appl. Math. Lett.* **2013**, *26*, 842–848. [CrossRef]
19. Cordero, A.; Torregrosa, J.R.; Vindel, P. Dynamics of a family of Chebyshev-Halley type methods. *Appl. Math. Comput.* **2013**, *219*, 8568–8583. [CrossRef]
20. Cordero, A.; Torregrosa, J.R. Variants of Newton's method using fifth-order quadrature formulas. *Appl. Math. Comput.* **2007**, *190*, 686–698. [CrossRef]
21. Ezquerro, J.A.; Hernández, A.M. On the R-order of the Halley method. *J. Math. Anal. Appl.* **2005**, *303*, 591–601. [CrossRef]
22. Ezquerro, J.A.; Hernández, M.A. New iterations of R-order four with reduced computational cost. *BIT Numer. Math.* **2009**, *49*, 325–342. [CrossRef]
23. Grau-Sánchez, M.; Noguera, M.; Gutiérrez, J.M. On some computational orders of convergence. *Appl. Math. Lett.* **2010**, *23*, 472–478. [CrossRef]
24. Gutiérrez, J.M.; Hernández, M.A. Recurrence relations for the super-Halley method. *Comput. Math. Appl.* **1998**, *36*, 1–8. [CrossRef]
25. Herceg, D.; Herceg, D. Sixth-order modifications of Newton's method based on Stolarsky and Gini means. *J. Comput. Appl. Math.* **2014**, *267*, 244–253. [CrossRef]
26. Hernández, M.A. Chebyshev's approximation algorithms and applications. *Comput. Math. Appl.* **2001**, *41*, 433–455. [CrossRef]
27. Hernández, M.A.; Salanova, M.A. Sufficient conditions for semilocal convergence of a fourth order multipoint iterative method for solving equations in Banach spaces. *Southwest J. Pure Appl. Math.* **1999**, *1*, 29–40.
28. Homeier, H.H.H. On Newton-type methods with cubic convergence. *J. Comput. Appl. Math.* **2005**, *176*, 425–432. [CrossRef]
29. Jarratt, P. Some fourth order multipoint methods for solving equations. *Math. Comput.* **1966**, *20*, 434–437. [CrossRef]
30. Kou, J. On Chebyshev–Halley methods with sixth-order convergence for solving non-linear equations. *Appl. Math. Comput.* **2007**, *190*, 126–131. [CrossRef]
31. Kou, J.; Wang, X. Semilocal convergence of a modified multi-point Jarratt method in Banach spaces under general continuity conditions. *Numer. Algorithms* **2012**, *60*, 369–390.
32. Li, D.; Liu, P.; Kou, J. An improvement of the Chebyshev-Halley methods free from second derivative. *Appl. Math. Comput.* **2014**, *235*, 221–225. [CrossRef]
33. Magreñán, Á.A. Different anomalies in a Jarratt family of iterative root-finding methods. *Appl. Math. Comput.* **2014**, *233*, 29–38.
34. Magreñán, Á.A. A new tool to study real dynamics: The convergence plane. *Appl. Math. Comput.* **2014**, *248*, 215–224. [CrossRef]
35. Neta, B. A sixth order family of methods for nonlinear equations. *Int. J. Comput. Math.* **1979**, *7*, 157–161. [CrossRef]
36. Ozban, A.Y. Some new variants of Newton's method. *Appl. Math. Lett.* **2004**, *17*, 677–682. [CrossRef]
37. Parhi, S.K.; Gupta, D.K. Recurrence relations for a Newton-like method in Banach spaces. *J. Comput. Appl. Math.* **2007**, *206*, 873–887.

38. Parhi, S.K.; Gupta, D.K. A sixth order method for nonlinear equations. *Appl. Math. Comput.* **2008**, *203*, 50–55. [CrossRef]
39. Ren, H.; Wu, Q.; Bi, W. New variants of Jarratt's method with sixth-order convergence. *Numer. Algorithms* **2009**, *52*, 585–603. [CrossRef]
40. Wang, X.; Kou, J.; Gu, C. Semilocal convergence of a sixth-order Jarratt method in Banach spaces. *Numer. Algorithms* **2011**, *57*, 441–456. [CrossRef]
41. Weerakoon, S.; Fernando, T.G.I. A variant of Newton's method with accelerated third order convergence. *Appl. Math. Lett.* **2000**, *13*, 87–93. [CrossRef]
42. Zhou, X. A class of Newton's methods with third-order convergence. *Appl. Math. Lett.* **2007**, *20*, 1026–1030. [CrossRef]
43. Amat, S.; Busquier, S.; Plaza, S. Dynamics of the King and Jarratt iterations. *Aequ. Math.* **2005**, *69*, 212–223. [CrossRef]
44. Amat, S.; Busquier, S.; Plaza, S. Chaotic dynamics of a third-order Newton-type method. *J. Math. Anal. Appl.* **2010**, *366*, 24–32. [CrossRef]
45. Amat, S.; Hernández, M.A.; Romero, N. A modified Chebyshev's iterative method with at least sixth order of convergence. *Appl. Math. Comput.* **2008**, *206*, 164–174. [CrossRef]
46. Bagchi, S. Computational Analysis of Network ODE Systems in Metric Spaces: An Approach. *J. Comput. Sci.* **2017**, *13*, 1–10. [CrossRef]

© 2019 by the authors. Licensee MDPI, Basel, Switzerland. This article is an open access article distributed under the terms and conditions of the Creative Commons Attribution (CC BY) license (http://creativecommons.org/licenses/by/4.0/).

MDPI
St. Alban-Anlage 66
4052 Basel
Switzerland
Tel. +41 61 683 77 34
Fax +41 61 302 89 18
www.mdpi.com

Symmetry Editorial Office
E-mail: symmetry@mdpi.com
www.mdpi.com/journal/symmetry

www.ingramcontent.com/pod-product-compliance
Lightning Source LLC
LaVergne TN
LVHW071946080526
838202LV00064B/6685